Frontiers in Mathematics

Matej Brešar
Mikhail A. Chebotar
Wallace S. Martindale 3rd

Functional
Identities

Birkhäuser Verlag
Basel · Boston · Berlin

Authors:

Matej Brešar
Department of Mathematics
and Computer Science
FNM
University of Maribor
2000 Maribor
Slovenia
e-mail: bresar@uni-mb.si

Wallace S. Martindale 3rd
Department of Mathematics
University of Massachusetts
Amherst, MA 01003
USA
e-mail: jerrywsm@comcast.net

Mikhail A. Chebotar
Department of Mathematical Sciences
Kent State University
Kent, OH 44242
USA
e-mail: mchebotar@gmail.com

2000 Mathematical Subject Classification: 15A04, 16N60, 16R50, 16W20, 16W10,
16W25, 17B60, 17C50, 47B47

Library of Congress Control Number: 2007924802

Bibliographic information published by Die Deutsche Bibliothek
Die Deutsche Bibliothek lists this publication in the Deutsche Nationalbibliografie;
detailed bibliographic data is available in the Internet at <http://dnb.ddb.de>.

ISBN 978-3-7643-7795-3 Birkhäuser Verlag, Basel - Boston - Berlin

© 2007 Birkhäuser Verlag, P.O. Box 133, CH-4010 Basel, Switzerland
Part of Springer Science+Business Media
Cover design: Birgit Blohmann, Zürich, Switzerland
Printed on acid-free paper produced from chlorine-free pulp. TCF ∞
Printed in Germany

ISBN 978-3-7643-7795-3   e-ISBN 978-3-7643-7796-0

9 8 7 6 5 4 3 2 1   www.birkhauser.ch

To Konstantin I. Beidar (1951–2004)

Contents

Preface

A *functional identity* (FI) can be informally described as an identical relation involving (arbitrary) elements in a ring together with ("unknown") functions; more precisely, elements are multiplied by values of functions. The goal of the general FI theory is to determine the form of these functions, or, when this is not possible, to determine the structure of the ring admitting the FI in question. This theory has turned out to be a powerful tool for solving a variety of problems in different areas. It is not always easy to recognize that the problem in question can be interpreted through some FI; often this is the most intriguing part of the process. But once one succeeds in discovering an FI that fits into the general theory, this abstract theory then as a rule yields the desired conclusions at a high level of generality.

Among classical algebraic concepts, the one of a polynomial identity (PI) seems to be, at least on the surface, the closest one to the concept of an FI. In fact, a PI is formally just a very special example of an FI (where functions are polynomials). However, the theory of PI's has quite different goals than the theory of FI's. One could say, especially from the point of view of applications, that the two theories are complementary to each other. Under some natural restrictions, PI theory deals with rings that are close to algebras of low dimensions, while FI theory gives definitive answers in algebras of sufficiently large or infinite dimensions.

FI theory is a relatively new one. Its roots lie in the Ph.D. Thesis of the first author from 1990, which was followed in the ensuing years by a series of papers in which he studied some basic FI's, in particular those concerning the so-called commuting maps, and also found first applications. The central figure in establishing the foundations of the general theory was Konstantin I. Beidar. This theory culminated around 2000 in the works by Beidar and the second author on d-free sets. Over the last few years, the emphasis has been mainly on applications of the general theory. A notable example are complete solutions of Herstein's conjectures on Lie homomorphisms and Lie derivations in associative rings (posed in Herstein's 1961 "AMS hour talk"). They were obtained in a series of papers by Beidar and the authors of this book which ended in 2002. Practically all of the advanced FI theory was used in these solutions. In fact, the main motivation for developing this theory was searching for tools that would settle these conjectures.

So far no book has been written exclusively on FI's. Commuting maps and some of their applications were treated in the book *Rings with generalized identities*

(Marcel Dekker, 1996) by Beidar, Martindale and Mikhalev, as well as in the book *Local multipliers of C*-algebras* (Springer, 2003) by Ara and Mathieu, in both cases in some rather special settings so that the overlap with the present book is very small. Up until now the general FI theory has existed only in research journals. We believe that the theory has reached maturity and that a monograph presenting FI's and their applications in a comprehensive and comprehensible manner shall be useful for the interested mathematical community.

Through their applications FI's have connections to different areas, say to Lie algebras, Jordan and other nonassociative algebras, linear algebra, operator theory, functional analysis, and mathematical physics. However, the basic setting of the book is (noncommutative) ring theory. Including non-algebraic topics in the book would make the exposition too diverse. Still, different parts of the book should be of interest to mathematicians having different basic backgrounds. While writing we kept in mind that the reader might not be a ring theorist. The prerequisites needed to follow the exposition are therefore carefully explained. The basic ones are surveyed at the very beginning of the book, while the somewhat more demanding ones appear in four appendices at the end. They consider topics which are more or less widely known and treated in other books. We concentrate on those aspects that are important for this book. Some results in appendices are proved, and some not; in any case we try to explain to the reader, not necessarily in a rigorous manner, the background of the results and concepts that are treated. The book is therefore fairly self-contained and suitable for self-study.

The book consists of three parts. Each of them has its own purpose. Part I is an introductory one, Part II gives a full account of the general FI theory, and Part III is devoted to applications of the general theory. Parts are divided into chapters which are further divided into sections. Chapters end with comments about the literature and the history.

Part I has two chapters. The purpose of Chapter 1 is to introduce the newcomer into the subject and to explain in an informal manner, mainly through examples, what this theory is all about. Chapter 2 contains results that are already of some interest in their own right, and they make it possible for one to use FI theory at a basic level. The concept of *d-freeness* is introduced in its simplest form, and the connections to the concept of the *strong degree* are discussed. The results of Chapter 2 are only partially superseded in Part II; namely, the setting in which they are obtained is somewhat different, and in a certain direction even more general than the setting considered later. In principle, a graduate student should be able to understand Part I without difficulties. One could use some sections as a part of a course in ring theory.

Part II is the core of the book. The general theory has so far been exposed only in numerous papers, which are often very technical and long. They depend upon each other, so it is not always easy to read them. The goal of Part II is to extract from these papers what is, in our opinion, most important and applicable, that is, to show what this theory has to offer. The basic concept upon which everything is based is that of a *d-free set*. Chapter 3 discusses basic properties and constructions of these sets. Chapter 4 is devoted to the study of FI's on *d*-free sets; the notions of *quasi-polynomials* and *core functions* are of central importance in this context. The purpose of Chapter 5 is to prove the existence of some important classes of *d*-free sets in prime and semiprime rings, and also to consider some special FI's in these rings. The exposition in Part II is necessarily technical, and in spite of our efforts to be as intelligible as possible, the reader studying the text in detail shall need some patience. It is impossible to avoid long formulas and somewhat complicated notation, so that "love at first sight" with this theory seems a bit unlikely; some effort and time is needed in order to appreciate it and find it enjoyable. On the other hand, the main results have clear statements and so hopefully Part II will serve as a useful source of references.

The meaning of the results of Part II becomes evident in Part III. Applications are the main reason for existence of the general theory. We believe that while this theory is already mature (at least in the direction in which it has been mainly developed), the possibilities for its applications are still far from being exploited. Our intention in Part III is therefore to show through several relevant examples the way this theory can be used. We do not try to present the results in utmost generality, but rather point out various areas where FI's are applicable and demonstrate the methods needed in applications. Chapter 6 deals with *Lie homomorphisms* and other "nonassociative" maps. In particular, solutions of Herstein's Lie map conjectures are discussed. Chapter 7 considers some *linear preserver problems*, particularly commutativity and normality preserving maps. Chapter 8 discusses miscellaneous topics which all, however, belong to the *Lie algebra* framework.

We are grateful to a number of colleagues for fruitful discussions on the subject of the book. It would be impossible to list all whose influence is felt in our writing. But we can not omit mentioning a man we can not thank enough and without whom this book would not exist. Unfortunately he is no longer with us. Konstantin (Kostia) I. Beidar passed away in March 2004. His impact on FI theory was decisive. He wrote more than 20 papers on the subject. Some of them were path-breaking. Kostia himself considered his fundamental results on FI's as some of his best mathematical achievements. It is just incredible that while being so productive in FI theory, at the same time he was also heavily involved in the work in many other fields. But his mathematical strength and knowledge was a kind of a legend among his numerous coauthors and colleagues. Solving problems and inventing new areas seemed so easy when working with Kostia. All three of us had the privilege of cooperating with him on several projects. Our main mathematical tie, however, was FI theory. The plan to write this book was made together with him. In several of our frequent visits to Tainan in Taiwan, where he was working after leaving Moscow State University in 1994, we discussed with him possible concepts of the book. A small portion of the book, a draft of Part I and some other preliminary notes, were written while he was still alive. Arriving at Part II, technically the most entangled part of the book, it was somehow expected that he would become involved in the actual writing. But his early and unexpected death has forced us to continue without him. We miss him; as a friend and as a mathematician. We hope that this book appropriately represents an important part of Kostia's mathematical legacy.

Maribor, Kent, Philadelphia   MATEJ BREŠAR
February 2007   MIKHAIL A. CHEBOTAR
                                                        WALLACE S. MARTINDALE 3RD

Part I

An Introductory Course

Chapter 1

What is a Functional Identity?

An exposition on a mathematical subject usually starts with basic definitions. We feel, however, that it is more appropriate to introduce functional identities through examples. We shall therefore present various simple examples of functional identities, so that the reader may guess which conclusions could be derived when facing these identities. These examples have been selected in order to illustrate the general theory, and not all of them are of great importance in their own right. Their consideration will be rather elementary; anyhow, many of the arguments that we shall present here will be used, sometimes in a hidden way, in much more general situations considered in further chapters. Examples will be followed by some basic definitions and notation, but even these will be given in a somewhat informal fashion. The last objective of this preliminary chapter is to point out a few instances where functional identities appear naturally. That is, we wish to indicate, without many details, why and where the theory of functional identities is applicable. So, in summary, the goal of this chapter is to give an informal introduction to functional identities which should be of help to a newcomer to the subject.

At the very beginning, however, we will survey the ring-theoretic notions and tools that will be needed throughout the book. They are very basic.

1.1 Prerequisites

Most readers are probably already familiar with the basic terminology in ring theory. However, some parts of the book are supposed to be of interest not exclusively to ring theorists. We shall therefore state and briefly comment on a few very fundamental definitions and results. Details can be found in many standard graduate level texts on algebra. Simultaneously we shall also explain some conventions and fix the notation that will be most frequently used. Thereby the reader can find below all the information needed to follow Part I. Occasionally, however, we shall also mention some facts about polynomial identities and generalized polynomial identities in Part I. In such instances the reader may consult appendices C and D or just continue reading without too much loss. A knowledge of the theory of (generalized) polynomial identities should certainly help the reader to understand Part I better, but it is not a necessity for following the essence of the exposition. Parts II and III require some more prerequisites which are surveyed in appendices.

Throughout the text, by a *ring* (*algebra*) we shall mean an associative ring (algebra over a commutative ring) possibly without unity and not necessarily commutative. As a matter of fact, many of our results are meaningless in the commutative setting: they are either trivially true or trivially false. When the existence of a unity will be required, we shall speak about a *unital ring* (*unital algebra*). We will also consider some nonassociative rings and algebras, but this will always be pointed out (say, a Lie ring will never be referred to as just a ring). We remark that a ring (algebra) is nonassociative if the associativity of multiplication is not required, which does not necessarily mean that it does not hold. For most of the time we shall denote rings and algebras by letters $\mathcal{A}, \mathcal{B}, \ldots$. The centers of rings will often play an important role, and even when not explicitly indicated we shall regard rings as algebras over their centers.

Let us list the definitions of some fundamental types of rings. By a *field* we mean a commutative division ring, while a *division ring* is a nonzero unital ring all of whose nonzero elements are invertible. More generally, a ring $\mathcal{A}$ is said to be *simple* if its only ideals are 0 and $\mathcal{A}$, and $ab \neq 0$ for some $a, b \in \mathcal{A}$. As usual, by an ideal of a ring we shall always mean a two-sided ideal. A ring $\mathcal{A}$ is called (*left*) *primitive* if it has a faithful simple (left) module $\mathcal{M}$. The simplicity of a module means that 0 and $\mathcal{M}$ are the only submodules of $\mathcal{M}$ and $\mathcal{A}\mathcal{M} \neq 0$, and faithfulness means that $a\mathcal{M} \neq 0$ for every nonzero $a \in \mathcal{A}$. A common generalization of all types mentioned so far is the concept of a prime ring: a ring $\mathcal{A}$ is called *prime* if the product of two nonzero ideals of $\mathcal{A}$ is always nonzero. Equivalently, $a\mathcal{A}b = 0$, where $a, b \in \mathcal{A}$, implies $a = 0$ or $b = 0$. Finally, a ring $\mathcal{A}$ containing no nonzero nilpotent ideals is said to be *semiprime*. Equivalently, $a\mathcal{A}a = 0$, where $a \in \mathcal{A}$, implies $a = 0$. Division, simple, primitive, prime, and semiprime algebras are defined analogously.

This is not a book about these types of rings; these definitions just provide an appropriate setting for studying functional identities. So we give just a few simple-minded examples to help a non-specialist to clarify these notions. Examples of fields are widely known. We assume the reader is familiar with the ring of quaternions which is the simplest, the most famous and historically the first example of a noncommutative division ring. It is well-known and easy to see that every unital simple ring is primitive. If $\mathcal{V}$ is a vector space over a field $\mathbb{F}$, then $\mathrm{End}_{\mathbb{F}}(\mathcal{V})$, the ring of all endomorphisms of $\mathcal{V}$, is always primitive, and is simple only when $\mathcal{V}$ is finite dimensional. In that case it is isomorphic to the matrix ring $M_n(\mathbb{F})$ where n is the dimension of $\mathcal{V}$. This notation is standard: for any ring $\mathcal{A}$, $M_n(\mathcal{A})$ denotes the ring of $n \times n$ matrices with entries from $\mathcal{A}$. It turns out that a unital ring $\mathcal{A}$ is simple, (left) primitive, prime, and semiprime, respectively, if and only if the same holds true for the ring $M_n(\mathcal{A})$. The ring $\mathbb{Z}$ of integers is the simplest example of a prime ring which is not primitive, and so, more generally, also $M_n(\mathbb{Z})$ is prime but not primitive. The direct product $\mathcal{A}_1 \times \mathcal{A}_2$ of two nonzero prime rings $\mathcal{A}_1$ and $\mathcal{A}_2$ is semiprime but not prime. The ring of all upper triangular $n \times n$ matrices over a field is not even semiprime since the strictly upper triangular matrices form a nilpotent ideal.

There are many ways to construct new rings from a given ring $\mathcal{A}$. Let us mention a few that shall be of particular importance for us.

First of all, for any n in $\mathbb{N}$, the set of all positive integers, we have the matrix ring $M_n(\mathcal{A})$. When $\mathcal{A}$ is unital, $M_n(\mathcal{A})$ contains the so-called *matrix units*, that is, matrices in which exactly one entry equals 1 and all other entries are 0. A matrix unit with entry 1 in the position (i,j) will be denoted by e_{ij}.

By $\mathrm{End}(\mathcal{A})$ we denote the ring of endomorphisms of the additive group of $\mathcal{A}$, that is, the set of additive maps of $\mathcal{A}$ into itself which forms a ring under pointwise addition and composition as multiplication. For any $a, b \in \mathcal{A}$ we define a *two-sided multiplication* $_a M_b \in \mathrm{End}(\mathcal{A})$ by $_a M_b(x) = axb$. By $\mathcal{M}(\mathcal{A})$ we denote the set of all elements in $\mathrm{End}(\mathcal{A})$ that can be written as finite sums of two-sided multiplications $_a M_b$. A typical element $f \in \mathcal{M}(\mathcal{A})$ is thus of the form $f : x \mapsto \sum_i a_i x b_i$. Note that $\mathcal{M}(\mathcal{A})$ is a subring of $\mathrm{End}(\mathcal{A})$. It will be called the *multiplication ring* of $\mathcal{A}$.

The notation $_{\mathcal{A}}\mathcal{M}$ means that $\mathcal{M}$ is a left module over a ring $\mathcal{A}$, and $\mathrm{End}_{\mathcal{A}}(\mathcal{M})$ denotes the ring of all $\mathcal{A}$-module endomorphisms of $\mathcal{M}$. Suppose that the module $\mathcal{M}$ is simple. Since the kernel and the range of every endomorphism $f \in \mathrm{End}_{\mathcal{A}}(\mathcal{M})$ are clearly submodules of $\mathcal{M}$, f is either 0 or it is invertible. That is to say, $\mathrm{End}_{\mathcal{A}}(\mathcal{M})$ is a division ring. This observation is known as *Schur's lemma*. A simple module $\mathcal{M}$ can therefore also be regarded as a vector space over a division ring $D = \mathrm{End}_{\mathcal{A}}(\mathcal{M})$. The simplicity of $\mathcal{M}$ implies that $\mathcal{A}x = \mathcal{M}$ for every nonzero $x \in \mathcal{M}$. That is to say, for every $0 \neq x \in \mathcal{M}$ and $y \in \mathcal{M}$ there exists $a \in \mathcal{A}$ such that $ax = y$. The celebrated *Jacobson density theorem* states that much more is true: If $x_1, \ldots, x_n$, where $n \in \mathbb{N}$ is arbitrary, are elements in $\mathcal{M}$ which are linearly independent over D, and $y_1, \ldots, y_n$ are any elements in $\mathcal{M}$, then there exist $a \in \mathcal{A}$ such that $ax_i = y_i$, $i = 1, \ldots, n$. One usually refers to this property by saying that $\mathcal{A}$ acts *densely* on the vector space $\mathcal{M}$.

Let $\mathcal{C}$ be a commutative ring and $X = \{x_1, x_2, \ldots\}$ be a countable set. By $\mathcal{C}\langle X \rangle$ we denote the *free algebra* on the set X over $\mathcal{C}$. The elements of $\mathcal{C}\langle X \rangle$ are polynomials in noncommuting variables from the set X, whose coefficients in $\mathcal{C}$ commute with indeterminates. Their multiplication is defined in the obvious way by juxtaposition. We remark here that the construction of the free algebra can be made rigorous by various methods (e.g., forming the vector space over $\mathcal{C}$ with basis the free semigroup on the set X or using the more high powered notion of tensor products), but we will not elaborate further in this regard. Considering an element $f \in \mathcal{C}\langle X \rangle$ as a polynomial, the notions such as the *degree of f* or *monomial of f* are self-explanatory. Let $f = f(x_1, \ldots, x_m)$; this notation indicates that no x_i with $i > m$ occurs in f. Further, let $\mathcal{A}$ be a $\mathcal{C}$-algebra, let $\mathcal{R}$ be a nonempty subset of $\mathcal{A}$, and let $\mathcal{R}^m = \mathcal{R} \times \mathcal{R} \times \ldots \times \mathcal{R}$ denote the Cartesian product (m times). Then f determines a function $F : \mathcal{R}^m \to \mathcal{A}$ given in the obvious way of substituting $r_1, \ldots, r_m$ for $x_1, \ldots, x_m$ in f, where $r_i \in \mathcal{R}$, $i = 1, \ldots, m$. Such a function F will be called the *polynomial function determined by f*. One should carefully note the distinction between a polynomial f and the polynomial function it determines, e.g., if $\mathcal{A}$ is commutative, then $f = x_1 x_2 - x_2 x_1$ is a nonzero element in $\mathcal{C}\langle X \rangle$, whereas F is the zero function on $\mathcal{A} \times \mathcal{A}$.

For elements x and y in a ring $\mathcal{A}$ we shall write $[x, y] = xy - yx$ and $x \circ y = xy + yx$. The element $[x, y]$ is called the *Lie product* (or the *commutator*) of elements x and y, and $x \circ y$ is called the *Jordan product* of x and y. In the case when $2\mathcal{A} = 0$ (i.e., $2a = 0$ for every $a \in \mathcal{A}$), the Lie and the Jordan product coincide, which makes their treatment rather muddled. Therefore we shall usually restrict our attention to rings which are 2-torsion free; we say that a ring (or just an additive group, for that matter) $\mathcal{A}$ is *n-torsion free*, where $n \in \mathbb{N}$, if $na \neq 0$ for every nonzero $a \in \mathcal{A}$. In case $\mathcal{A}$ is a prime ring, it is more customary to express its *n*-torsion freeness through the notion of the *characteristic* (char($\mathcal{A}$)): $\mathcal{A}$ is *n*-torsion free if and only if char($\mathcal{A}$) $\neq n$. We also remark that the characteristic of a prime ring is always a prime number or 0.

Let us fix some more notation. For a nonempty subset $\mathcal{T}$ of a ring $\mathcal{A}$ we shall write $\langle \mathcal{T} \rangle$ for the subring of $\mathcal{A}$ generated by $\mathcal{T}$, and $(\mathcal{T})$ for the ideal of $\mathcal{A}$ generated by $\mathcal{T}$. The set $\mathcal{C}(\mathcal{T}) = \{a \in \mathcal{A} \mid [a, \mathcal{T}] = 0\}$ is called the *centralizer* of $\mathcal{T}$ in $\mathcal{A}$. The *center* of $\mathcal{A}$ is of course the centralizer of $\mathcal{A}$ in $\mathcal{A}$; however, instead of $\mathcal{C}(\mathcal{A})$ we will usually denote it by $\mathcal{Z}(\mathcal{A})$. When $\mathcal{T} = \{t\}$, we shall write $\mathcal{C}(t)$, $\langle t \rangle$ and (t) instead of $\mathcal{C}(\{t\})$, $\langle\{t\}\rangle$ and $(\{t\})$. If $\mathcal{R}$ and $\mathcal{S}$ are additive subgroups of $\mathcal{A}$, we denote by $\mathcal{R}\mathcal{S}$, $[\mathcal{R}, \mathcal{S}]$, and $\mathcal{R} \circ \mathcal{S}$ the additive subgroup of $\mathcal{A}$ generated by all rs, $[r, s]$, and $r \circ s$, respectively, where $r \in \mathcal{R}$ and $s \in \mathcal{S}$ are arbitrary elements. Similarly we define $\mathcal{R}_1 \mathcal{R}_2 \ldots \mathcal{R}_n$, $[\mathcal{R}_1, [\mathcal{R}_2, [\ldots, \mathcal{R}_n] \ldots]]$ etc. A warning is necessary: sometimes $\mathcal{R}^n$ will denote the additive subgroup generated by all $r_1 r_2 \ldots r_n$, $r_i \in \mathcal{R}$, but more often it will denote the Cartesian product of n copies of $\mathcal{R}$. But from the context it should be clear in each case what we have in mind.

An additive subgroup $\mathcal{L}$ of $\mathcal{A}$ such that $[\mathcal{L}, \mathcal{L}] \subseteq \mathcal{L}$ is called a *Lie subring* of a ring $\mathcal{A}$. If an additive subgroup $\mathcal{I}$ of a Lie subring $\mathcal{L}$ satisfies $[\mathcal{I}, \mathcal{L}] \subseteq \mathcal{I}$, then $\mathcal{I}$ is said to be a *Lie ideal* of $\mathcal{L}$. For example, $[\mathcal{L}, \mathcal{L}]$ is a Lie ideal of $\mathcal{L}$, and so is $\mathcal{Z}(\mathcal{L}) = \mathcal{C}(\mathcal{L}) \cap \mathcal{L} = \{l \in \mathcal{L} \mid [l, \mathcal{L}] = 0\}$, the *center of $\mathcal{L}$*. We shall be usually interested in *noncentral* Lie ideals of $\mathcal{L}$, i.e., those which are not contained in $\mathcal{Z}(\mathcal{L})$. Of course, a Lie ideal of $\mathcal{L}$ is also a Lie subring of $\mathcal{A}$. *Jordan subrings* and their *Jordan ideals* are defined similarly. Important examples of Lie and Jordan subrings are provided in rings with involution. An *involution*, usually denoted by $*$, on a ring $\mathcal{A}$ is an antiautomorphism of $\mathcal{A}$ of period 1 or 2, i.e., $*$ satisfies $(x + y)^* = x^* + y^*$, $(xy)^* = y^* x^*$ and $(x^*)^* = x$ for all $x, y \in \mathcal{A}$. Note that the set $\mathcal{S}(\mathcal{A}) = \{x \in \mathcal{A} \mid x^* = x\}$ of *symmetric elements* of $\mathcal{A}$ is a Jordan subring of $\mathcal{A}$, and the set $\mathcal{K}(\mathcal{A}) = \{x \in \mathcal{A} \mid x^* = -x\}$ of *skew elements* of $\mathcal{A}$ is a Lie subring of $\mathcal{A}$. When the context is clear we shall write $\mathcal{S}$ and $\mathcal{K}$ instead of $\mathcal{S}(\mathcal{A})$ and $\mathcal{K}(\mathcal{A})$, and similarly we shall write $\mathcal{Z}$ instead of $\mathcal{Z}(\mathcal{A})$.

Let $\mathcal{B}$ and $\mathcal{A}$ be rings and let $\mathcal{L}$ be a Lie subring of $\mathcal{B}$. An additive map $\alpha : \mathcal{L} \to \mathcal{A}$, $x \mapsto x^{\alpha}$, is called a *Lie homomorphism* if $[x,y]^{\alpha} = [x^{\alpha}, y^{\alpha}]$ for all $x, y \in \mathcal{L}$. A *Jordan homomorphism* from a Jordan subring of $\mathcal{B}$ into $\mathcal{A}$ is defined analogously, i.e., it is an additive map that preserves the Jordan product. There is another type of maps that will often appear. Let $\mathcal{A}$ be a ring and let $\mathcal{M}$ be an $(\mathcal{A}, \mathcal{A})$-bimodule. An additive map $\delta : \mathcal{A} \to \mathcal{M}$ is called a *derivation* if $(xy)^{\delta} = x^{\delta}y + xy^{\delta}$ for all $x, y \in \mathcal{A}$. For example, for any fixed $a \in \mathcal{A}$ the map $x \mapsto [a, x]$ is a derivation of $\mathcal{A}$ into itself. Such derivations are called *inner derivations*. Further, an additive map $\delta : \mathcal{L} \to \mathcal{M}$, where $\mathcal{L}$ is a Lie subring of $\mathcal{A}$ is called a *Lie derivation* if $[x, y]^{\delta} = [x^{\delta}, y] + [x, y^{\delta}]$ for all $x, y \in \mathcal{L}$. The definition of a *Jordan derivation* is self-explanatory.

There is a simple explanation for the terminology introduced above. Lie subrings of associative rings are examples of Lie rings, and Jordan subrings are examples of Jordan rings. These are defined as follows. A nonassociative ring (resp. algebra) $\mathcal{L}$ with multiplication $(x, y) \mapsto \{x, y\}$ is called a *Lie ring* (resp. *Lie algebra*) if $\{x, x\} = 0$ for all $x \in \mathcal{L}$ and $\{x, \{y, z\}\} + \{y, \{z, x\}\} + \{z, \{x, y\}\} = 0$ for all $x, y, z \in \mathcal{L}$. The latter identity is called the *Jacobi identity*. A nonassociative ring (resp. algebra) $\mathcal{J}$ with multiplication $(x, y) \mapsto x \cdot y$ is called a *Jordan ring* (resp. *Jordan algebra*) if $x \cdot y = y \cdot x$ and $(x^2 \cdot y) \cdot x = x^2 \cdot (y \cdot x)$ for all $x, y \in \mathcal{J}$ (here, of course, x^2 stands for $x \cdot x$). One can verify that every associative ring can be turned into a Lie ring (resp. Jordan ring) by defining a new product as a Lie product $[x, y] = xy - yx$ (resp. Jordan product $x \circ y = xy + yx$). Actually, even in an abstract Lie algebra it is more common to denote the product by $[.\,,.]$, rather than by $\{.\,,.\}$. However, we introduced this notation to avoid confusion, so $[.\,,.]$ will always denote the Lie product arising from an associative ring.

The reader should be familiar with the process of *linearization*. In a loose manner this can be described as obtaining new relations from a given relation by replacing, iteratively if necessary, an arbitrary element satisfying this relation by the sum of two arbitrary elements. Let us explain this by an example. The reader shall easily find out when the method presented in this example also works in some related situations. Let $\mathcal{G}$ and $\mathcal{H}$ be additive groups and let $B : \mathcal{G}^n = \mathcal{G} \times \mathcal{G} \times \ldots \times \mathcal{G} \to \mathcal{H}$ be an n-*additive map*, that is, a map additive in each argument. The map $x \mapsto B(x, x, \ldots, x)$ will be called the *trace* of B. Suppose that the trace of B is zero. We claim that in this case

$$\sum_{\pi \in S_n} B(x_{\pi(1)}, x_{\pi(2)}, \ldots, x_{\pi(n)}) = 0$$

for all $x_1 \ldots, x_n \in \mathcal{G}$ where S_n denotes the symmetric group of order n. One usually says that the latter identity is obtained by *linearizing* $B(x, x, \ldots, x) = 0$ for all $x \in \mathcal{G}$. When $n = 2$, that is when B is a *biadditive map*, we get $B(x_1, x_2) + B(x_2, x_1) = B(x_1 + x_2, x_1 + x_2) - B(x_1, x_1) - B(x_2, x_2) = 0$ immediately. Now consider the case when $n = 3$. Replacing x by $x_1 + x_2$ in $B(x, x, x) = 0$ we get

$$B(x_1, x_1, x_2) + B(x_1, x_2, x_1) + B(x_2, x_1, x_1)$$
$$+ B(x_2, x_2, x_1) + B(x_2, x_1, x_2) + B(x_1, x_2, x_2) = 0$$

for all $x_1, x_2 \in \mathcal{G}$. Incidentally we remark that substituting x_2 for $-x_2$ in this last identity yields $B(x_1, x_1, x_2) + B(x_1, x_2, x_1) + B(x_2, x_1, x_1) = 0$ provided that $\mathcal{H}$ is 2-torsion free. In any case, however, replacing x_2 by $x_2 + x_3$ we arrive at $\sum_{\pi \in S_3} B(x_{\pi(1)}, x_{\pi(2)}, x_{\pi(3)}) = 0$ for all $x_1, x_2, x_3 \in \mathcal{G}$, proving our claim. The idea of this argument still works for any $n > 3$, but writing the proof in detail is rather tedious and we omit it.

1.2   Simple Examples of Functional Identities

We shall use the abbreviation FI for a functional identity.

Let $\mathcal{A}$ be a ring and let $E, F : \mathcal{A} \to \mathcal{A}$ be functions such that

$$E(x)y + F(y)x = 0 \quad \text{for all } x, y \in \mathcal{A}. \tag{1.1}$$

This is a very basic example of an FI. Let us stress that we do not assume any further conditions on functions E and F besides that they satisfy (1.1). The FI theory deals with set-theoretic functions satisfying certain identities. These functions should be considered as "unknowns".

A trivial possibility when (1.1) is fulfilled is when $E = F = 0$. Are there any other possibilities? If $\mathcal{A}$ is commutative, then there certainly are: just take, for example, E to be the identity map and $F = -E$. More generally, suppose that $\mathcal{A}$ contains a nonzero central ideal $\mathcal{I}$, i.e., an ideal contained in $\mathcal{Z}(\mathcal{A})$. Given any $c \in \mathcal{I}$ we can define E and F by $E(x) = -F(x) = cx$ and (1.1) is fulfilled. Now let $\mathcal{A}$ be any ring. Note that (1.1) implies

$$(E(x)yz)w = -F(yz)xw = (E(xw)y)z = -(F(y)x)wz = E(x)ywz$$

for all $x, y, z, w \in \mathcal{A}$. That is,

$$E(\mathcal{A})\mathcal{A}[\mathcal{A}, \mathcal{A}] = 0.$$

If $\mathcal{A}$ is prime and noncommutative, it follows immediately that $E = 0$. In that case also $F = 0$ by (1.1). Now suppose that $\mathcal{A}$ is semiprime and $E \neq 0$. Let $\mathcal{I} = (E(\mathcal{A}))$, i.e., $\mathcal{I}$ is the ideal generated by the range of E. Note that $[\mathcal{I}, \mathcal{A}]\mathcal{A}[\mathcal{I}, \mathcal{A}] = 0$ and hence $[\mathcal{I}, \mathcal{A}] = 0$ by the semiprimeness of $\mathcal{A}$. That is, $\mathcal{I}$ is a central ideal of $\mathcal{A}$. Finally, suppose that $\mathcal{A} = M_n(\mathcal{C})$, $n \geq 1$, where $\mathcal{C}$ is a commutative unital ring. If $n \geq 2$, then it is easy to see that $([\mathcal{A}, \mathcal{A}])$, the ideal generated by $[\mathcal{A}, \mathcal{A}]$, contains the identity matrix, which in turn implies that $E = 0$, and hence also $F = 0$.

One way of looking at functional identities is that these are equations with functions appearing as unknowns. Accordingly, we will often refer to functions satisfying a certain FI as solutions of that FI. We shall say that $E = F = 0$ is the *standard solution* of the FI (1.1). We have seen that the existence of a nonstandard solution implies that the ring is very special. More precisely, we can summarize our observations as follows:

Example 1.1. Let $\mathcal{A}$ be a semiprime (resp. prime) ring. Then there exists a nonstandard solution of (1.1) if and only if $\mathcal{A}$ contains a nonzero central ideal (resp. $\mathcal{A}$ is commutative). Further, if $\mathcal{A} = M_n(\mathcal{C})$ where $\mathcal{C}$ is a commutative unital ring, then there exists a nonstandard solution of (1.1) if and only if $n = 1$.

The next FI which we are going to consider is only slightly more complicated than (1.1). Instead of requiring that the expression $E(x)y + F(y)x$ is always zero we assume that it is central, that is, we consider the FI

$$E(x)y + F(y)x \in \mathcal{Z} = \mathcal{Z}(\mathcal{A}) \quad \text{for all } x, y \in \mathcal{A}. \tag{1.2}$$

One might say that (1.2) does not look like an identity, but rather as an inclusion of one set in another; anyway, we can rewrite (1.2) as $[E(x)y + F(y)x, z] = 0$ for all $x, y, z \in \mathcal{A}$ which is certainly an identity.

As in the previous case, we define $E = F = 0$ to be the standard solution of (1.2). However, nonstandard solutions of (1.2) exist not only in any commutative ring $\mathcal{C}$, but also in the ring $\mathcal{A} = M_2(\mathcal{C})$. Indeed, define $E, F : \mathcal{A} \to \mathcal{A}$ by $E(x) = F(x) = x - \text{tr}(x)1$, where $\text{tr}(x)$ denotes the trace of x and 1 is the identity matrix, and check that this gives a nonstandard solution of (1.2). On the other hand, if $\mathcal{A} = M_n(\mathcal{C})$ where $n \geq 3$ and $\mathcal{C}$ is a commutative unital ring, then (1.2) has only the standard solution. Indeed, set $\pi(x, y) = E(x)y + F(y)x$ for $x, y \in \mathcal{A}$. Our assumption is that $\pi(x, y)$ lies in $\mathcal{Z}$. Therefore, for any $x, y, t \in \mathcal{A}$ the element $\pi(xt, y) - \pi(x, y)t = E(xt)y - E(x)yt$ lies in $\mathcal{C}(t)$. Therefore, commuting it with t we get

$$-E(x)yt^2 + (tE(x) + E(xt))yt - tE(xt)y = 0.$$

Let $1 \leq i \leq n$, set $t = e_{12} + e_{23}$, $y = e_{i1}$ in the last relation, and then multiply the identity so obtained from the right by e_{3i}. Hence we arrive at $E(x)e_{ii} = 0$ for all $x \in \mathcal{A}$, $1 \leq i \leq n$. This clearly yields $E = 0$, and hence also $F = 0$. Now we can state

Example 1.2. Let $\mathcal{A} = M_n(\mathcal{C})$ where $\mathcal{C}$ is a commutative unital ring. Then there exists a nonstandard solution of (1.2) if and only if $n \leq 2$.

How does one proceed from Examples 1.1 and 1.2? In the next step it seems natural to seek FI's where the ring $M_3(\mathcal{C})$ would also play an exceptional role. Let us first reveal that the example in $M_2(\mathcal{C})$ was derived from the Cayley–Hamilton theorem. The reader might be familiar with this theorem only for matrices over (some) fields, so let us point out that *"Every matrix is a zero of its characteristic polynomial"* is still true for matrices over any commutative ring $\mathcal{C}$ (see e.g.,

[187, p. 18]). Thus, for every matrix $x \in \mathcal{A} = M_n(\mathcal{C})$ we have

$$x^n + \alpha_1(x)x^{n-1} + \alpha_2(x)x^{n-2} + \ldots + \alpha_{n-1}(x)x + \alpha_n(x)1 = 0$$

where $\alpha_i(x) \in \mathcal{C}$, $i = 1, \ldots, n$, are coefficients of the characteristic polynomial $\det(\lambda 1 - x)$. These coefficients can be expressed by Newton's formulas, but for our purposes it suffices to consider them in a less precise manner. Expanding the characteristic polynomial we first see that $\alpha_1(x) = -tr(x)$ and so $\alpha_1 : \mathcal{A} \to \mathcal{C}$ is an additive map (actually, even a $\mathcal{C}$-linear map, but this is not important for our immediate goal). Secondly, $\alpha_2(x)$ can be expressed as the sum of the terms each of which is of the form $\pm x_{i_1 i_2} x_{j_1 j_2}$, where x_{ij}'s are entries of x. But then $\alpha_2 : \mathcal{A} \to \mathcal{C}$ is the trace of a biadditive map $\zeta_2 : \mathcal{A}^2 \to \mathcal{C}$. Indeed, one can define $\zeta_2(x, y)$ in the obvious way, as the sum of the terms of the form $\pm x_{i_1 i_2} y_{j_1 j_2}$. Further, the expression for $\alpha_3(x)$ consists of summands of the form $\pm x_{i_1 i_2} x_{j_1 j_2} x_{k_1 k_2}$, and so $\alpha_3 : \mathcal{A} \to \mathcal{C}$ is the trace of a 3-additive map $\zeta_3 : \mathcal{A}^3 \to \mathcal{C}$. In this way we find out that the Cayley–Hamilton theorem implies that

$$x^n + \zeta_1(x)x^{n-1} + \zeta_2(x, x)x^{n-2} + \ldots + \zeta_{n-1}(x, x, \ldots, x)x + \zeta_n(x, \ldots, x)1 = 0$$

for all $x \in \mathcal{A}$, where each $\zeta_k : \mathcal{A}^k \to \mathcal{C}$ is a k-additive map. The last term is a scalar matrix (incidentally, it is equal to $(-1)^n \det(x)1$), and so we have

$$x^n + \zeta_1(x)x^{n-1} + \zeta_2(x, x)x^{n-2} + \ldots + \zeta_{n-1}(x, \ldots, x)x \in \mathcal{Z}$$

(note that the center $\mathcal{Z}$ of $\mathcal{A}$ is equal to $\mathcal{C}1$). For $n = 2$ this reads as $x^2 + \zeta_1(x)x \in \mathcal{Z}$. A linearization gives rise to a nonstandard solution of (1.2) mentioned above. Now let $n = 3$. Then a linearization gives

$$E(x, y)z + E(x, z)y + E(y, z)x \in \mathcal{Z}$$

for all $x, y, z \in \mathcal{A}$, where $E : \mathcal{A}^2 \to \mathcal{A}$ is given by

$$E(x, y) = (x + \zeta_1(x))y + (y + \zeta_1(y))x + (\zeta_2(x, y) + \zeta_2(y, x))1. \qquad (1.3)$$

Note that $E \neq 0$ (after all, this also follows from Example 1.2).

The conclusion from the above paragraph can be interpreted as follows. The FI

$$E(x, y)z + F(x, z)y + G(y, z)x \in \mathcal{Z} \qquad \text{for all } x, y, z \in \mathcal{A} \qquad (1.4)$$

has a nonstandard solution on $\mathcal{A} = M_3(\mathcal{C})$, namely, the one where $E = F = G$ is the function defined according to (1.3). By the standard solution of (1.4) we of course mean that $E = F = G = 0$. It is clear from the above discussion that nonstandard solutions of (1.4) also exist in $M_n(\mathcal{C})$ for $n = 1, 2$.

Let us, for a while, treat (1.4) in any ring $\mathcal{A}$. It should already be self-explanatory that $E, F, G : \mathcal{A}^2 \to \mathcal{A}$ are considered to be arbitrary functions. Set $\pi(x, y, z) = E(x, y)z + F(x, z)y + G(y, z)x \in \mathcal{Z}$. A direct computation shows that

$$\pi(xt, yt, z) - \pi(xt, y, z)t - \pi(x, yt, z)t + \pi(x, y, z)t^2$$
$$= E(xt, yt)z - (E(xt, y) + E(x, yt))zt + E(x, y)zt^2.$$

Obviously, this element lies in $\mathcal{C}(t)$ for any $t \in \mathcal{A}$. Therefore, commuting it with t we arrive at

$$E(x,y)zt^3 + E_1(x,y,t)zt^2 + E_2(x,y,t)zt + E_3(x,y,t)z = 0$$

for all $x, y, z, t \in \mathcal{A}$. Here, the E_i's are some new functions which can be easily expressed in terms of E; however, we omit this calculation since it is of no importance for the proceeding discussion.

Assume now that $\mathcal{A} = M_n(\mathcal{C})$ with $n \geq 4$ and set $t = e_{12} + e_{23} + e_{34}$ and $z = e_{i1}$ in the last relation, and then multiply the identity so obtained from the right by e_{4i}. Then we get $E(x,y)e_{ii} = 0$ for all $x, y \in \mathcal{A}$, $1 \leq i \leq n$, so that $E = 0$. In a similar fashion we show that $F = 0$ and $G = 0$. Thus, we have established

Example 1.3. Let $\mathcal{A} = M_n(\mathcal{C})$ where $\mathcal{C}$ is a commutative unital ring. Then there exists a nonstandard solution of (1.4) if and only if $n \leq 3$.

Let us mention that appropriate modifications of the above arguments give similar results in rings quite different from the matrix rings. But to show this we need to introduce some other concepts which will be done in subsequent chapters.

Now the reader can probably guess how the examples given so far could be unified and extended. Nevertheless, we stop this line of investigation at this point since we have no intention of proving general statements in this chapter. Instead we consider a different kind of an extension of the basic FI (1.1). Let a be a fixed nonzero element in a ring $\mathcal{A}$ and consider the identity

$$E(x)ya + F(y)xa = 0 \quad \text{for all } x, y \in \mathcal{A}. \tag{1.5}$$

So, besides involving functions this identity also involves a fixed element of the ring. Such identities will be called *generalized functional identities*. The reasons for this terminology will become more clear later. Again we say that $E = F = 0$ is the standard solution of (1.5). In the case when $a = 1$, or equivalently when a is invertible, (1.5) has already been treated in Example 1.1 and we have seen that nonstandard solutions can exist only very exceptionally. In the case when a is not invertible, or better when a is "far" from being invertible, this is no longer true. For instance, for any $n \in \mathbb{N}$ there exists a nonstandard solution in $\mathcal{A} = M_n(\mathcal{C})$ given by $a = e_{11}$, $E(x) = -F(x) = x_{11}e_{11}$. More generally, if a is an element in a ring $\mathcal{A}$ such that

$$a\mathcal{A}a = \mathcal{Z}a \neq 0, \tag{1.6}$$

then one can verify that $E(x) = -F(x) = axa$ gives a nonstandard solution of (1.5). Now assume that $\mathcal{A}$ is a simple unital ring and let us prove a kind of a converse to these examples. Assuming (1.5) we have, on the one hand, $E(x)yaza = -F(yaz)xa$, and on the other hand, $E(x)yaza = -F(y)xaza$. Hence

$$F(yaz)xa = F(y)xaza \quad \text{for all } x, y, z \in \mathcal{A}. \tag{1.7}$$

Suppose that $a \neq 0$ and $F(y_0) \neq 0$ for some $y_0 \in \mathcal{A}$. Since $\mathcal{A}$ is simple, there are $x_i, y_i \in \mathcal{A}$ such that $\sum_i x_i F(y_0) y_i = 1$. Consequently, using (1.7) we get

$$xaza = \sum_i x_i (F(y_0) y_i xaza) = \sum_i x_i F(y_0 az) y_i xa \quad \text{for all } x, z \in \mathcal{A}.$$

This shows that for any $z \in \mathcal{A}$ there is $c = \sum_i x_i F(y_0 az) y_i \in \mathcal{A}$ such that $xaza = cxa$ for all $x \in \mathcal{A}$. This implies that $cyxa = yxaza = ycxa$ for all $x, y \in \mathcal{A}$, that is, $[c, \mathcal{A}]\mathcal{A}a = 0$ which clearly yields $c \in \mathcal{Z}$. But then $aza = ca$. Thus we proved $a\mathcal{A}a \subseteq \mathcal{Z}a$. But then actually (1.6) holds true since $\mathcal{Z}$, as the center of a unital simple ring, is a field (indeed, just note that for every central element c the set $c\mathcal{A}$ is an ideal).

Which are the simple unital rings $\mathcal{A}$ that contain elements a satisfying (1.6)? As already indicated, $M_n(\mathcal{C})$ with $\mathcal{C}$ a field is such an example: even any matrix unit satisfies (1.6). In fact, matrix rings are the only examples. This is well-known; let us show this by using the Jacobson density theorem. So assume that $a \in \mathcal{A}$ satisfies (1.6). We claim that $\mathcal{A}a$ is a *minimal left ideal* of $\mathcal{A}$. This means that $\mathcal{A}a$ does not properly contain any nonzero left ideal of $\mathcal{A}$. Indeed, assume that $\mathcal{L}$ is a nonzero left ideal of $\mathcal{A}$ such that $\mathcal{L} \subseteq \mathcal{A}a$. Then $0 \neq a\mathcal{L} \subseteq a\mathcal{A}a = \mathcal{Z}a$, whence $0 \neq au = ca$ for some $u \in \mathcal{L}$, $c \in \mathcal{Z}$. Thus $a = c^{-1}au \in \mathcal{L}$ and so $\mathcal{L} = \mathcal{A}a$, proving our claim. Now regard $\mathcal{A}a$ as a left $\mathcal{A}$-module. To each element $x \in \mathcal{A}$ we can associate a $\mathcal{Z}$-linear operator $L_x : \mathcal{A}a \to \mathcal{A}a$ given by $L_x(ya) = xya$. Since $\mathcal{A}$ is simple, the map $x \mapsto L_x$ is an embedding of the ring $\mathcal{A}$ into the ring $\mathrm{End}_{\mathcal{Z}}(\mathcal{A}a)$. We may therefore view $\mathcal{A}$ as a subring of $\mathrm{End}_{\mathcal{Z}}(\mathcal{A}a)$. Since $\mathcal{A}a$ is minimal as a left ideal of $\mathcal{A}$ it follows that it is simple as an $\mathcal{A}$-module. Pick $f \in \mathrm{End}_{\mathcal{A}}(\mathcal{A}a)$. In view of (1.6) there is $b \in \mathcal{A}$ such that $aba = a$, so that $f(a) = f(aba) = abf(a) \subseteq a\mathcal{A}a$. Therefore, $f(a) = ca$ for some $c \in \mathcal{Z}$ and hence $f(xa) = xf(a) = cxa$ for every $x \in \mathcal{A}$. That is, every $f \in \mathrm{End}_{\mathcal{A}}(\mathcal{A}a)$ is a multiplication with an element from $\mathcal{Z}$. Conversely, every such multiplication is an element of $\mathrm{End}_{\mathcal{A}}(\mathcal{A}a)$. In this way we can therefore identify $\mathrm{End}_{\mathcal{A}}(\mathcal{A}a)$ with $\mathcal{Z}$. By the Jacobson density theorem the ring $\mathcal{A}$ acts densely on the vector space $_{\mathcal{Z}}(\mathcal{A}a)$. We claim that $_{\mathcal{Z}}(\mathcal{A}a)$ is finite dimensional. Since $\mathcal{A}$ is simple and unital, there are $x_i, y_i \in \mathcal{A}$ such that $\sum_i x_i a y_i = 1$. Using (1.6) it follows that

$$\mathcal{A}a = \left(\sum_i x_i a y_i \right) \mathcal{A}a \subseteq \sum_i \mathcal{Z} x_i a$$

which proves our claim. Because of the dense action $\mathcal{A}$ must therefore contain all linear operators on $\mathcal{A}a$. Accordingly, $\mathcal{A}$ is isomorphic to $M_n(\mathcal{Z})$ where n is the dimension of $\mathcal{A}a$ over $\mathcal{Z}$. We can now state

Example 1.4. Let $\mathcal{A}$ be a simple unital ring. Then there exists $a \in \mathcal{A}$ such that (1.5) has a nonstandard solution if and only if $\mathcal{A} \cong M_n(\mathcal{Z})$ for some $n \in \mathbb{N}$ and some field $\mathcal{Z}$.

We continue with an FI whose standard solutions are not as trivial as in the previous cases. Let $F : \mathcal{A} \to \mathcal{A}$ be such that

$$F(x)y + F(y)x = yF(x) + xF(y) \quad \text{for all } x, y \in \mathcal{A}. \tag{1.8}$$

Obvious examples are the identity map and every map with its range in $\mathcal{Z}$. Accordingly, every map of the form

$$F(x) = \lambda x + \mu(x), \quad \lambda \in \mathcal{Z}, \quad \mu : \mathcal{A} \to \mathcal{Z} \tag{1.9}$$

is a solution of (1.8), and these are the solutions that will be called standard in the present case.

The FI (1.8) is closely related to the notion of a commuting map which will be often considered in the sequel. A map F is said to to be *commuting* on a set $\mathcal{S} \subseteq \mathcal{A}$ if $[F(x), x] = 0$ for all $x \in \mathcal{S}$. Incidentally we also mention a somewhat more general notion of a map that is *centralizing* on $\mathcal{S}$; by definition such a map satisfies $[F(x), x] \in \mathcal{Z}$ for all $x \in \mathcal{S}$. If F is additive and commuting on $\mathcal{A}$, then linearizing $[F(x), x] = 0$ we get (1.8).

The most convenient way to consider (1.8) is via the following concept: A biadditive map $\Delta : \mathcal{A}^2 \to \mathcal{A}$ is called a *biderivation* if it is a derivation in each argument, that is, for every $y \in \mathcal{A}$ the maps $x \mapsto \Delta(x, y)$ and $x \mapsto \Delta(y, x)$ are derivations. For example, for any $\lambda \in \mathcal{Z}$, $(x, y) \mapsto \lambda[x, y]$ is a biderivation. We shall call such maps *inner biderivations*. It is easy to construct non-inner biderivations on commutative rings, say, take the ring of polynomials over a field and define $\Delta(f, g) = f'g'$ where f' is the formal derivative of the polynomial f. In noncommutative rings, however, it happens quite often that all biderivations are inner. In such rings (1.8) can have only a standard solution. Namely, (1.8) can be written as $[F(x), y] = [x, F(y)]$. This implies that the map $\Delta(x, y) = [F(x), y]$ is (even an inner) derivation in each argument, and so it is a biderivation. Therefore, by assumption there exists $\lambda \in \mathcal{Z}$ such that $[F(x), y] = \lambda[x, y] = [\lambda x, y]$ for all $x, y \in \mathcal{Z}$, meaning that $\mu(x) = F(x) - \lambda x$ lies in $\mathcal{Z}$.

So let us consider an arbitrary biderivation Δ on a ring $\mathcal{A}$. Since Δ is a derivation in the first argument, we have

$$\Delta(xu, yv) = \Delta(x, yv)u + x\Delta(u, yv),$$

and since it is also a derivation in the second argument it follows that

$$\Delta(xu, yv) = \Delta(x, y)vu + y\Delta(x, v)u + x\Delta(u, y)v + xy\Delta(u, v).$$

On the other hand, first using the derivation law in the second and after that in the first argument we get

$$\Delta(xu, yv) = \Delta(xu, y)v + y\Delta(xu, v)$$
$$= \Delta(x, y)uv + x\Delta(u, y)v + y\Delta(x, v)u + yx\Delta(u, v).$$

Comparing both relations we obtain $\Delta(x, y)[u, v] = [x, y]\Delta(u, v)$ for all $x, y, u, v \in \mathcal{A}$. Replacing v by zv and using $[u, zv] = [u, z]v + z[u, v]$, $\Delta(u, zv) = \Delta(u, z)v + z\Delta(u, v)$ we obtain

$$\Delta(x, y)z[u, v] = [x, y]z\Delta(u, v)$$

for all $x, y, z, u, v \in \mathcal{A}$. This is the crucial identity. To illustrate its utility, assume for example that $\mathcal{A}$ is unital and $([\mathcal{A}, \mathcal{A}]) = \mathcal{A}$. Thus, there are $z_i, u_i, v_i, w_i \in \mathcal{A}$ such that $\sum_i z_i [u_i, v_i] w_i = 1$, hence

$$\Delta(x, y) = \sum_i \Delta(x, y) z_i [u_i, v_i] w_i = \sum_i [x, y] z_i \Delta(u_i, v_i) w_i.$$

That is, $\Delta(x, y) = [x, y]\lambda$ for all $x, y \in \mathcal{A}$ where $\lambda = \sum_i z_i \Delta(u_i, v_i) w_i \in \mathcal{A}$. We claim that $\lambda \in \mathcal{Z}$. Indeed, we have

$$[x, y]z\lambda + y[x, z]\lambda = [x, yz]\lambda = \Delta(x, yz)$$
$$= \Delta(x, y)z + y\Delta(x, z) = [x, y]\lambda z + y[x, z]\lambda,$$

showing that $[x, y][z, \lambda] = 0$ for all $x, y, z \in \mathcal{A}$. Replacing z by zw and using $[zw, \lambda] = [z, \lambda]w + z[w, \lambda]$ it follows that $[\mathcal{A}, \mathcal{A}]\mathcal{A}[\lambda, \mathcal{A}] = 0$, which in turn implies $\lambda \in \mathcal{Z}$ in view of our assumption.

Example 1.5. Let $\mathcal{A}$ be a unital ring such that $([\mathcal{A}, \mathcal{A}]) = \mathcal{A}$. Then every biderivation on $\mathcal{A}$ is inner. Consequently, every solution of the FI (1.8) is a standard solution (1.9). In particular, every additive commuting map on $\mathcal{A}$ is of the form (1.9).

For instance, in a simple ring $\mathcal{A}$, or in the ring $\mathcal{A} = M_n(\mathcal{C})$ with $\mathcal{C}$ a commutative unital ring, we can now say that (1.8) has only standard solutions. Here one does not exclude the case when $\mathcal{A}$ is commutative because every map on a commutative ring is trivially of the form (1.9).

In our final example we consider a generalized FI

$$E(x)ya = axF(y) \quad \text{for all } x, y \in \mathcal{A}. \tag{1.10}$$

Here of course a is a fixed element in a ring $\mathcal{A}$ and $E, F : \mathcal{A} \to \mathcal{A}$. Suppose for a moment that a is invertible. Setting $y = a^{-1}$ in (1.10) we then get $E(x) = axq$ where $q = F(a^{-1}) \in \mathcal{A}$. This yields $ax(F(y) - qya) = 0$ and hence $F(y) = qya$. So we see which solution we can naturally expect. We shall say

$$E(x) = axq, \quad F(y) = qya, \tag{1.11}$$

where $q \in \mathcal{A}$, is a standard solution of (1.10). Let us modify the above argument to show that (1.10) has only standard solutions also under the milder assumption that $\mathcal{A}$ is unital and $(a) = \mathcal{A}$. So let $\sum_i x_i a y_i = 1$ hold true for some $x_i, y_i \in \mathcal{A}$. Then we have $E(x) = \sum_i E(x)x_i a y_i = \sum_i axF(x_i)y_i = axq$ where $q = \sum_i F(x_i)y_i \in \mathcal{A}$. As above this implies $F(y) = qya$, so we can state

Example 1.6. Let $\mathcal{A}$ be a unital ring. Suppose that $a \in \mathcal{A}$ is such that $(a) = \mathcal{A}$. Then every solution of (1.10) is a standard solution (1.11). In particular, in a simple unital ring (1.10) has only standard solutions provided that $a \neq 0$.

Now consider (1.10) in the ring $\mathbb{Z}$ and let, for example, $a = 2$. Taking E and F to be both identity maps we certainly get a solution of (1.10). However, there is no

$q \in \mathbb{Z}$ satisfying (1.11). On the other hand, if we change slightly the definition of a standard solution by allowing that q lies in $\mathbb{Q}$, then such q certainly exists: $q = \frac{1}{2}$. Though trivial, this example already indicates that it is sometimes appropriate to involve rings bigger than the original ring when studying functional identities. In Parts II and III we shall indeed deal with such rings. In Part I, however, we confine our attention to rings such that the presence of another bigger ring can be avoided.

1.3 Basic Concepts

Examples from the preceding section indicate how one might investigate FI's. One first has to find all "obvious" solutions of a given FI, that is, the solutions which do not depend on some structural properties of the ring but are merely consequences of a formal calculation. Such solutions are called standard solutions. The eventual existence of a nonstandard solution usually implies that the ring has a very special structure. So, intuitively it should already be clear what functional identities are about. But we still have to place these observations in some general framework.

Let us first explain what we formally mean by a (generalized) functional identity. The definitions that we shall give are very general and admittedly will not have much practical meaning in this book. We just want to point out what is the "home" of the theory in the broadest sense. The meaning of the definition is thus more or less informative, and besides, it just does not seem right in mathematics to speak about notions without making precise what they are. However, if one is unfamiliar with some of the notions that will be mentioned in the next paragraphs, then one may simply ignore them since their understanding is not of essential importance for the book.

Some of the readers may have already guessed that FI's could be viewed as generalizations of polynomial identities (PI's). Let us give a concrete hint. A ring $\mathcal{A}$ satisfies a multilinear PI of degree 3 if $\sum_{\pi \in S_3} n_\pi x_{\pi(1)} x_{\pi(2)} x_{\pi(3)} = 0$ for all $x_1, x_2, x_3 \in \mathcal{A}$, where the n_π's are integers, with at least one n_π equal to 1, and S_3 is the symmetric group of order 3. This can be written as $E(x_1, x_2)x_3 + F(x_1, x_3)x_2 + G(x_2, x_3)x_1 = 0$ where $E(x_1, x_2) = n_1 x_1 x_2 + n_{(12)} x_2 x_1$, and F and G are defined analogously. But this is just a special case of the FI (1.4) where E, F, G are supposed to be just entirely arbitrary functions. This example also illustrates the earlier remark that nonstandard solutions usually imply that the ring is of a special nature: here the standard solution should just be the zero functions and PI-rings are of a special nature.

We shall define the concept of an FI by modifying appropriately the definition of a PI. Let $\mathcal{A}$ be a ring, let $\mathcal{R}$ be a nonempty subset of $\mathcal{A}$, and let $F_i : \mathcal{R}^m \to \mathcal{A}$, $i = 1, \dots, n$, be functions. Actually, we could assume more generally that the F_i's map $\mathcal{R}^m$ into a ring containing $\mathcal{A}$, or even into an $(\mathcal{A}, \mathcal{A})$-bimodule, but let us not bother about various possible generalizations right now. Next, let $X = \{x_1, y_1, x_2, y_2, \dots\}$ be a countable set, let $\mathbb{Z}\langle X \rangle$ be the free algebra on X over $\mathbb{Z}$,

and let
$$f = f(x_1, \ldots, x_m, y_1, \ldots, y_n) \in \mathbb{Z}\langle X \rangle, \quad m \geq 1, n \geq 0$$
be a polynomial such that at least one of its monomials of highest degree has coefficient 1. We shall say that f is a *functional identity on* $\mathcal{R}$ *with functions* $F_1, \ldots, F_n$ if
$$f(r_1, \ldots, r_m, F_1(r_1, \ldots, r_m), \ldots, F_n(r_1, \ldots, r_m)) = 0$$
for all $r_1, \ldots, r_m \in \mathcal{R}$. In this case we shall also say that the functions $F_1, \ldots, F_n$ are *solutions* of this functional identity.

For instance, the identity (1.4) is clearly equivalent to $[E(x, y)z + F(x, z)y + G(y, z)x, u] = 0$ for all $x, y, z, u \in \mathcal{A}$, and so it can now be described by saying that
$$f(x_1, x_2, x_3, x_4, y_1, y_2, y_3) = [y_1 x_1 + y_2 x_2 + y_3 x_3, x_4]$$
is an FI on $\mathcal{A}$ with functions F_1, F_2, F_3 defined in the obvious way, that is, so that F_i does not depend on the i-th variable (e.g., $F_1(a_1, a_2, a_3) = G(a_2, a_3)$ etc.). Similarly, the identity (1.8) can be expressed via the polynomial $[y_1, x_1] + [y_2, x_2]$ and functions F_1, F_2 given by $F_1(x_1, x_2) = F(x_2)$, $F_2(x_1, x_2) = F(x_1)$, while saying that a map F is commuting is the same as saying that $[y_1, x_1]$ is an FI on $\mathcal{A}$ with the map F.

What is the connection between FI's and PI's? One could, of course, say that a PI is just a special case of an FI with $n = 0$ (i.e., there are no functions) in the definition of an FI. But this superficial remark says nothing about the substantive connection between FI's and PI's such as the one hinted at in the example on multilinear PI's of degree 3. Strictly speaking the notion of a PI is not quite comparable with that of an FI: a PI is an element of the free algebra whereas the notion of an FI involves functions. But, this small technicality aside, any PI can be shown to be in essence an FI as we now proceed to indicate. Let $\mathcal{R}$ be a nonempty subset of a ring $\mathcal{A}$ and let $g = g(x_1, \ldots, x_m) \in \mathbb{Z}\langle X \rangle$ be a PI on $\mathcal{R}$. We write $g = f_1 x_1 + \ldots + f_m x_m$, $f_i = f_i(x_1, \ldots, x_m)$, and let F_i be the polynomial function $\mathcal{R}^m \to \mathcal{A}$ determined by f_i, $i = 1, \ldots, m$. Then $f = y_1 x_1 + \ldots + y_m x_m$ with functions $F_1, \ldots, F_m$ is the appropriate FI on $\mathcal{R}$. In this sense the PI g can be viewed as the FI f. Of course there are other ways of decomposing g (e.g., $g = x_1 h_1 + \ldots + x_m h_m$) which could equally well have been used.

Similarly we define a *generalized functional identity* (GFI). Actually, the definition is literally the same as that of an FI with one change: we require that f, instead of belonging to $\mathbb{Z}\langle X \rangle$, is an element of the coproduct of $\mathcal{A}$ and $\mathbb{Z}\langle X \rangle$ (when $n = 0$, this definition basically reduces to the definition of a generalized polynomial identity (GPI), cf. appendix D). For example, (1.10) can be expressed as that $f(x_1, x_2, y_1, y_2) = y_2 x_2 a - a x_1 y_1$ is a GFI on $\mathcal{A}$ with maps F_1, F_2 given by $F_1(x_1, x_2) = F(x_2)$, $F_2(x_1, x_2) = F(x_1)$.

Anyway, we shall use the phrases a "functional identity" and a "generalized functional identity" rather informally in this book. Roughly speaking, those identities that in addition to functions also involve some fixed elements in a ring (such as (1.5) and (1.10)) should be in principle called generalized functional identities. As already indicated, the reason for this terminology is that the concept of an FI extends the concept of a PI, while the concept of a GFI extends the concept of a GPI. However, while the theories of PI's and GPI's are quite different and to some extent independent of each other, the goals, the methods and the results of the theories of FI's and GFI's are very similar. In fact, in order to obtain some result on FI's one is often forced to deal with certain GFI's first. We shall therefore use the adjective "generalized" only when we shall want to stress that they extend the concept of GPI's. On the other hand, for some identities that are closely related to FI's but are formally GFI's we shall rather use the term *identities with coefficients* where a coefficient refers to fixed ring elements appearing in the identity.

In contrast to PI theory one cannot expect in general that some (or any!) reasonable conclusions can be derived when considering just an arbitrary (G)FI from these definitions. Furthermore, it seems almost impossible (at this stage?) to classify those (G)FI's that "make some sense" and those that do not. We shall therefore confine ourselves to some special (G)FI's in this book, which themselves are already quite general. We shall now describe the most fundamental type of identities that shall be thoroughly analyzed. They correspond to the polynomial $\sum_i y_{1i}x_i + \sum_j x_j y_{2j}$. To describe them more precisely we first need to introduce some additional notation.

Let $m \in \mathbb{N}$. For elements $x_1, x_2, \ldots, x_m$ in a ring $\mathcal{A}$ we shall write

$$\overline{x}_m = (x_1, \ldots, x_m) \in \mathcal{A}^m.$$

Here of course $\mathcal{A}^m$ denotes the Cartesian product of m copies of $\mathcal{A}$. For convenience we also define $\mathcal{A}^0 = \{0\}$, i.e., $\mathcal{A}^0$ contains only the zero element of $\mathcal{A}$. Further, for any $1 \leq i \leq m$ we set

$$\overline{x}_m^i = (x_1, \ldots, x_{i-1}, x_{i+1}, \ldots, x_m) \in \mathcal{A}^{m-1},$$

and for $1 \leq i < j \leq m$ we set

$$\overline{x}_m^{ij} = \overline{x}_m^{ji} = (x_1, \ldots, x_{i-1}, x_{i+1}, \ldots, x_{j-1}, x_{j+1} \ldots, x_m) \in \mathcal{A}^{m-2}.$$

In particular, in view of our convention we have $\overline{x}_1^1 = \overline{x}_2^{12} = 0$.

Now let $\mathcal{I}$ and $\mathcal{J}$ be finite subsets of $\mathbb{N}$ and let $m \in \mathbb{N}$ be such that $\mathcal{I}, \mathcal{J} \subseteq \{1, \ldots, m\}$ (it may well be the case that $\mathcal{I} \cup \mathcal{J}$ is a proper subset of $\{1, \ldots, m\}$). Further, let $E_i, F_j : \mathcal{A}^{m-1} \to \mathcal{A}$, $i \in \mathcal{I}$, $j \in \mathcal{J}$, be arbitrary functions (a map defined on $\mathcal{A}^0 = \{0\}$ should be considered as a fixed element in $\mathcal{A}$) . We will be interested in FI's involving expressions

$$\sum_{i \in \mathcal{I}} E_i(\overline{x}_m^i)x_i = \sum_{i \in \mathcal{I}} E_i(x_1, \ldots, x_{i-1}, x_{i+1}, \ldots, x_m)x_i$$

and

$$\sum_{j \in \mathcal{J}} x_j F_j(\overline{x}_m^j) = \sum_{j \in \mathcal{J}} x_j F_j(x_1, \ldots, x_{j-1}, x_{j+1}, \ldots, x_m).$$

More precisely, the basic FI's we are going to consider are

$$\sum_{i \in \mathcal{I}} E_i(\overline{x}_m^i)x_i + \sum_{j \in \mathcal{J}} x_j F_j(\overline{x}_m^j) = 0 \quad \text{for all } \overline{x}_m \in \mathcal{A}^m, \tag{1.12}$$

$$\sum_{i \in \mathcal{I}} E_i(\overline{x}_m^i)x_i + \sum_{j \in \mathcal{J}} x_j F_j(\overline{x}_m^j) \in \mathcal{Z} \quad \text{for all } \overline{x}_m \in \mathcal{A}^m. \tag{1.13}$$

Here, of course, $\mathcal{Z} = \mathcal{Z}(\mathcal{A})$ is the center of $\mathcal{A}$. The case when $\mathcal{I}$ or $\mathcal{J}$ is $\emptyset$, the empty set, is not excluded. In such case it should be understood that *a sum over $\emptyset$ is 0*. So, for example, when $\mathcal{J} = \emptyset$ (1.12) reduces to

$$\sum_{i \in \mathcal{I}} E_i(\overline{x}_m^i)x_i = 0 \quad \text{for all } \overline{x}_m \in \mathcal{A}^m. \tag{1.14}$$

Examples of these basic FI's are provided in the previous section. For instance (1.8) is an example of (1.12) and (1.4) is an example of (1.13) (with $\mathcal{J} = \emptyset$). What can we say concerning the solutions (i.e., the functions E_i and F_j) of (1.12) and (1.13)? A perusal of the examples in the previous section indicates that there are "obvious" solutions (already called "standard") and sometimes special (i.e., "nonstandard") solutions due to the peculiarities of the particular ring in question.

Right now we want to focus further, albeit still in an informal fashion, on the meaning of standard solutions of (1.12) and (1.13), since this notion is central to the theme of this book. A rigorous definition will be given in the next chapter, but for now we simply want to get a better feel for the subject. The definition of a standard solution of (1.13) is the same as of that of (1.12); we shall discuss the situation concerning (1.13) a little later, since the reader might understandably wonder how the same (formal) solution can simultaneously satisfy (1.12) and (1.13)!

The reader should first ask the question: what form should the functions E_i and F_j have so that (1.12) will *always* be satisfied, no matter what the ring $\mathcal{A}$ is? Put another way, treating (1.12) in a purely formal way without regard to the ring in question, what can one always say about the form which E_i and F_j must assume?

The following special case of (1.12) is illustrative for our purposes, and our analysis of it should be a helpful stepping stone to making the reader feel at ease with the formal definition of a standard solution given in the next chapter. Let

$$E_1(x_2, x_3)x_1 + E_2(x_1, x_3)x_2 + x_2 F_2(x_1, x_3) + x_3 F_3(x_1, x_2) = 0 \tag{1.15}$$

for all $\overline{x}_3 \in \mathcal{A}^3$. Here $\mathcal{I} = \{1, 2\}$, $\mathcal{J} = \{2, 3\}$ and $m = 3$. The idea is to make E_1 and E_2 as general as the limitations imposed by (1.15) will allow, meanwhile showing that the natural choices are available for F_2 and F_3 so that (1.15) is satisfied. Certainly the function $x_2 p_{12}(x_3)$, where $p_{12} : \mathcal{A} \to \mathcal{A}$ is an arbitrary set-theoretic function should be included as a summand of E_1, since it can be "canceled" immediately by making sure that the function $-p_{12}(x_3)x_1$ is a summand of F_2. Similarly $x_3 p_{13}(x_2)$, where p_{13} is arbitrary, must be included as a summand of E_1 since it can be "canceled" by taking $-p_{13}(x_2)x_1$ as a summand of F_3. The reader already sees that E_2 must have $x_3 p_{23}(x_1)$ as a summand (with p_{23} arbitrary), since it can be canceled by letting F_3 have the summand $-p_{23}(x_1)x_2$. Furthermore, E_2 must have a central summand $\lambda_2(x_1, x_3)$, where $\lambda_2 : \mathcal{A}^2 \to \mathcal{Z}$ is an arbitrary set-theoretic map, since it can be balanced out by letting F_2 have the central summand $-\lambda_2(x_1, x_3)$. However, in this example E_1 should not have any nonzero central summand $\lambda_1 : \mathcal{A}^2 \to \mathcal{Z}$, since $\lambda_1(x_2, x_3)x_1$ cannot be canceled. In summary we have then shown that the functions

$$\begin{aligned}
E_1(x_2, x_3) &= x_2 p_{12}(x_3) + x_3 p_{13}(x_2), \\
E_2(x_1, x_3) &= x_3 p_{23}(x_1) + \lambda_2(x_1, x_3), \\
F_2(x_1, x_3) &= -p_{12}(x_3)x_1 - \lambda_2(x_1, x_3), \\
F_3(x_1, x_2) &= -p_{13}(x_2)x_1 - p_{23}(x_1)x_2
\end{aligned} \tag{1.16}$$

(where $p_{12}, p_{13}, p_{23} : \mathcal{A} \to \mathcal{A}$ and $\lambda_2 : \mathcal{A}^2 \to \mathcal{Z}$ are arbitrary) *always* form a solution of (1.15). We note that the central component λ_i only appears for $i \in \mathcal{I} \cap \mathcal{J}$ (in this example $\mathcal{I} \cap \mathcal{J} = \{2\}$). Accordingly (1.16) will be called a *standard solution of* (1.15).

Considering the FI (1.14) in a similar manner one notices that besides the trivial possibility when all maps are 0 there is no other natural choice that (1.14) is satisfied. Therefore we define that $E_i = 0$, $i \in \mathcal{I}$, is the *standard solution of* (1.14).

One notices that the functions E_i and F_j involved in the definition of the basic FI's (1.12) and (1.13) are allowed to be arbitrary set-theoretic maps from $\mathcal{A}^{m-1}$ to $\mathcal{A}$. There are in fact applications of the theory in which it is useful not to make further assumptions on these functions. However, for more complicated FI's which may not be "multilinear" in the sense that (1.12) and (1.13) are, it is usually necessary to impose the condition that the functions involved be multiadditive. In this way the linearization process may be invoked to transform the given FI to a more tractable "multilinear" one. As a simple illustration consider a commuting map $F : \mathcal{A} \to \mathcal{A}$ (already encountered in Example 1.5), which by definition satisfies the functional identity $[F(x), x] = 0$ for all $x \in \mathcal{A}$. If F is additive, we have seen that F may often be of the very specific form (1.9). But without the assumption of additivity F could be any one of a myriad of strange looking maps, i.e., *any* set-theoretic map sending each x into its centralizer will suffice.

We are now ready to introduce, for now in a loose informal way, the central theme of this book. With reference to the FI (1.12) (and to the notation therein involved) two general problems present themselves.

First, given a ring $\mathcal{A}$, suppose there exists $d \in \mathbb{N}$ such that

(a) whenever $\max\{|\mathcal{I}|, |\mathcal{J}|\} \leq d$, (1.12) has only a standard solution.

What can be said about the structure of such rings? For example, for $d = 2$ it is immediate that $\mathcal{A}$ cannot be commutative since $x \cdot y - y \cdot x = 0$ would provide a nonstandard solution of the FI of the type (1.14). More generally, it is easy to see (cf. Lemma 2.9 below) that for an arbitrary $d \in \mathbb{N}$, (a) implies that $\mathcal{A}$ cannot satisfy a polynomial identity of degree $\leq d$.

Secondly, given a ring $\mathcal{A}$, can one show for an appropriate $d \in \mathbb{N}$ that the condition (a) holds? For every unital ring it is a simple exercise to show that (a) holds for $d = 1$ (this is basically also proved later, in the discussion before Lemma 2.8). In general, however, this would appear to be a formidable problem. Nevertheless it is important to obtain positive results in this regard since otherwise the theory to which the first problem addresses itself would be a rather empty one.

Now let us bring the FI (1.13) into the picture (we have seen that (1.13) may occur in a natural way, e.g., see the discussion preceding Example 1.3 concerning the Cayley–Hamilton theorem). Let $d \in \mathbb{N}$ and let $\mathcal{A}$ satisfy (a). It is tempting to try to show that

(b) whenever $\max\{|\mathcal{I}|, |\mathcal{J}|\} \leq d - 1$, (1.13) has only a standard solution.

Indeed, one might think that this could be shown by commuting (1.13) with a new variable y and using the fact that the resulting FI of the form (1.12) has only a standard solution. However, an example (for $d = 2$) has been produced by Brešar [65] showing that (a) holds but (b) does not. Similarly, (a) also does not always follow from (b), so (a) and (b) in general are independent of each other [65].

On the other hand, it turns out that in some important classes of rings, the conditions (a) and (b) are equivalent. We shall elaborate more fully on this in the sequel. For now let us illustrate the remark that (a) (sometimes) implies (b) with a sample argument in the case where $\mathcal{A}$ is a simple unital ring and $d = 2$. The particular FI we will consider is

$$E(y)x + yF(x) = \lambda(x, y) \in \mathcal{Z} \tag{1.17}$$

for all $x, y \in \mathcal{A}$. Assume that (a) holds. Then $\mathcal{A}$ is not commutative, and thus there is an element $t \in \mathcal{A}$ of degree ≥ 2 over $\mathcal{Z}$, i.e., $1, t$ are $\mathcal{Z}$-independent. Replacing y by ty in (1.17) we have

$$E(ty)x + tyF(x) = \alpha, \quad \alpha = \lambda(x, ty). \tag{1.18}$$

Multiplying (1.17) on the left by t we have

$$tE(y)x + tyF(x) = \beta t, \quad \beta = \lambda(x, y). \tag{1.19}$$

Subtracting (1.19) from (1.18) yields

$$G(y)x = \alpha - \beta t, \quad G(y) = E(ty) - tE(y). \tag{1.20}$$

Commuting (1.20) with t gives

$$G(y)xt - tG(y)x = 0. \tag{1.21}$$

Since $\mathcal{A}$ is simple it is easy to see that $\mathcal{A}$ is a simple left module over the multiplication ring $\mathcal{M}(\mathcal{A})$, and moreover, that the associated division ring $\mathrm{End}(_{\mathcal{M}(\mathcal{A})}\mathcal{A})$

is just $\mathcal{Z}$. In particular, by the Jacobson density theorem there exists $\mathcal{E} = \mathcal{M}(\mathcal{A})$, $\mathcal{E} : x \mapsto \sum_k a_k x b_k$, such that $\mathcal{E}(t) = 1$ and $\mathcal{E}(1) = 0$. Replacing x by a_k in (1.21), multiplying the identity so obtained on the right by b_k, and summing over k, we are left with $G(y) = 0$, whence (1.20) becomes $\alpha - \beta t = 0$. In particular $\beta = 0$ and so (1.17) reduces to $E(y)x + yF(x) = 0$, which by (a) has the standard solution $E(y) = yp$ and $F(x) = -px$, where $p \in \mathcal{A}$.

For $d \in \mathbb{N}$ we shall say that a ring $\mathcal{A}$ is *d-free* if both (a) and (b) hold; in fact, using the terminology precisely we should say that "$\mathcal{A}$ is a *d-free subset of itself*", but at this point we shall just superficially call such rings *d*-free. The concept of *d*-freeness will be of crucial importance in this book. In Parts II and III we shall deal with FI's on arbitrary sets and accordingly we shall define *d*-free sets as well as some more general notions.

Our next illustration is a sample argument involved in showing that a ring $\mathcal{A}$ is *d*-free. Let $\mathcal{A}$ be a simple unital ring and suppose there exists an element t in $\mathcal{A}$ which is of degree ≥ 3 over $\mathcal{Z}$. We shall show that the FI

$$E(x, y)z + F(x, z)y + G(y, z)x = 0 \qquad (1.22)$$

has only the standard solution $E = F = G = 0$; this is just the easiest of various calculations one must make in order to show that $\mathcal{A}$ is 3-free. The FI (1.22) is just a special case of the FI (1.4) and so the same calculations as in the consideration of Example 1.3 now yield

$$E(xt, yt)z - (E(xt, y) + E(x, yt))zt + E(x, y)zt^2 = 0. \qquad (1.23)$$

Since $1, t, t^2$ are by assumption $\mathcal{Z}$-independent there exists $\mathcal{E} \in \mathcal{M}(\mathcal{A})$ such that $\mathcal{E}(1) = \mathcal{E}(t) = 0$ and $\mathcal{E}(t^2) = 1$. Proceeding as in the previous illustration, (1.23) reduces to $E(x, y) = 0$. Similarly $F = G = 0$.

What hints do we gather from the preceding two illustrations which will shed light on some of the concepts and methods of proof which the reader will encounter in the sequel? It is hoped that these examples will help to pave the way so that the theory in its full generality, with its necessarily complicated terminology and sometimes lengthy proofs, will not come as such a surprise.

At the forefront perhaps is the appearance of a fixed element t of $\mathcal{A}$ of sufficiently high degree over the center $\mathcal{Z}$, whether we are assuming the existence of such an element or trying to prove its existence. The simple device of replacing a variable x by tx or xt, multiplying the original FI on the left or right by t, and comparing these two identities enabled us (very loosely speaking) to get rid of one of the variables. This strongly suggests that many of the proofs in the general theory are going to be inductive in nature. Of course, one pays the price of removing one of the variables by creating a new FI involving powers of t. But the independence of these powers of t may allow one to apply the Jacobson density theorem to $\mathcal{M}(\mathcal{A})$ acting on $\mathcal{A}$ over $\mathcal{Z}$ and thereby to get rid of all but one summand. We saw this phenomenon occur for simple unital rings, and later, in Part II, we shall see that a similar situation will prevail for the more general (semi)prime rings. At any rate we see that in a natural way problems concerning d-freeness, originally involving (1.12) and (1.13), may be transferred to FI's of the form

$$\sum_{u,\,i} E_{iu}(\bar{x}_m^i) x_i t^u + \sum_{v,\,j} t^v x_j F_{jv}(\bar{x}_m^j) = 0. \tag{1.24}$$

Accordingly, the reader should expect to see a more general definition of d-freeness given (it will be called $(t;d)$-freeness) based on FI's of the form (1.24).

1.4   Finding Functional Identities in Different Areas

It is our aim now to list a few mathematical concepts which give rise to certain FI's, and thereby indicate why and where FI theory is applicable. We shall give only definitions of these concepts and show how FI's can be produced out of them, and will discuss neither the historic background concerning them nor the new results about them that can be obtained using FI's. This will be done in detail in Part III.

It is already clear that FI's can be viewed as generalizations of PI's. However, here we wish to discuss those areas where the connections with FI's are not so obvious as in PI (and GPI) theory.

The development of FI theory has been intimately connected with solving problems on Lie maps in associative rings. Therefore we start our discussion with the Lie homomorphism problem. Let $\mathcal{A}$ and $\mathcal{B}$ be rings, let $\mathcal{L}$ be a Lie subring of $\mathcal{B}$, and let $\alpha : \mathcal{L} \to \mathcal{A}$ be a Lie homomorphism. The basic question is whether it is possible to extend α to the subring $\langle \mathcal{L} \rangle$ generated by $\mathcal{L}$ so that this extension is, roughly speaking, "close" to a ring (anti)homomorphism (let us not worry about details right now). Suppose first that $\mathcal{L} = \mathcal{B}$ is already a ring and also, for simplicity suppose that α is a Lie isomorphism (i.e., a bijection) of $\mathcal{B}$ onto $\mathcal{A}$. The goal is to find out how α acts on the product xy of two elements in $\mathcal{B}$. Since we know how it acts on their Lie product $[x, y]$, it suffices (as long as the rings are 2-torsion free) to describe the action of α on their Jordan product $x \circ y$ (as $2xy = [x, y] + x \circ y$). The main idea of our approach is very simple. Since y^2 commutes with y we have $[(y^2)^\alpha, y^\alpha] = 0$ for every $y \in \mathcal{B}$. Setting $x = y^\alpha$ we can write this as

$$\left[\left(\left(x^{\alpha^{-1}}\right)^2\right)^\alpha, x\right] = 0 \quad \text{for all } x \in \mathcal{A}. \tag{1.25}$$

Defining $F : \mathcal{A}^2 \to \mathcal{A}$ by

$$F(x, z) = \left(x^{\alpha^{-1}} \circ z^{\alpha^{-1}}\right)^\alpha$$

and assuming that $\mathcal{A}$ is 2-torsion free, we thus have

$$[F(x, x), x] = 0 \quad \text{for all } x \in \mathcal{A}. \tag{1.26}$$

That is to say, the trace of the biadditive map F is a commuting map. Note that F is symmetric, i.e., $F(x,z) = F(z,x)$. Therefore, linearizing (1.26) we get

$$\sum_{i=1}^{3} F(\overline{x}_3^i)x_i - \sum_{j=1}^{3} x_j F(\overline{x}_3^j) = 0 \quad \text{for all } \overline{x}_3 \in \mathcal{A}^3. \tag{1.27}$$

This is the identity of the type (1.12) (with $m = 3$ and $\mathcal{I} = \mathcal{J} = \{1,2,3\}$). In general d-freeness only allows one to say a little about "unknown" functions, as illustrated in (1.16) where nothing further can be said about the p_{ij}'s. The current example (1.26), however, involves just a single function, and in fact, assuming $\mathcal{A}$ is 3-free, we can explicitly solve for F. We did not give the precise definition of d-freeness yet, but in view of the above informal explanations we believe that the following outline of arguing will be understandable to the reader. From the first summation in (1.16) we have

$$F(x,z) = xp(z) + zq(x) + \nu(x,z),$$

where $p, q : \mathcal{A} \to \mathcal{A}$ are additive and $\nu : \mathcal{A}^2 \to \mathcal{Z}$ is biadditive. Since F is symmetric, we see, using 3-freeness again, that $p = q$ and so we have

$$F(x,z) = xp(z) + zp(x) + \nu(x,z). \tag{1.28}$$

From the second summation of (1.27) we have

$$F(x,z) = r(z)x + r(x)z + \nu(x,z). \tag{1.29}$$

Comparing (1.28) and (1.29) and using 3-freeness, it can be easily concluded that $p(z) = \lambda z + \mu(z)$ where $\lambda \in \mathcal{Z}$ and $\mu : \mathcal{A} \to \mathcal{Z}$ is additive. Hence it follows from (1.28) that

$$F(x,z) = \lambda x \circ z + \mu(x)z + \mu(z)x + \nu(x,z). \tag{1.30}$$

We remark that the expression obtained in (1.30) is an example of what we shall call a *quasi-polynomial* (such expressions have also shown up in previous examples); when we come to expand on the general theory in Chapter 4 we shall see that one of the main goals when treating FI's is to show that the functions are indeed quasi-polynomials.

Obtaining (1.30) goes a long way to showing how α acts on the Jordan product. Of course there is more to the proof, but the main breakthrough has been made.

Later we shall see that the outlined procedure also works in a considerably more general setting. It is essential, however, that $\mathcal{A}$ is at least 3-free (possibly in a more general sense than outlined above, but this will be discussed later). For example, in the case when $\mathcal{A} = M_n(\mathbb{F})$ with $\mathbb{F}$ a field, the general theory successfully produces the definitive result as long as $n \geq 3$. The $n = 1$ case is trivial, while in the case $n = 2$ the general theory fails though the final result is actually the same as for any other n. The problem is that the ring $M_2(\mathbb{F})$ is not 3-free which is clear already from Example 1.3. Therefore, in this concrete case another approach is necessary. This matrix example is quite illuminating since it illustrates both the power and the limitations of FI theory. Often the theory is not applicable to rings which are "too close" to commutative ones. In these, usually very concrete rings one needs to apply different methods; often the classical PI theory then turns out to be useful.

Now consider a more entangled situation when $\mathcal{A}$ and $\mathcal{B}$ are rings with involution and α is a Lie isomorphism from the set of skew elements $\mathcal{K}(\mathcal{B})$ of $\mathcal{B}$ onto the set of skew elements $\mathcal{K}(\mathcal{A})$ of $\mathcal{A}$. Since $\mathcal{K}(\mathcal{B})$ is not closed under the Jordan product, the same trick as above does not work; however, the cube of a skew element is clearly skew again, and so the obvious modification of the above argument gives rise to the FI

$$[F(k, k, k), k] = 0 \quad \text{for all } k \in \mathcal{K}(\mathcal{A}),$$

where $F : \mathcal{K}(\mathcal{A})^3 \to \mathcal{K}(\mathcal{A})$ is defined by

$$F(k_1, k_2, k_3) = \left(\sum_{\pi \in S_3} (k_{\pi(1)})^{\alpha^{-1}} (k_{\pi(2)})^{\alpha^{-1}} (k_{\pi(3)})^{\alpha^{-1}} \right)^{\alpha}.$$

A linearization gives

$$\sum_{i=1}^{4} F(\overline{x}_4^i) x_i - \sum_{i=1}^{4} x_i F(\overline{x}_4^i) = 0 \quad \text{for all } \overline{x}_4 \in \mathcal{K}(\mathcal{A})^4.$$

This is an FI on $\mathcal{K}(\mathcal{A})$ and not on the whole ring $\mathcal{A}$. Therefore, the assumption of d-freeness of $\mathcal{A}$ is not directly applicable in this situation. However, in Part II we shall consider FI's on subsets of rings and find out that under some natural conditions on $\mathcal{A}$, $\mathcal{K}(\mathcal{A})$ is d-free.

But even this more sophisticated approach is not (directly) applicable to the study of Lie isomorphisms of many other important Lie subrings of an associative ring $\mathcal{A}$, such as for example $[\mathcal{A}, \mathcal{A}]$ or $[\mathcal{K}(\mathcal{A}), \mathcal{K}(\mathcal{A})]$. Namely, these Lie subrings are not closed under any powers and so there is no such intimate relation between them and the associative structure of $\mathcal{A}$. Nevertheless, there are ways to produce FI's in such cases as well. The problem, however, is considerably more difficult and the main ideas cannot be explained just in a few lines and so will not be discussed here. Let us just mention that one has to face several rather different FI's in these problems, not only those that arise from commuting maps. Also, some of these FI's involve a rather large number of variables which somehow justifies the need to create the general theory via the concept of d-freeness (just as an illustration we mention that when dealing with Lie isomorphisms of $[\mathcal{K}(\mathcal{A}), \mathcal{K}(\mathcal{A})]$ where $\mathcal{A}$ is a simple unital ring with involution, the general theory produces the definitive result as long as $\mathcal{A}$ is 21-free, which is equivalent to the condition that the dimension of $\mathcal{A}$ over its center is > 400).

The Lie homomorphism problem is just the first and the most thoroughly studied problem among several Lie theoretic problems for which FI's have turned out to be applicable. We will now indicate, admittedly rather superficially, a few more.

Let δ be a Lie derivation of a ring $\mathcal{A}$ into itself. Note that δ satisfies

$$[(x^2)^\delta - x^\delta x - x x^\delta, x] = 0 \quad \text{for all } x \in \mathcal{A},$$

which can be interpreted as the FI (1.26) (which in turn implies (1.27)). So one may expect that the consideration of Lie homomorphisms and Lie derivations should be similar. This is true indeed, and moreover, via the concept of d-freeness the problems on Lie derivations can be as a rule reduced to analogous problems on Lie homomorphisms. Let us also mention in this context that FI's have been used in some related but entirely analytic problems, namely, the ones concerning the so-called *automatic continuity* of Lie isomorphisms and Lie derivations on Banach algebras.

Let $\mathcal{A}$ be an associative algebra, and let there exist an additional multiplication $* : \mathcal{A}^2 \to \mathcal{A}$ such that $(\mathcal{A}, +, *)$ is a nonassociative algebra. Suppose that $*$ is *third-power associative*, meaning that $(x * x) * x = x * (x * x)$ for every $x \in \mathcal{A}$, and suppose further that $*$ is connected with the associative multiplication through the formula $y * x - x * y = [y, x]$ for all $x, y \in \mathcal{A}$. Is it possible to describe $*$? This problem appears in the theory of the so-called *Lie-admissible algebras*. The basic idea of the FI approach to this problem is very simple. Substituting $x * x$ for y in $y * x - x * y = [y, x]$ we obtain

$$[x * x, x] = 0 \quad \text{for all } x \in \mathcal{A},$$

that is, again we have arrived at the FI (1.26), and hence (1.27).

We say that $(\mathcal{P}, +, \cdot, \{., .\})$ is a *Poisson algebra* if $(\mathcal{P}, +, \cdot)$ is an associative algebra, $(\mathcal{P}, +, \{., .\})$ is a Lie algebra, and $\{x \cdot y, z\} = x \cdot \{y, z\} + \{x, z\} \cdot y$ for all $x, y, z \in \mathcal{P}$. Now if $(\mathcal{P}, +, \{., .\})$ is a Lie subalgebra of some associative algebra $\mathcal{A}$ so that $\{x, y\} = [x, y]$ ($= xy - yx$ where xy denotes the product of x and y in $\mathcal{A}$), then setting $x = y = z$ we get

$$[x \cdot x, x] = 0 \quad \text{for all } x \in \mathcal{P}.$$

So again we have arrived at an FI of the type (1.26). This is the starting point of investigating the connection between the products involved.

We have thereby shown how various Lie theoretic concepts give rise to the FI (1.26), that is, commuting traces of biadditive maps appear. In all these instances, just as in the Lie homomorphism problem, a more thorough analysis of these concepts yields more complicated FI's. But for now we stop at this point.

There are many parallels between Lie and Jordan structures in associative rings. So one may wonder whether FI's are also applicable to Jordan theory. This is true indeed. Let us outline only one example. Let α be a Jordan isomorphism of a ring $\mathcal{B}$ onto a ring $\mathcal{A}$. Now we are interested in the action of α on the Lie product of elements, so we are searching for some identical relation connecting the Lie and the Jordan product (i.e., an appropriate analogy of $[x \circ x, x] = 0$ which led to an FI in the Lie homomorphism case). We shall derive one such identity by expressing $[[x, y], [z, w]]$ in two different ways. On the one hand we have

$$[[x, y], [z, w]] = (y \circ [z, w]) \circ x - (x \circ [z, w]) \circ y,$$

and on the other hand,

$$[[x, y], [z, w]] = (z \circ [x, y]) \circ w - (w \circ [x, y]) \circ z.$$

Comparing we obtain

$$(y \circ [z, w]) \circ x - (x \circ [z, w]) \circ y = (z \circ [x, y]) \circ w - (w \circ [x, y]) \circ z.$$

Now set $F(u, v) = [u^{\alpha^{-1}}, v^{\alpha^{-1}}]^{\alpha}$ for $u, v \in \mathcal{A}$, and note that the above identity yields

$$(y \circ F(z, w)) \circ x - (x \circ F(z, w)) \circ y = (z \circ F(x, y)) \circ w - (w \circ F(x, y)) \circ z$$

for all $x, y, z, w \in \mathcal{A}$. Expanding the terms, we can rewrite this as

$$[[x, y], F(z, w)] = [F(x, y), [z, w]].$$

This is clearly an FI of the type (1.12). Assuming that $\mathcal{A}$ is 4-free, one can then describe F and thereby find out how α acts on the Lie product.

One can pose a considerably more general problem concerning the structure of maps that preserve an arbitrary multilinear polynomial in noncommuting variables. More precisely, given rings $\mathcal{A}$ and $\mathcal{B}$ and an arbitrary multilinear polynomial $f(x_1, \ldots, x_n) \in \mathbb{Z}\langle x_1, x_2, \ldots \rangle$, we consider an additive map $\alpha : \mathcal{B} \to \mathcal{A}$ satisfying

$$f(x_1, \ldots, x_n)^{\alpha} = f(x_1^{\alpha}, \ldots, x_n^{\alpha}) \quad \text{for all } x_1, \ldots, x_n \in \mathcal{B}. \tag{1.31}$$

This concept clearly extends and unifies the classical concepts of Lie homomorphisms (the polynomial $f(x_1, x_2) = x_1 x_2 - x_2 x_1$) and Jordan homomorphisms (the polynomial $f(x_1, x_2) = x_1 x_2 + x_2 x_1$). It is not obvious how to create a suitable FI when facing (1.31), but it can be done. Under reasonable assumptions one can then describe the form of these maps.

Finally we mention connections of FI theory with some so-called *linear preserver problems*. By a linear preserver we mean a linear map between algebras which, roughly speaking, preserves certain properties of some algebra elements. The goal is to describe the form of such a map. Most of the results on linear preserving are entirely linear algebraic, that is, they are concerned with algebras of matrices. On the other hand, various linear preserver problems have also been considered in algebras of bounded linear operators as well as on some other algebras appearing in functional analysis. Using FI's we can obtain ring-theoretic generalizations of some of them.

One of the most well-known linear preserver problems is the one concerning *commutativity preservers*. That is, we want to find the form of a linear map α from an algebra $\mathcal{B}$ to an algebra $\mathcal{A}$ with the property that elements x^α and y^α in $\mathcal{A}$ commute whenever x and y in $\mathcal{B}$ commute. Of course, every Lie homomorphism satisfies this condition, and in fact the same idea that works out for Lie homomorphisms is applicable in this more general problem. Again we have $[(y^2)^\alpha, y^\alpha] = 0$ for every $y \in \mathcal{B}$, and so under the assumption of bijectivity of α we arrive at (1.25). That is to say, again we have to deal with the FI's (1.26) and (1.27). One can consider this problem in a greater generality, say, assuming that only the commutativity of symmetric elements in algebras with involution is preserved.

Another class of problems where FI's are applicable is the one on *normality preservers*. Let $\mathcal{A}$ and $\mathcal{B}$ be algebras with involution. An element x is said to be *normal* if it commutes with x^*. Consider a linear map $\alpha : \mathcal{B} \to \mathcal{A}$ such that $x^\alpha \in \mathcal{A}$ is normal whenever $x \in \mathcal{B}$ is normal. Assume further that α is $*$-linear, meaning that $(x^*)^\alpha = (x^\alpha)^*$ for each $x \in \mathcal{B}$. Pick $s \in \mathcal{S}(\mathcal{B})$ and $k \in \mathcal{K}(\mathcal{B})$ such that $[s, k] = 0$. Then $x = s + k$ is normal, hence $x^\alpha = s^\alpha + k^\alpha$ is normal, i.e., it commutes with $(x^\alpha)^* = s^\alpha - k^\alpha$. But then $[s^\alpha, k^\alpha] = 0$ provided that $\mathcal{A}$ is 2-torsion free. In particular, for each $k \in \mathcal{K}(\mathcal{B})$ we have that the element k^2 lies in $\mathcal{S}(\mathcal{B})$ and it commutes with $k, k^3 \in \mathcal{K}(\mathcal{B})$. Consequently,

$$[(k^2)^\alpha, k^\alpha] = 0 \quad \text{and} \quad [(k^2)^\alpha, (k^3)^\alpha] = 0$$

for all $k \in \mathcal{K}(\mathcal{B})$. Under certain additional assumptions these two identities can be interpreted as FI's.

So one can see that there are various FI's that deserve special attention. In particular, commuting traces of biadditive maps often naturally appear in various situations. One way of looking at a biadditive map on a ring $\mathcal{A}$ is that this is a new product transforming $\mathcal{A}$ into another (nonassociative) ring. The condition that the trace of this map is commuting can be read as that the square (with respect to the new nonassociative product) of each element commutes (with respect to the original associative product) with the element itself. This point of view perhaps gives some better insight into why commuting traces of biadditive maps arise in problems concerning nonassociative structures in associative rings.

We will not go any further at present. Part III will be entirely devoted to a realization of the ideas that were outlined in a very loose manner here.

Literature and Comments. The first functional identity was discovered at the beginning of the 1990s by accident, as an attempt to unify several existing results on centralizing maps. In 1957 Posner [182] proved that the existence of a nonzero centralizing derivation on a prime ring implies that the ring is commutative. This result was then extended in different directions. In particular, analogous results for some other maps, for instance for nontrivial centralizing automorphisms [159], were obtained (see [66] for more details and references). When trying to discover some general law behind these different but strikingly similar results, Brešar found out that actually *any* centralizing additive map of a prime ring of characteristic not 2 must be of the form (1.9) (however, with λ and $\mu(x)$ possibly belonging to the so-called extended centroid rather than to the center). This result appeared in his 1990 Ph.D. Thesis, and was somewhat later, in 1993, published in the paper [56] (by chance two related subsequent papers [54, 55] of Brešar were published somewhat earlier). Soon after Brešar considered commuting traces of biadditive maps (i.e., the FI (1.26)) and applied the result obtained to the Lie map and commutativity preserver problems [58]. These results initiated the series of papers on various FI's and their applications, written by numerous authors: [6, 12, 35, 39, 57, 59, 60, 62, 63, 74, 75, 77, 78, 79, 80, 81, 87, 137, 139, 140, 141, 142, 192]. Although the results obtained in these papers were related, each of them was proved by a slightly different method. The lack of a systematic approach thus became apparent. In 1998 Beidar [16], motivated by a result of Chebotar on generalized functional identities [88], introduced FI's (1.12) and (1.13) and proved a result which covered and unified a number of results obtained in the above list of papers. In this book we mostly consider the results that were obtained after this fundamental paper of Beidar.

Most of the examples from Section 1.2 were taken from the survey article of Brešar [64]. The notion of a biderivation was introduced and studied in noncommutative rings independently by different authors [77, 107, 189], but it seems the first one was Skosyrskii [189]. The definition of a *d*-free set was introduced in the paper [29] by Beidar and Chebotar.

Chapter 2

The Strong Degree and the FI-Degree

In Section 2.1 we will introduce the concept of the strong degree of a unital ring. The definition involves a condition which is rather technical, but we shall see that the strong degree can be rather easily computed for certain classes of rings. The main reason for dealing with this concept is its connection with functional identities - this will be the topic of Section 2.4. Before that, in Sections 2.2 and 2.3, we will consider certain versions of the concept of d-freeness (called strong d-freeness and strong $(t; d)$-freeness). Unlike in Chapter 1, we shall now consider these notions in a rigorous manner.

Roughly speaking, the goal of this chapter is to show that FI's of fundamental types can be handled in rings whose strong degree is sufficiently large. This will yield definitive results for certain classes of rings, in particular for simple unital rings and rings of $n \times n$ matrices over commutative unital rings. It should be mentioned, however, that the strong degree approach to FI's is satisfactory only at a basic level of generality. For instance, in order to pass from simple to prime rings we shall have to introduce some other, more complicated notions (in Chapter 5). In various directions the results obtained here will be largely superseded in Part II; however, not in every direction. The advantage of studying FI's via the strong degree is that this makes it possible for one to treat maps having their ranges in an arbitrary bimodule over a ring $\mathcal{A}$, while in Part II (in particular in Chapter 5), we shall have to confine ourselves to the situation where the role of a bimodule is replaced by a fixed ring containing $\mathcal{A}$ and attached to $\mathcal{A}$ in a particularly nice way.

At any rate, the strong degree approach to FI's is self-contained, easy to follow, and it can also be viewed as a good illustration for FI methods. That is why it is included in Part I.

Throughout this chapter we assume that $\mathcal{A}$ is a unital ring.

2.1   The Strong Degree

Recall that $\mathcal{M}(\mathcal{A})$ denotes the multiplication ring of $\mathcal{A}$. For convenience we define $t^0 = 1$ for any $t \in \mathcal{A}$.

Definition 2.1. The *strong degree* of a nonzero element $t \in \mathcal{A}$ is greater than n (notation $s\text{-deg}(t) > n$), where $n \geq 0$ is an integer, if for any $0 \leq i \leq n$ there exists $\mathcal{E}_i \in \mathcal{M}(\mathcal{A})$ such that
$$\mathcal{E}_i(t^j) = \delta_{ij}$$
(the "Kronecker delta") for each $j = 0, 1, \ldots, n$. Further, if $s\text{-deg}(t) > n-1$ but $s\text{-deg}(t) \not> n$, we say that the strong degree of t is n ($s\text{-deg}(t) = n$). If $s\text{-deg}(t) > n$ for any positive integer n, then we write $s\text{-deg}(t) = \infty$. Finally, the *strong degree* of $\mathcal{A}$ is defined as
$$s\text{-deg}(\mathcal{A}) = \sup\{s\text{-deg}(t) \mid t \in \mathcal{A}\}.$$

Our main aim in this section is to compute $s\text{-deg}(\mathcal{A})$ for some rings $\mathcal{A}$. We begin with a few simple observations. First of all, if $\mathcal{A}_0 \subseteq \mathcal{A}$ are rings with the same unity, then clearly
$$s\text{-deg}(\mathcal{A}_0) \leq s\text{-deg}(\mathcal{A}).$$

It is obvious that every nonzero element $t \in \mathcal{A}$ has strong degree at least 1. Note that $\mathcal{E}(t) = 1$ for some $\mathcal{E} \in \mathcal{M}(\mathcal{A})$ if and only if (t), the ideal of $\mathcal{A}$ generated by t, is equal to $\mathcal{A}$. This is therefore a necessary condition for t to have strong degree > 1. But it is certainly not a sufficient condition, just consider $t = 1$. More generally, it is clear that the strong degree of central elements cannot be > 1. Thus

$$s\text{-deg}(z) = 1 \quad \text{for every } z \in \mathcal{Z}. \tag{2.1}$$

Let $\mathcal{M}$ be a unital $(\mathcal{A}, \mathcal{A})$-bimodule. This means that $\mathcal{M}$ is both a left $\mathcal{A}$-module and a right $\mathcal{A}$-module, and moreover $(xm)y = x(my)$ and $1m = m1 = m$ holds for all $x, y \in \mathcal{A}$ and $m \in \mathcal{M}$. We set $\mathcal{Z}(\mathcal{M}) = \{\lambda \in \mathcal{M} \mid \lambda x = x\lambda$ for all $x \in \mathcal{A}\}$ for the *center of* $\mathcal{M}$. The next lemma will be of crucial importance in the consideration of functional identities.

Lemma 2.2. *Let $\mathcal{M}$ be a unital $(\mathcal{A}, \mathcal{A})$-bimodule, let $u_i, v_j \in \mathcal{M}$, $i = 0, 1, \ldots, m$, $j = 0, 1, \ldots, n$, and let $t \in \mathcal{A}$.*

  (i) *If $s\text{-deg}(t) > m$ and $\sum_{i=0}^{m} u_i x t^i = 0$ for all $x \in \mathcal{A}$, then each $u_i = 0$.*

 (ii) *If $s\text{-deg}(t) > n$ and $\sum_{j=0}^{n} t^j x v_j = 0$ for all $x \in \mathcal{A}$, then each $v_j = 0$.*

(iii) *If $s\text{-deg}(t) > \max\{n, m\}$ and $\sum_{i=0}^{m} u_i x t^i + \sum_{j=0}^{n} t^j x v_j = 0$ for all $x \in \mathcal{A}$, then $u_i \in \sum_{j=0}^{n} \mathcal{Z}(\mathcal{M}) t^j$ and $v_j \in \sum_{i=0}^{m} \mathcal{Z}(\mathcal{M}) t^i$ for all i, j.*

Proof. To prove (i) let $\mathcal{E}_i$, $i = 0, 1, \ldots, m$, be as in Definition 2.1. We fix $0 \leq j \leq m$, write $\mathcal{E}_j(x) = \sum_{k=1}^{p} a_k x b_k$, and note that $\sum_{i=0}^{m} u_i a_k t^i = 0$. Multiplying on the right by b_k and then summing over k, we see that $0 = \sum_{i=0}^{m} u_i \mathcal{E}_j(t^i) = u_j$. Similarly we prove (ii).

Now assume that the conditions of (iii) are fulfilled. By symmetry we only need to show that each $v_j \in \sum_{i=0}^{m} \mathcal{Z}(\mathcal{M})t^i$. Let $\mathcal{E}_i \in \mathcal{M}(\mathcal{A})$, $0 \leq i \leq \max\{n, m\}$ be as in Definition 2.1. We fix $0 \leq k \leq n$, write $\mathcal{E}_k(x) = \sum_{l=1}^{q} a_l x b_l$, note that $\sum_{j=0}^{n} t^j b_l x v_j = -\sum_{i=0}^{m} u_i b_l x t^i$, and conclude (as we did in the proof of (i) above) that $x v_k = \sum_{j=0}^{n} \mathcal{E}_k(t^j) x v_j = \sum_{i=0}^{m} z_i x t^i$ for all $x \in \mathcal{A}$, where $z_i = -\mathcal{E}_k(u_i)$. By taking $x = 1$ we see that it suffices to show that each z_i lies in $\mathcal{Z}(\mathcal{M})$. To this end we fix $0 \leq l \leq m$ and (by a now familiar method) obtain $x\mathcal{E}_l(v_k) = \sum_{i=0}^{m} z_i x \mathcal{E}_l(t^i) = z_l x$ for all $x \in \mathcal{A}$. Setting $x = 1$ we see that $z_l = \mathcal{E}_l(v_k)$, whence $z_l x = x z_l$, i.e., $z_l \in \mathcal{Z}(\mathcal{M})$. $\qquad\square$

For brevity we write $\mathcal{Z}$ for $\mathcal{Z}(\mathcal{A})$. We say that $t \in \mathcal{A}$ is *algebraic over* $\mathcal{Z}$ if there exist $z_0, \ldots, z_n \in \mathcal{Z}$ such that

$$z_0 + z_1 t + \ldots + z_n t^n = 0 \quad \text{and} \quad z_n \neq 0. \tag{2.2}$$

Moreover, in this case we say that t is algebraic of degree at most n. The *degree of algebraicity* of t over $\mathcal{Z}$ (its definition should be self-explanatory) will be denoted simply by $\deg(t)$. We shall write $\deg(t) = \infty$ in the case when t is not algebraic over $\mathcal{Z}$. The definition of the degree of algebraicity of $\mathcal{A}$, $\deg(\mathcal{A})$, is analogous to that of the strong degree of $\mathcal{A}$.

Assuming that $\deg(t) \leq n$, i.e. that (2.2) holds, it follows that $\sum_{i=0}^{n} z_i x t^i = 0$ for all $x \in \mathcal{A}$, and hence $s\text{-}\deg(t) \leq n$ by Lemma 2.2 (i) (applied for $\mathcal{M} = \mathcal{A}$). This means that

$$s\text{-}\deg(t) \leq \deg(t) \quad \text{for every } t \in \mathcal{A}. \tag{2.3}$$

This inequality may be strict. For example, let $\mathcal{A} = \mathbb{F}\langle X \rangle$ be the free algebra over a field $\mathbb{F}$ on a set X containing at least two elements. Pick $t \in \mathcal{A} \setminus \mathbb{F}$, i.e., t is not a constant polynomial. Then clearly t is not algebraic so that $\deg(t) = \infty$. If $s\text{-}\deg(t)$ was > 1, there would be $a_i, b_i \in \mathcal{A}$ such that $\sum a_i b_i = 0$ and $\sum a_i t b_i = 1$. However, writing $t = \lambda + t_0$ where $\lambda \in \mathbb{F}$ and t_0 has constant term 0, it follows that $\sum a_i t b_i = \lambda \sum a_i b_i + \sum a_i t_0 b_i = \sum a_i t_0 b_i$ should have constant term 0, a contradiction. Therefore, $s\text{-}\deg(t) = 1$. In fact, this implies that $s\text{-}\deg(\mathcal{A}) = 1$ while $\deg(\mathcal{A}) = \infty$.

As will become apparent later, the strong degree approach to FI's is not efficient in the case when the strong degree of the ring is only 1. We remark that $\mathbb{F}\langle X \rangle$ is a primitive ring, so that this chapter will not give any sufficient answer about FI's in primitive (not to say prime) rings. At present we have to confine ourselves to the narrower class of simple rings.

Lemma 2.3. *Let $\mathcal{A}$ be a simple unital ring. Then $s\text{-}\deg(t) = \deg(t)$ for every nonzero $t \in \mathcal{A}$.*

Proof. We have to show that $s\text{-}\deg(t) \geq \deg(t)$ for every $t \in \mathcal{A}$. Assume therefore that $t \in \mathcal{A}$ is such that $\deg(t) > n$. We must show that $s\text{-}\deg(t) > n$. According to our assumption, the elements $1, t, \ldots, t^n$ are linearly independent over $\mathcal{Z}$. The simplicity of the ring $\mathcal{A}$ implies the simplicity of $\mathcal{A}$ regarded as a left $\mathcal{M}(\mathcal{A})$-module. Noting that the associated division ring $\mathrm{End}(_{\mathcal{M}(\mathcal{A})}\mathcal{A})$ is just $\mathcal{Z}$, the desired conclusion now follows at once from the Jacobson density theorem.   □

Theorem 2.4. *Let $\mathcal{R}$ be a unital ring. Then*

$$s\text{-deg}(M_n(\mathcal{R})) \geq n \cdot s\text{-deg}(\mathcal{R}).$$

Proof. Set $\mathcal{A} = M_n(\mathcal{R})$. We identify $\mathcal{R}$ with the subring of all scalar matrices of $\mathcal{A}$ and respectively consider $\mathcal{M}(\mathcal{R})$ as a subring of $\mathcal{M}(\mathcal{A})$. Assume that s-$\deg(\mathcal{R}) \geq m$. The theorem will be proved by showing that s-$\deg(\mathcal{A}) > mn - 1$. Pick $a \in \mathcal{R}$ with s-$\deg(a) \geq m$ and set $t = e_{12} + e_{23} + \ldots + e_{n-1,n} + ae_{n1}$. Our goal is to show that s-$\deg(t) > nm - 1$.

Note that $t^n = a$ and hence

$$t^{np} = a^p \quad \text{for all } p = 0, 1, \ldots. \tag{2.4}$$

Next, given $0 \leq i, j \leq n - 1$, we have

$$e_{k1} M_{e_{i+1,k}}(t^j) = e_{k1} t^j e_{i+1,k} = \delta_{ij} e_{kk} \quad \text{for all } 1 \leq k \leq n.$$

Setting $\mathcal{U}_i = \sum_{k=1}^{n} e_{k1} M_{e_{i+1,k}}$, $0 \leq i \leq n - 1$, we see that

$$\mathcal{U}_i(t^j) = \delta_{ij} \quad \text{for all } 0 \leq i, j \leq n - 1. \tag{2.5}$$

Since s-$\deg(a) \geq m$, there exist $\mathcal{V}_q \in \mathcal{M}(\mathcal{R}) \subseteq \mathcal{M}(\mathcal{A})$, $0 \leq q \leq m - 1$, such that $\mathcal{V}_q(a^p) = \delta_{qp}$ for all $0 \leq p, q \leq m-1$. Given $0 \leq k \leq mn-1$, write $k = qn+i$ where $0 \leq i \leq n - 1$. Clearly $0 \leq q \leq m - 1$. Set $\mathcal{E}_k = \mathcal{V}_q \mathcal{U}_i$. Now let $0 \leq \ell \leq mn - 1$, $\ell = pn + j$ where $0 \leq j \leq n - 1$. We have that $t^\ell = t^{np}t^j = a^p t^j$. Taking into account that the left multiplication by a^p commutes with each $\mathcal{U}_i$, we infer from both (2.4) and (2.5) that

$$\mathcal{E}_k(t^\ell) = (\mathcal{V}_q \mathcal{U}_i)(a^p t^j) = \mathcal{V}_q(a^p)\mathcal{U}_i(t^j) = \delta_{qp}\delta_{ij} = \delta_{k\ell}$$

for all $0 \leq k, \ell \leq mn - 1$. Therefore s-$\deg(t) > mn - 1$.   □

Corollary 2.5. *Let $\mathcal{C}$ be a commutative unital ring. Then*

$$s\text{-deg}(M_n(\mathcal{C})) = n.$$

Proof. By Theorem 2.4, s-$\deg(M_n(\mathcal{C})) \geq n$. On the other hand, (2.3) and the Cayley–Hamilton theorem show that s-$\deg(M_n(\mathcal{C})) \leq n$.   □

Corollary 2.6. *Let $\mathcal{V}$ be a vector space over a field $\mathbb{F}$. Then*

$$s\text{-deg}(\mathrm{End}_{\mathbb{F}}(\mathcal{V})) = \dim_{\mathbb{F}}(\mathcal{V}).$$

Proof. Set $\mathcal{A} = \mathrm{End}_{\mathbb{F}}(\mathcal{V})$. If $\dim_{\mathbb{F}}(\mathcal{V}) = \infty$, then $\mathcal{V} \cong \mathcal{V} \oplus \mathcal{V}$, hence $\mathcal{A} \cong \mathrm{End}_{\mathbb{F}}(\mathcal{V} \oplus \mathcal{V}) \cong M_2(\mathrm{End}_{\mathbb{F}}(\mathcal{V})) = M_2(\mathcal{A})$, and so s-$\deg(\mathcal{A}) \geq 2s$-$\deg(\mathcal{A})$ by Theorem 2.4. But then s-$\deg(\mathcal{A}) = \infty$. If $\dim_{\mathbb{F}}(\mathcal{V}) = n < \infty$, then $\mathcal{A} \cong M_n(\mathbb{F})$ and so the result follows from Corollary 2.5.   □

2.2 Strongly d-Free Rings and the FI-Degree

Our first aim in this section is to give a rigorous definition of the concept of a strongly d-free ring. We shall do this rather concisely, so we advise the reader to consult Chapter 1 for an intuitive explanation concerning d-freeness. Before starting with a formal discussion we also mention that unlike before we shall now consider FI's involving functions with their range in an $(\mathcal{A}, \mathcal{A})$-bimodule $\mathcal{M}$. Fortunately, this greater level of generality does not cause any complications in our arguments. If we would confine ourselves only to the fundamental case when $\mathcal{M} = \mathcal{A}$, the proofs would be practically the same.

Throughout this section $\mathcal{A}$ will be a unital ring and $\mathcal{M}$ will be an arbitrary unital $(\mathcal{A}, \mathcal{A})$-bimodule with center $\mathcal{Z}(\mathcal{M})$. We write $\mathcal{Z}$ for $\mathcal{Z}(\mathcal{A})$. Further, let $\mathcal{I}$ and $\mathcal{J}$ be finite subsets of $\mathbb{N}$, let $m \in \mathbb{N}$ be such that $\mathcal{I} \cup \mathcal{J} \subseteq \{1, \ldots, m\}$, and let $E_i, F_j : \mathcal{A}^m \to \mathcal{M}$, $i \in \mathcal{I}$, $j \in \mathcal{J}$, be arbitrary functions. Recall that the basic FI's we shall deal with are

$$\sum_{i \in \mathcal{I}} E_i(\bar{x}_m^i) x_i + \sum_{j \in \mathcal{J}} x_j F_j(\bar{x}_m^j) = 0 \quad \text{for all } \bar{x}_m \in \mathcal{A}^m, \tag{2.6}$$

and a slightly more general one,

$$\sum_{i \in \mathcal{I}} E_i(\bar{x}_m^i) x_i + \sum_{j \in \mathcal{J}} x_j F_j(\bar{x}_m^j) \in \mathcal{Z}(\mathcal{M}) \quad \text{for all } \bar{x}_m \in \mathcal{A}^m. \tag{2.7}$$

In view of our convention that a sum over $\emptyset$ is zero, the FI's

$$\sum_{i \in \mathcal{I}} E_i(\bar{x}_m^i) x_i = 0 \quad \text{for all } \bar{x}_m \in \mathcal{A}^m, \tag{2.8}$$

$$\sum_{j \in \mathcal{J}} x_j F_j(\bar{x}_m^j) = 0 \quad \text{for all } \bar{x}_m \in \mathcal{A}^m \tag{2.9}$$

are particular cases of (2.6), and

$$\sum_{i \in \mathcal{I}} E_i(\bar{x}_m^i) x_i \in \mathcal{Z}(\mathcal{M}) \quad \text{for all } \bar{x}_m \in \mathcal{A}^m, \tag{2.10}$$

$$\sum_{j \in \mathcal{J}} x_j F_j(\bar{x}_m^j) \in \mathcal{Z}(\mathcal{M}) \quad \text{for all } \bar{x}_m \in \mathcal{A}^m \tag{2.11}$$

are particular cases of (2.7).

Suppose there exist maps

$$p_{ij} : \mathcal{A}^{m-2} \to \mathcal{M}, \ i \in \mathcal{I}, \ j \in \mathcal{J}, \ i \neq j,$$
$$\lambda_k : \mathcal{A}^{m-1} \to \mathcal{Z}(\mathcal{M}), \ k \in \mathcal{I} \cup \mathcal{J},$$

such that

$$E_i(\overline{x}_m^i) = \sum_{\substack{j \in \mathcal{J}, \\ j \neq i}} x_j p_{ij}(\overline{x}_m^{ij}) + \lambda_i(\overline{x}_m^i), \quad i \in \mathcal{I},$$

$$F_j(\overline{x}_m^j) = -\sum_{\substack{i \in \mathcal{I}, \\ i \neq j}} p_{ij}(\overline{x}_m^{ij})x_i - \lambda_j(\overline{x}_m^j), \quad j \in \mathcal{J}, \tag{2.12}$$

$$\lambda_k = 0 \quad \text{if} \quad k \notin \mathcal{I} \cap \mathcal{J}.$$

One can readily check that (2.12) is a solution of (2.6) and hence also of (2.7). We shall say that every solution of the form (2.12) is a *standard solution* of (2.6), as well as of (2.7). Considering the case when $\mathcal{J} = \emptyset$ we see that $E_i = 0$ for each i is the (only) standard solution of the FI's (2.8) and (2.10).

The case when $|\mathcal{I} \cup \mathcal{J}| \leq 2$ perhaps needs some additional explanation. In view of our convention that $\mathcal{A}^0 = \{0\}$, a map defined on $\mathcal{A}^0$ is a constant, and can therefore be identified with a fixed element in $\mathcal{M}$. So, for example, when $\mathcal{I} = \mathcal{J} = \{1\}$ and $m = 1$, (2.6) reads as that elements $E_1, F_1 \in \mathcal{M}$ satisfy $E_1 x_1 + x_1 F_1 = 0$ for all $x_1 \in \mathcal{A}$, and the standard solution of this identity can be, in view of the convention that the sum over $\emptyset$ is 0, simply expressed as $E_1 = -F_1 \in \mathcal{Z}(\mathcal{M})$. When $\mathcal{I} = \{1\}$, $\mathcal{J} = \emptyset$ and $m = 1$, (2.6) means that $E_1 \in \mathcal{M}$ satisfies $E_1 x_1 = 0$ for all $x_1 \in \mathcal{A}$, and the standard solution is of course $E_1 = 0$. When $|\mathcal{I} \cup \mathcal{J}| = 2$ and $m = 2$, the (constant) maps p_{ij} from the definition of the standard solution can be, as already indicated above, regarded as elements in $\mathcal{M}$.

Definition 2.7. A ring $\mathcal{A}$ is said to be *strongly d-free*, where $d \in \mathbb{N}$, if for every unital $(\mathcal{A}, \mathcal{A})$-bimodule $\mathcal{M}$, for all $m \in \mathbb{N}$ and all $\mathcal{I}, \mathcal{J} \subseteq \{1, 2, \ldots, m\}$, the following two conditions are satisfied:

(a) If $\max\{|\mathcal{I}|, |\mathcal{J}|\} \leq d$, then (2.6) implies (2.12).

(b) If $\max\{|\mathcal{I}|, |\mathcal{J}|\} \leq d - 1$, then (2.7) implies (2.12).

If $\mathcal{A}$ is strongly d-free for any $d \in \mathbb{N}$, then we say that $\mathcal{A}$ is ∞-*free*.

The necessity of requiring both conditions (a) and (b) has been already discussed in Chapter 1.

Let us point out that (2.12) trivially implies (2.6). Thus, if $\mathcal{A}$ is strongly d-free and (2.7) holds with $\max\{|\mathcal{I}|, |\mathcal{J}|\} \leq d - 1$, then the right-hand side of (2.7) must be 0.

The least we can say is that every unital ring $\mathcal{A}$ is strongly 1-free. To prove this, one just has to verify (a) with $\max\{|\mathcal{I}|, |\mathcal{J}|\} = 1$. There are only four FI's that have to be considered: $E(\overline{x}_m^1)x_1 = 0$ (i.e., $\mathcal{I} = \{1\}$ and $\mathcal{J} = \emptyset$), $x_1 F(\overline{x}_m^1) = 0$ (i.e., $\mathcal{I} = \emptyset$ and $\mathcal{J} = \{1\}$), $E(\overline{x}_m^1)x_1 + x_1 F(\overline{x}_m^1) = 0$ (i.e., $\mathcal{I} = \mathcal{J} = \{1\}$), and $E(\overline{x}_m^1)x_1 + x_2 F(\overline{x}_m^2) = 0$ (i.e., $\mathcal{I} = \{1\}$, $\mathcal{J} = \{2\}$). Substituting 1 for x_1 in each of these FI's (and also 1 for x_2 in the last one) it follows easily that all of them have only standard solutions.

In a loose manner one can say that a ring $\mathcal{A}$ is strongly d-free if the FI's on $\mathcal{A}$ of the types (2.6) and (2.7) in "not too many" variables (e.g., when $m < d$) have only standard solutions. Are these standard solutions unique, that is, are the functions p_{ij}'s and λ_k's appearing in (2.12) uniquely determined? Should the functions p_{ij}'s and λ_k's be multiadditive in the case when the E_i's and F_j's are? We answer these questions and simultaneously gather together a few simple but useful observations in

Lemma 2.8. *Let $\mathcal{A}$ be a strongly d-free ring. Then:*

(i) *$\mathcal{A}$ is d'-free for every $d' < d$.*

(ii) *If $|\mathcal{I}| \leq d$, then (2.8) implies that each $E_i = 0$.*

(iii) *If $|\mathcal{J}| \leq d$, then (2.9) implies that each $F_j = 0$.*

(iv) *If $|\mathcal{I}| \leq d - 1$, then (2.10) implies that each $E_i = 0$.*

(v) *If $|\mathcal{J}| \leq d - 1$, then (2.11) implies that each $F_j = 0$.*

(vi) *If $\max\{|\mathcal{I}|, |\mathcal{J}|\} \leq d$, then the p_{ij}'s and λ_i's (from (2.12)) are unique.*

(vii) *If $\max\{|\mathcal{I}|, |\mathcal{J}|\} \leq d$ and all E_i's and F_j's are $(m-1)$-additive, then all p_{ij}'s and λ_i's (from (2.12)) are $(m-2)$-additive and $(m-1)$-additive, respectively.*

Proof. (i) is trivial. (ii) and (iv) follow immediately from the definition by choosing $\mathcal{J} = \emptyset$, while (iii) and (v) follow by choosing $\mathcal{I} = \emptyset$.

Next, suppose that $\max\{|\mathcal{I}|, |\mathcal{J}|\} \leq d$ and that we have two standard solutions, that is, there exist maps $p_{ij}, q_{ij} : \mathcal{A}^{m-2} \to \mathcal{M}$, $i \in \mathcal{I}$, $j \in \mathcal{J}$, $i \neq j$, and $\lambda_k, \mu_k : \mathcal{A}^{m-1} \to \mathcal{Z}(\mathcal{M})$, $k \in \mathcal{I} \cup \mathcal{J}$, $\lambda_k = \mu_k = 0$ if $k \notin \mathcal{I} \cap \mathcal{J}$, such that

$$E_i(\overline{x}_m^i) = \sum_{\substack{j \in \mathcal{J}, \\ j \neq i}} x_j p_{ij}(\overline{x}_m^{ij}) + \lambda_i(\overline{x}_m^i) = \sum_{\substack{j \in \mathcal{J}, \\ j \neq i}} x_j q_{ij}(\overline{x}_m^{ij}) + \mu_i(\overline{x}_m^i)$$

for all $\overline{x}_m \in \mathcal{A}^m$ and each $i \in \mathcal{I}$. Accordingly

$$\sum_{\substack{j \in \mathcal{J}, \\ j \neq i}} x_j [p_{ij}(\overline{x}_m^{ij}) - q_{ij}(\overline{x}_m^{ij})] = \mu_i(\overline{x}_m^i) - \lambda_i(\overline{x}_m^i) \in \mathcal{Z}(\mathcal{M}).$$

If $i \in \mathcal{J}$, then (v) implies that $p_{ij} = q_{ij}$ for all $j \in \mathcal{J} \setminus \{i\}$ because $|\mathcal{J} \setminus \{i\}| \leq d - 1$, and hence also $\lambda_i = \mu_i$. If $i \notin \mathcal{J}$, then $\lambda_i = \mu_i = 0$ and so we have

$$\sum_{j \in \mathcal{J}} x_j [p_{ij}(\overline{x}_m^{ij}) - q_{ij}(\overline{x}_m^{ij})] = 0.$$

But now (iii) yields $p_{ij} = q_{ij}$ for all $j \in \mathcal{J}$. Thus $p_{ij} = q_{ij}$ and $\lambda_k = \mu_k$ for all $i \in \mathcal{I}$, $j \in \mathcal{J}$, $k \in \mathcal{I} \cup \mathcal{J}$, proving (vi).

Proving (vii) is easy, basically it is just an exercise in notation. Assume that the E_i's and F_j's are $(m-1)$-additive and that (2.12) holds. Replacing x_j by $y_j + z_j$ in the first identity of (2.12) we arrive at the situation where clearly (v) is applicable, and hence (vii) can be easily inferred.

It has been indicated earlier that PI's may be regarded as special cases of
FI's. However, if one takes the point of view that the desired goal in FI theory is to
try to show that the only solutions of a particular FI are the standard ones, then PI
theory must be regarded as complementary to FI theory: a PI is an FI with highly
nonstandard solutions. Let us explain this more specifically. Suppose that a ring
$\mathcal{A}$ satisfies a polynomial identity of degree n. It is well known and easy to see (linearization!) that then $\mathcal{A}$ also satisfies a multilinear polynomial identity of degree
$\leq n$. Therefore, there exists a multilinear polynomial $f = f(x_1, \ldots, x_m)$ of degree
$m \leq n$ such that $\mathcal{A}$ satisfies f and $\mathcal{A}$ does not satisfy a polynomial identity of degree $m - 1$. Because of multilinearity we can write $f(x_1, \ldots, x_m) = \sum_{i=1}^{m} f_i(\overline{x}_m^i)x_i$
where at least one of the f_i's is a nonzero polynomial of degree $m - 1$. But then
it follows from Lemma 2.8 (ii) (with $\mathcal{M} = \mathcal{A}$) that $\mathcal{A}$ can be strongly d-free only
for $d < m(\leq n)$. Thus, a ring cannot be simultaneously strongly d-free and satisfy
a PI of degree d. Let us record this observation as

Lemma 2.9. *A strongly d-free ring does not satisfy a polynomial identity of degree*
$\leq d$.

As observed above, every unital ring is strongly 1-free. With this and Lemma
2.8 (i) in mind we introduce the next concept.

Definition 2.10. Let $\mathcal{A}$ be a unital ring. We say that $\mathcal{A}$ has FI-*degree* d (notation
FI-deg$(\mathcal{A}) = d$) if $\mathcal{A}$ is strongly d-free and is not strongly $(d+1)$-free. In the case
when $\mathcal{A}$ is ∞-free, we say that $\mathcal{A}$ has FI-*degree* ∞ (FI-deg$(\mathcal{A}) = \infty$).

For example, it is immediate from the above discussion that FI-deg$(\mathcal{C}) = 1$
for any commutative unital ring $\mathcal{C}$. This trivial observation is a special case (when
$d = 1$) of the next lemma. In order to understand its background we advise the
reader to consult the discussion following Example 1.2.

Lemma 2.11. *Let $\mathcal{A}$ be a strongly d-free unital ring. If there exist traces of k-
additive maps $\alpha_k : \mathcal{A} \to \mathcal{Z}$, $k = 1, \ldots, d$, such that*

$$x^d + \alpha_1(x)x^{d-1} + \ldots + \alpha_{d-1}(x)x + \alpha_d(x) = 0 \qquad (2.13)$$

for all $x \in \mathcal{A}$, then FI-deg$(\mathcal{A}) = d$.

Proof. By assumption for every $k = 1, \ldots, d$ there is a k-additive map $\zeta_k : \mathcal{A}^k \to$
$\mathcal{Z}$, such that $\alpha_k(x) = \zeta_k(x, x, \ldots, x)$. Consider the k-additive functions $T_k : \mathcal{A}^k \to$
$\mathcal{A}$, $1 \leq k \leq d$, defined by

$$= \sum \Big(x_{\pi(1)}x_{\pi(2)} \cdots x_{\pi(k)} + \zeta_1(x_{\pi(1)})x_{\pi(2)} \cdots x_{\pi(k)} + \ldots$$

$$, x_{\pi(2)}, \ldots, x_{\pi(k-1)})x_{\pi(k)} + \zeta_k(x_{\pi(1)}, x_{\pi(2)}, \ldots, x_{\pi(k)}) \Big).$$

We define $T_0 = 1$. First, the linearization of (2.13) is just $T_d(\overline{x}_d) = 0$. Second, for $1 \leq k \leq d$, T_k enjoys the recursive property

$$T_k(\overline{x}_k) = \sum_{i=1}^{k} T_{k-1}(\overline{x}_k^i)x_i + \sum_{\pi \in S_k} \zeta_k(x_{\pi(1)}, x_{\pi(2)}, \ldots, x_{\pi(k)}). \qquad (2.14)$$

If $T_k = 0$ for some $k < d$, then (2.14) takes the form (2.10) and since $\mathcal{A}$ is strongly $(k+1)$-free (Lemma 2.8 (i)) it follows that $T_{k-1} = 0$ (Lemma 2.8 (iv)), ultimately leading to the contradiction $T_0 = 0$. Thus $T_k \neq 0$ for all $k < d$ and in particular $T_{d-1} \neq 0$. However, in view of $T_d = 0$ and (2.14), T_{d-1} satisfies $\sum_{i=1}^{d} T_{d-1}(\overline{x}_d^i)x_i \in \mathcal{Z}$. But then $\mathcal{A}$ cannot be strongly $(d+1)$-free by Lemma 2.8 (iv). $\qquad \square$

2.3 Strongly $(t; d)$-Free Rings

As already indicated in the previous chapter, in order to establish d-freeness of certain rings it is necessary first to analyze some more general identities with coefficients which are powers of some fixed element t. This is the main reason for introducing the concept of a strongly $(t; d)$-free ring in this section.

We keep the notation from the preceding section and add some new notation.

Throughout this section t will be a fixed element in $\mathcal{A}$, and we set $\mathcal{C}(t) = \{\mu \in \mathcal{M} \,|\, \mu t = t\mu\}$. Further, $a, b \geq 0$ will be integers and

$$E_{iu}, F_{jv} : \mathcal{A}^{m-1} \to \mathcal{M}, \ i \in \mathcal{I}, \ j \in \mathcal{J}, \ 0 \leq u \leq a, \ 0 \leq v \leq b$$

will be arbitrary maps. We consider the FI

$$\sum_{i \in \mathcal{I}} \sum_{u=0}^{a} E_{iu}(\overline{x}_m^i)x_i t^u + \sum_{j \in \mathcal{J}} \sum_{v=0}^{b} t^v x_j F_{jv}(\overline{x}_m^j) = 0 \qquad (2.15)$$

for all $\overline{x}_m \in \mathcal{A}^m$. We define a *standard solution* of (2.15) as

$$E_{iu}(\overline{x}_m^i) = \sum_{\substack{j \in \mathcal{J}, \ v=0 \\ j \neq i}}^{b} t^v x_j p_{iujv}(\overline{x}_m^{ij}) + \sum_{v=0}^{b} \lambda_{iuv}(\overline{x}_m^i)t^v,$$

$$F_{jv}(\overline{x}_m^j) = -\sum_{\substack{i \in \mathcal{I}, \ u=0 \\ i \neq j}}^{a} p_{iujv}(\overline{x}_m^{ij})x_i t^u - \sum_{u=0}^{a} \lambda_{juv}(\overline{x}_m^j)t^u, \qquad (2.16)$$

$$\lambda_{kuv} = 0 \quad \text{if} \quad k \notin \mathcal{I} \cap \mathcal{J},$$

for all $\overline{x}_m \in \mathcal{A}^m$, $i \in \mathcal{I}$, $j \in \mathcal{J}$, $0 \leq u \leq a$, $0 \leq v \leq b$, where

$$p_{iujv} : \mathcal{A}^{m-2} \to \mathcal{M}, \ i \in \mathcal{I}, \ j \in \mathcal{J}, \ i \neq j, \ 0 \leq u \leq a, \ 0 \leq v \leq b,$$
$$\lambda_{kuv} : \mathcal{A}^{m-1} \to \mathcal{Z}(\mathcal{M}), \ k \in \mathcal{I} \cup \mathcal{J}, \ 0 \leq u \leq a, \ 0 \leq v \leq b.$$

Definition 2.12. A ring $\mathcal{A}$ is said to be *strongly $(t; d)$-free*, where $t \in \mathcal{A}$ and $d \in \mathbb{N}$, if for every unital $(\mathcal{A}, \mathcal{A})$-bimodule $\mathcal{M}$, all $m \in \mathbb{N}$, all $\mathcal{I}, \mathcal{J} \subseteq \{1, 2, \ldots, m\}$, and all integers $a, b \geq 0$ the following condition is satisfied: If $\max\{|\mathcal{I}| + a, |\mathcal{J}| + b\} \leq d$, then (2.15) implies (2.16).

So, unlike Definition 2.7, this definition requires only one condition to be fulfilled. One might of course wonder why there is no analogy of the condition (b), that is, why this definition does not also consider the FI

$$\sum_{i \in \mathcal{I}} \sum_{u=0}^{a} E_{iu}(\overline{x}_m^i) x_i t^u + \sum_{j \in \mathcal{J}} \sum_{v=0}^{b} t^v x_j F_{jv}(\overline{x}_m^j) \in \mathcal{Z}(\mathcal{M}) \tag{2.17}$$

for all $\overline{x}_m \in \mathcal{A}^m$. The reason is simple: an appropriate version of the condition concerning (2.17) automatically follows from the required one. We will establish this at the end of this section (Corollary 2.16), and from this we will then be able to conclude that a strongly $(t; d)$-free ring is also d-free (Corollary 2.17). First we need some auxiliary results.

When one of the sets $\mathcal{I}$ and $\mathcal{J}$ is empty we get two particular cases of (2.15):

$$\sum_{i \in \mathcal{I}} \sum_{u=0}^{a} E_{iu}(\overline{x}_m^i) x_i t^u = 0 \quad \text{for all } \overline{x}_m \in \mathcal{A}^m, \tag{2.18}$$

$$\sum_{j \in \mathcal{J}} \sum_{v=0}^{b} t^v x_j F_{jv}(\overline{x}_m^j) = 0 \quad \text{for all } \overline{x}_m \in \mathcal{A}^m. \tag{2.19}$$

We also state two somewhat more general relations

$$\sum_{i \in \mathcal{I}} \sum_{u=0}^{a} E_{iu}(\overline{x}_m^i) x_i t^u \in \mathcal{C}(t) \quad \text{for all } \overline{x}_m \in \mathcal{A}^m, \tag{2.20}$$

$$\sum_{j \in \mathcal{J}} \sum_{v=0}^{b} t^v x_j F_{jv}(\overline{x}_m^j) \in \mathcal{C}(t) \quad \text{for all } \overline{x}_m \in \mathcal{A}^m. \tag{2.21}$$

Let us record an analogue of Lemma 2.8.

Lemma 2.13. *Let $\mathcal{A}$ be a strongly $(t; d)$-free ring. Then:*

(i) *$\mathcal{A}$ is strongly $(t; d')$-free for every $d' < d$.*

(ii) *If $|\mathcal{I}| + a \leq d$, then (2.18) implies that each $E_{iu} = 0$.*

(iii) *If $|\mathcal{J}| + b \leq d$, then (2.19) implies that each $F_{jv} = 0$.*

(iv) *If $|\mathcal{I}| + a \leq d - 1$, then (2.20) implies that each $E_{iu} = 0$.*

(v) *If $|\mathcal{J}| + b \leq d - 1$, then (2.21) implies that each $F_{jv} = 0$.*

(vi) *If* $\max\{|\mathcal{I}| + a, |\mathcal{J}| + b\} \leq d$, *then the* p_{iujv}'s *and* λ_{iuv}'s *(from* (2.16)*) are unique.*

(vii) *If* $\max\{|\mathcal{I}|+a, |\mathcal{J}|+b\} \leq d$ *and all* E_{iu}'s *and* F_{jv}'s *are* $(m-1)$-*additive, then all* p_{iujv}'s *and* λ_{iuv}'s *(from* (2.16)*) are* $(m-2)$-*additive and* $(m-1)$-*additive, respectively.*

Proof. As in the proof of Lemma 2.8 we note that (i), (ii), and (iii) are trivial. To prove (iv), assume $|\mathcal{I}|+a \leq d-1$ and (2.20). Commuting $\sum_{i \in \mathcal{I}} \sum_{u=0}^{a} E_{iu}(\bar{x}_m^i)x_i t^u$ with t we get

$$\sum_{i \in \mathcal{I}} \sum_{u=0}^{a+1} G_{iu}(\bar{x}_m^i)x_i t^u = 0$$

for all $\bar{x}_m \in \mathcal{A}^m$, where

$$G_{i0}(\bar{x}_m^i) = -tE_{i0}(\bar{x}_m^i),$$
$$G_{iu}(\bar{x}_m^i) = -tE_{iu}(\bar{x}_m^i) + E_{i,u-1}(\bar{x}_m^i), \quad u = 1, \ldots, a,$$
$$G_{i,a+1}(\bar{x}_m^i) = E_{ia}(\bar{x}_m^i)$$

for each $i \in \mathcal{I}$. Since $|\mathcal{I}| + (a+1) \leq d$ by our assumption, (ii) implies that $G_{iu} = 0$ for every $u = 0, 1, \ldots, a+1$. But this readily yields that each $E_{iu} = 0$. The proof of (vi) is similar. Finally, (vi) and (vii) can be proved similarly as analogous assertions in Lemma 2.8. $\qquad\square$

Let us point out a necessary condition that an element t yielding strong $(t; d)$-freeness must satisfy.

Lemma 2.14. *Suppose that* $t \in \mathcal{A}$ *is such that* $\mathcal{A}$ *is a strongly* $(t; d)$-*free ring. If* $\sum_{u=0}^{d-1} \gamma_u t^u = 0$ *for some* $\gamma_u \in \mathcal{Z}(\mathcal{M})$, *then each* $\gamma_u = 0$.

Proof. Our assumption implies that $\sum_{u=0}^{d-1} \gamma_u x_1 t^u = 0$ for all $x_1 \in \mathcal{A}$. This can be regarded as a special case of (2.18) and so Lemma 2.13 (ii) tells us that each $\gamma_u = 0$. $\qquad\square$

In the proof of the next lemma we shall for the first time use a certain method that will be henceforth frequently used when taking the inductive step in a proof (we remark that many of the proofs in the area of functional identities are of an inductive nature). Let us demonstrate this method by an example. Consider the familiar function

$$H(\bar{x}_m) = \sum_{i \in \mathcal{I}} \sum_{u=0}^{a} E_{iu}(\bar{x}_m^i)x_i t^u + \sum_{j \in \mathcal{J}} \sum_{v=0}^{b} t^v x_j F_{jv}(\bar{x}_m^j). \qquad (2.22)$$

Assume $1 \in \mathcal{J}$. Now form the function

$$H(tx_1, x_2, \ldots, x_m) - tH(\bar{x}_m). \qquad (2.23)$$

By expanding (2.23) and collecting terms in an obvious way one sees that (2.23) can be rewritten as

$$\sum_{i\in\mathcal{I}}\sum_{u=0}^{a}G_{iu}(\overline{x}_m^i)x_it^u + \sum_{\substack{j\in\mathcal{J},\\ j\neq 1}}\sum_{v=0}^{b+1}t^v x_j H_{jv}(\overline{x}_m^j)$$

for appropriate G_{iu}'s and H_{jv}'s. Thus the size of $|\mathcal{J}|$ has been decreased by at least 1, i.e., $\mathcal{J}$ has been replaced by $\overline{\mathcal{J}} = \mathcal{J}\setminus\{1\}$. In an informal and nonrigorous way, we shall sometimes refer to any process similar to the above (such as forming (2.23)) as a *t-substitution operation*.

Lemma 2.15. *Suppose there exist maps* $\mu_w : \mathcal{A}^m \to \mathcal{Z}(\mathcal{M})$, $0 \leq w \leq c$, *such that*

$$\sum_{i\in\mathcal{I}}\sum_{u=0}^{a}E_{iu}(\overline{x}_m^i)x_it^u + \sum_{j\in\mathcal{J}}\sum_{v=0}^{b}t^v x_j F_{jv}(\overline{x}_m^j) = \sum_{w=0}^{c}\mu_w(\overline{x}_m)t^w$$

for all $\overline{x}_m \in \mathcal{A}^m$. *If* $\mathcal{A}$ *is a strongly* $(t;d)$*-free ring and* $\max\{a+|\mathcal{I}|, c+|\mathcal{J}|\} \leq d-1$, *then each* $\mu_w = 0$.

Proof. We proceed by induction on $|\mathcal{J}|$. If $|\mathcal{J}| = 0$, then Lemma 2.13 (iv) implies that each $E_{iu} = 0$. Consequently

$$\sum_{w=0}^{c}\mu_w(\overline{x}_m)t^w = 0$$

and so Lemma 2.14 tells us that each $\mu_w = 0$.

Now let $|\mathcal{J}| \geq 1$, say $1 \in \mathcal{J}$. Now we can use a t-substitution operation. Define H by (2.22), now form (2.23) and note that this yields

$$\sum_{i\in\mathcal{I}}\sum_{u=0}^{a}G_{iu}(\overline{x}_m^i)x_it^u + \sum_{\substack{j\in\mathcal{J},\\ j\neq 1}}\sum_{v=0}^{b+1}t^v x_j H_{jv}(\overline{x}_m^j) = \mu_0(tx_1, x_2,\ldots, x_m)$$

$$+ \sum_{w=1}^{c}\{\mu_w(tx_1, x_2,\ldots, x_m) - \mu_{w-1}(\overline{x}_m)\}t^w - \mu_c(\overline{x}_m)t^{c+1}$$

for some G_{iu}'s and H_{jv}'s. Since $c + 1 + |\mathcal{J}\setminus\{1\}| = c + |\mathcal{J}|$, we are in a position to apply the induction assumption. Hence it follows that

$$\mu_0(tx_1, x_2,\ldots, x_m) = 0,$$
$$\mu_w(tx_1, x_2,\ldots, x_m) - \mu_{w-1}(\overline{x}_m) = 0,$$
$$-\mu_c(\overline{x}_m) = 0$$

for all $\overline{x}_m \in \mathcal{A}^m$. This clearly implies that each $\mu_w = 0$. $\qquad\qquad\square$

We are now in a position to state a result announced at the beginning of this section, i.e., the one related to the condition (b) from Definition 2.7.

Corollary 2.16. *Let $\mathcal{A}$ be a strongly $(t; d)$-free ring. If $\max\{|\mathcal{I}|+a, |\mathcal{J}|+b\} \leq d-1$, then (2.17) implies (2.16).*

Proof. Setting $c = 0$ in Lemma 2.15 we see that (2.17) together with $\max\{|\mathcal{I}|+a, |\mathcal{J}|\} \leq d-1$ implies (2.15). From (2.15) and $\max\{|\mathcal{I}|+a, |\mathcal{J}|+b\} \leq d-1 < d$ we then get (2.16). $\qquad\square$

Setting $a = b = 0$ in Definition 2.12 and Corollary 2.16 we immediately get

Corollary 2.17. *If a ring $\mathcal{A}$ is strongly $(t; d)$-free for some $t \in \mathcal{A}$, then $\mathcal{A}$ is d-free.*

This result is of great significance. Namely, the usual way of establishing the strong d-freeness of a ring is to prove its strong $(t; d)$-freeness for suitable t.

2.4 s-deg$(\mathcal{A}) \leq$ FI-deg$(\mathcal{A})$

The formula from the title will follow from a somewhat more general statement concerning strong $(t; d)$-freeness (Theorem 2.19). Therefore we keep all notation from the preceding section. We shall slightly abbreviate this notation by omitting writing the arguments of functions. More precisely, for maps $F : \mathcal{A}^{m-1} \to \mathcal{M}$, $G : \mathcal{A}^{m-2} \to \mathcal{M}$ we shall write

$$F^i \quad \text{for} \quad F(\bar{x}^i_m),$$
$$G^{ij} \quad \text{for} \quad G(\bar{x}^{ij}_m).$$

So, for example, (2.15) will be succinctly written as

$$\sum_{i \in \mathcal{I}} \sum_{u=0}^{a} E^i_{iu} x_i t^u + \sum_{j \in \mathcal{J}} \sum_{v=0}^{b} t^v x_j F^j_{jv} = 0,$$

and the first line of a standard solution (2.16) will be written as

$$E^i_{iu} = \sum_{\substack{j \in \mathcal{J}, \ v=0 \\ j \neq i}}^{b} t^v x_j p^{ij}_{iujv} + \sum_{v=0}^{b} \lambda^i_{iuv} t^v.$$

Moreover, for $H : \mathcal{A}^m \to \mathcal{M}$ we shall write

$$H(x_k t) \text{ for } H(x_1, \ldots, x_{k-1}, x_k t, x_{k+1}, \ldots, x_m)$$

and for $F : \mathcal{A}^{m-1} \to \mathcal{M}$ we shall write

$$F^i(x_k t) \text{ for}$$
$$F(x_1, \ldots, x_{i-1}, x_{i+1}, \ldots, x_{k-1}, x_k t, x_{k+1}, \ldots, x_m) \text{ if } i < k,$$
$$\text{and} \quad F(x_1, \ldots, x_{k-1}, x_k t, x_{k+1}, \ldots, x_{i-1}, x_{i+1}, \ldots, x_m) \text{ if } i > k.$$

General statements on FI's are usually proved progressively in a sense that one starts with simpler FI's and then proceeds step by step to more complicated ones. The proof of each step itself is usually based on induction on the number of variables involved.

Lemma 2.18. *If s-$\deg(t) \geq |\mathcal{I}| + a$, then (2.18) implies that each $E_{iu} = 0$. Similarly, if s-$\deg(t) \geq |\mathcal{J}| + b$, then (2.19) implies that each $F_{jv} = 0$.*

Proof. Because of the symmetry it suffices to prove only the first assertion. We proceed by induction on $|\mathcal{I}|$. For $|\mathcal{I}| = 1$, say $\mathcal{I} = \{1\}$, we have

$$\sum_{u=0}^{a} E_{1u}^{1} x_1 t^u = 0 \quad \text{for every } x_1 \in \mathcal{A}.$$

Since s-$\deg(t) \geq a + 1$, fixing $x_2, \ldots, x_m$ and applying Lemma 2.2 (i) we conclude that each $E_{1u} = 0$.

In the inductive case $|\mathcal{I}| > 1$, say $1, 2 \in \mathcal{I}$, we apply the t-substitution operation. Set

$$H(\bar{x}_m) = \sum_{i \in \mathcal{I}} \sum_{u=0}^{a} E_{iu}^{i} x_i t^u = 0$$

and note that

$$0 = H(x_1 t) - H(\bar{x}_m) t$$

$$= \sum_{\substack{i \in \mathcal{I}, \\ i \neq 1}} E_{i0}^{i}(x_1 t) x_i + \sum_{\substack{i \in \mathcal{I}, \\ i \neq 1}} \sum_{u=1}^{a} \{E_{iu}^{i}(x_1 t) - E_{i,u-1}^{i}\} x_i t^u - \sum_{\substack{i \in \mathcal{I}, \\ i \neq 1}} E_{ia}^{i} x_i t^{a+1}.$$

By the induction assumption it follows that $E_{ia} = 0$, $E_{iu}^{i}(x_1 t) - E_{i,u-1}^{i} = 0$ for all $i \neq 1$, $u = 0, 1, \ldots, a$, hence each $E_{iu} = 0$ for $i \neq 1$. Repetition of the preceding process, but with respect to x_2 enables us to obtain $E_{1u} = 0$ for all $u = 0, 1, \ldots, a$. □

Theorem 2.19. *Let $\mathcal{A}$ be a unital ring and let $t \in \mathcal{A}$. If $d \in \mathbb{N}$ is such that s-$\deg(t) \geq d$, then $\mathcal{A}$ is strongly $(t; d)$-free; in particular, $\mathcal{A}$ is strongly d-free.*

Proof. We may assume that (2.15) holds with s-$\deg(t) \geq \max\{|\mathcal{I}| + a, |\mathcal{J}| + b\}$. Our goal is to derive (2.16) from these assumptions.

Suppose we have already proved that each E_{iu} has the form given according to (2.16) (a part of this assumption is that $\lambda_{iuv} = 0$ if $i \notin \mathcal{J}$). We claim that then the F_{jv}'s are given according to (2.16) as well. Indeed, substituting the expressions for E_{iu}'s in (2.15) we obtain that

$$\sum_{j \in \mathcal{J}} \sum_{v=0}^{b} t^v x_j \left[F_{jv}^{j} + \sum_{\substack{i \in \mathcal{I}, \\ i \neq j}} \sum_{u=0}^{a} p_{iujv}^{ij} x_i t^u + \sum_{u=0}^{a} \lambda_{juv}^{j} t^u \right] = 0.$$

Our claim now follows from Lemma 2.18. Analogously, if all the F_{jv}'s are given according to (2.16), then all E_{iu}'s are given according to (2.16) as well.

If $|\mathcal{I}| = 0$ or $|\mathcal{J}| = 0$ then the result follows from Lemma 2.18.

We proceed by induction on $|\mathcal{I}| + |\mathcal{J}|$. In view of the preceding remark the first case to consider is when $|\mathcal{I}| = 1 = |\mathcal{J}|$. There are essentially just two subcases to consider.

The first one is when $\mathcal{I} = \{1\} = \mathcal{J}$. We have

$$\sum_{u=0}^{a} E_{1u}^1 x_1 t^u + \sum_{v=0}^{b} t^v x_1 F_{1v}^1 = 0$$

and, fixing $x_2, x_3, \ldots, x_m$, we then have by Lemma 2.2 (iii) that

$$E_{1u}(x_2, \ldots, x_m) = \sum_{v=0}^{b} \lambda_{1uv}(x_2, \ldots, x_m) t^v$$

for some $\lambda_{1uv}(r_2, \ldots, r_m) \in \mathcal{Z}(M)$, $0 \leq u \leq a$, $0 \leq v \leq b$. We see that the E_{1u}'s are given according to (2.16). This forces the F_{1v}'s to be given according to (2.16).

The second subcase is when $\mathcal{I} = \{2\}$ and $\mathcal{J} = \{1\}$. We have

$$\sum_{u=0}^{a} E_{2u}^2 x_2 t^u + \sum_{v=0}^{b} t^v x_1 F_{1v}^1 = 0. \tag{2.24}$$

Fix any v with $0 \leq v \leq b$. Since $s\text{-deg}(t) \geq \max\{|\mathcal{I}| + a, |\mathcal{J}| + b\} > b$, there exists $c_l, d_l \in \mathcal{A}$, $l = 1, 2, \ldots, n$, such that, for $w = 0, 1, \ldots, b$

$$\sum_{l=1}^{n} c_l t^w d_l = \begin{cases} 1 & \text{if } w = v, \\ 0 & \text{if } w \neq v. \end{cases}$$

Therefore

$$F_{1v}^1 = \sum_{w=0}^{b} \sum_{l=1}^{n} c_l t^w d_l F_{1w}^1 = \sum_{l=1}^{n} c_l \sum_{w=0}^{b} t^w d_l F_{1w}^1,$$

which in view of (2.24) yields

$$F_{1v}^1 = -\sum_{l=1}^{n} c_l \sum_{u=0}^{a} E_{2u}(d_l, x_3, \ldots, x_m) x_2 t^u = -\sum_{u=0}^{a} p_u(x_3, \ldots, x_m) x_2 t^u,$$

where $p_u(x_3, \ldots, x_m) = \sum_{l=1}^{n} c_l E_{2u}(d_l, x_3, \ldots, x_m)$. This means that all F_{1v}'s are given according to (2.16) and so the same holds true for all E_{2u}'s. The proof of the second subcase is complete.

We may now assume $|\mathcal{I}| + |\mathcal{J}| > 2$ and make the inductive step. We may assume $|\mathcal{J}| \geq 2$ and that $1, 2 \in \mathcal{J}$. Set

$$H(\bar{x}_m) = \sum_{i \in \mathcal{I}} \sum_{u=0}^{a} E_{iu}^i x_i t^u + \sum_{j \in \mathcal{J}} \sum_{v=0}^{b} t^v x_j F_{jv}^j$$

and apply the t-substitution operation. Our assumption is that $H = 0$. Computing $H(tx_1) - tH(\overline{x}_m)$ results in

$$\sum_{i \in \mathcal{I}} \sum_{u=0}^{a} G_{iu}^i x_i t^u + \sum_{\substack{j \in \mathcal{J}, \\ j \neq 1}} x_j F_{j0}^j(tx_1)$$

$$+ \sum_{\substack{j \in \mathcal{J}, \\ j \neq 1}} \sum_{v=1}^{b} t^v x_j \{F_{jv}^j(tx_1) - F_{j,v-1}^j\} - \sum_{\substack{j \in \mathcal{J}, \\ j \neq 1}} t^{b+1} x_j F_{jb}^j = 0$$

for appropriate G_{iu}'s. Note that $|\mathcal{J} \setminus \{1\}| + b + 1 = |\mathcal{J}| + b$ and so the degree condition on t again holds. By induction we have

$$F_{jb}^j = -\sum_{\substack{i \in \mathcal{I}, \\ i \neq j}} \sum_{u=0}^{a} q_{iuj\,b+1}^{ij} x_i t^u - \sum_{u=0}^{a} \mu_{ju\,b+1}^j t^u, \quad j \neq 1,$$

$$F_{jv}^j(tx_1) - F_{j,v-1}^j = -\sum_{\substack{i \in \mathcal{I}, \\ i \neq j}} \sum_{u=0}^{a} q_{iujv}^{ij} x_i t^u - \sum_{u=0}^{a} \mu_{juv}^j t^u,$$

$$j \neq 1, \quad v = 1, 2, \ldots, b,$$

and $\mu_{juv} = 0$ if $j \notin \mathcal{I}$. From this identity, beginning with F_{jb} and proceeding recursively, we see that

$$F_{jv}^j = -\sum_{\substack{i \in \mathcal{I}, \\ i \neq j}} \sum_{u=0}^{a} p_{iujv}^{ij} x_i t^u - \sum_{u=0}^{a} \lambda_{juv}^j t^u, \quad j \neq 1$$

for appropriate p_{iujv} and λ_{juv} with $\lambda_{juv} = 0$ if $j \notin \mathcal{I}$. Similarly, by considering $H(tx_2) - tH(\overline{x}_m)$ we see that

$$F_{1v}^1 = -\sum_{\substack{i \in \mathcal{I}, \\ i \neq 1}} \sum_{u=0}^{a} p_{iu1v}^{i1} x_i t^u - \sum_{u=0}^{a} \lambda_{1uv}^1 t^u,$$

$\lambda_{1uv} = 0$ if $1 \notin \mathcal{I}$, and so all F_{jv}'s are in the standard form. As we have seen before this forces the E_{iu}'s to be given according to (2.16). $\square$

The last assertion in Theorem 2.19 can be stated as

Corollary 2.20. *Let $\mathcal{A}$ be a unital ring. Then $s\text{-deg}(\mathcal{A}) \leq \text{FI-deg}(\mathcal{A})$.*

We now turn our attention to some concrete classes of rings.

Corollary 2.21. *Let $\mathcal{A}$ be a simple unital ring. Then*

$$\text{FI-}\deg(\mathcal{A}) = \sqrt{\dim_{\mathcal{Z}}(\mathcal{A})}.$$

In particular, $\text{FI-}\deg(\mathcal{A}) = \infty$ (i.e., $\mathcal{A}$ is ∞-free) if and only if $\mathcal{A}$ is not a PI-ring.

Proof. Let $d = \deg(\mathcal{A})$. We already know that $s\text{-}\deg(\mathcal{A}) = d$ (Lemma 2.3); furthermore, it is well-known that $\sqrt{\dim_{\mathcal{Z}}(\mathcal{A})} = d$ (see Corollary C.3 below). Corollary 2.20 tells us that $\text{FI-}\deg(\mathcal{A}) \geq d$. But then Lemma 2.11 together with Corollary C.3 implies that $\text{FI-}\deg(\mathcal{A}) = d$. The last statement follows from the fact that $\mathcal{A}$ is a PI-ring if and only if it is finite dimensional over its center (see Theorem C.1). $\qquad\square$

Corollary 2.20 and Theorem 2.4 give

Corollary 2.22. *Let $\mathcal{R}$ be any unital ring, and let $n \in \mathbb{N}$. Then*

$$\text{FI-}\deg(M_n(\mathcal{R})) \geq n.$$

When $\mathcal{R} = \mathcal{C}$ is a commutative ring, this inequality becomes equality. This follows from Lemma 2.11 and the Cayley–Hamilton theorem (cf. the discussion following Example 1.2).

Corollary 2.23. *Let $\mathcal{C}$ be a commutative unital ring, and let $n \in \mathbb{N}$. Then*

$$\text{FI-}\deg(M_n(\mathcal{C})) = n.$$

Finally, from Corollary 2.23, Corollary 2.20, and Corollary 2.6 we get

Corollary 2.24. *Let $\mathcal{V}$ be a vector space over a field $\mathbb{F}$. Then*

$$\text{FI-}\deg(\text{End}_{\mathbb{F}}(\mathcal{V})) = \dim_{\mathbb{F}}(\mathcal{V}).$$

In particular, $\text{FI-}\deg(\text{End}_{\mathbb{F}}(\mathcal{V})) = \infty$ (i.e., $\text{End}_{\mathbb{F}}(\mathcal{V})$ is ∞-free) if and only if $\mathcal{V}$ is infinite dimensional.

Literature and Comments. The concept of the strong degree was introduced by Beidar, Brešar and Chebotar in [19] and most of the results in this chapter are taken from this paper. It should be mentioned, however, that the proofs from Section 2.2 are only superficially outlined in [19] since they are just simple modifications of those given earlier by Beidar and Martindale in [38], where FI's in a somewhat different setting of prime rings are considered. Also, [19] does not consider bimodules. The necessity for involving bimodules was first observed in [1] where the consideration of Lie derivations from von Neumann algebras into their bimodules rests heavily on the results exposed above.

Part II

The General Theory

Chapter 3

Constructing d-Free Sets

In this chapter we will study d-free sets and some related (but more complicated) concepts such as $(t; d)$-freeness and $(*; t; d)$-freeness. Besides introducing these concepts and considering their formal properties, our main objective will be to present various constructions that yield new d-free (resp. $(t; d)$-free) sets from given ones. In Chapter 5 we shall actually establish d-freeness of certain particular classes of sets and then use the results of the present chapter to show that the list of d-free sets is really quite extensive.

3.1 Notation

The problem of notation plays an inordinately large role in FI theory. By the very nature of the subject there is the ever-present problem of finding the right balance between accurate but heavy-handed notation on the one hand and simple but nonrigorous notation on the other hand. Therefore, before embarking on the general theory, we shall first make some comments about the notation we shall be using. This notation extends the one introduced in Part I.

We first introduce the setting where the general theory takes place. Throughout, $\mathcal{Q}$ will be a unital ring with center $\mathcal{C}$. In applications $\mathcal{Q}$ will often be a ring of quotients of a given ring, and $\mathcal{C}$ will often be a field (e.g., the extended centroid in the prime ring case). This explains the background of our notation; however, in the general theory $\mathcal{Q}$ is just any unital ring and $\mathcal{C}$ is its center. We shall consider FI's involving functions mapping into $\mathcal{Q}$, so $\mathcal{Q}$ will play a similar role as did the bimodule $\mathcal{M}$ in Part I. Assuming that $\mathcal{Q}$ was a bimodule over some ring (instead of being a ring itself) would cause some technical difficulties since occasionally there will be a need to multiply elements in $\mathcal{Q}$ between themselves. That is why we have decided to work in the context of rings rather than bimodules.

By $\mathcal{C}^*$ we denote the group of invertible elements in $\mathcal{C}$. For a subset $\mathcal{R}$ of $\mathcal{Q}$ we write $\mathcal{C}(\mathcal{R})$ for the centralizer of $\mathcal{R}$ in $\mathcal{Q}$; for $t \in \mathcal{Q}$ we write $\mathcal{C}(t)$ instead of $\mathcal{C}(\{t\})$.

The notion of an *algebraic element* (over a commutative ring) and its degree is a standard one, and was already defined in the preceding chapter. For $x \in \mathcal{Q}$ we denote by $\deg(x)$ *the degree of algebraicity of x over* $\mathcal{C}$ provided of course that x is algebraic over $\mathcal{C}$. In case x is not algebraic over $\mathcal{C}$ we shall write $\deg(x) = \infty$. So $\deg(x) \geq d$ means that either x is not algebraic over $\mathcal{C}$ or its degree of algebraicity is $\geq d$. For a nonempty subset $\mathcal{R}$ of $\mathcal{Q}$ we define

$$\deg(\mathcal{R}) = \sup\{\deg(x) \mid x \in \mathcal{R}\}.$$

If $\mathcal{R}$ contains elements that are not algebraic over $\mathcal{C}$, then clearly $\deg(\mathcal{R}) = \infty$; the converse is not true in general.

We fix $m \in \mathbb{N}$. For nonempty subsets $\mathcal{R}_1, \mathcal{R}_2, \ldots, \mathcal{R}_m$ of $\mathcal{Q}$ we set

$$\widehat{\mathcal{R}} = \prod_{k=1}^{m} \mathcal{R}_k = \mathcal{R}_1 \times \mathcal{R}_2 \times \ldots \times \mathcal{R}_m.$$

Let us point out that the notation $\widehat{\mathcal{R}}$ will be exclusively associated to m, i.e., it will always denote the product of m sets $\mathcal{R}_1, \ldots, \mathcal{R}_m$. In case all the $\mathcal{R}_i$'s are equal, $\mathcal{R}_i = \mathcal{R}$ for every i, we shall write $\mathcal{R}^m$ instead of $\widehat{\mathcal{R}}$. For $1 \leq i \leq m$ we set

$$\widehat{\mathcal{R}}^i = \prod_{\substack{k=1, \\ k \neq i}}^{m} \mathcal{R}_k,$$

and for $1 \leq i < j \leq m$ we set

$$\widehat{\mathcal{R}}^{ij} = \widehat{\mathcal{R}}^{ji} = \prod_{\substack{k=1, \\ k \neq i,j}}^{m} \mathcal{R}_k.$$

For elements $x_1 \in \mathcal{R}_1$, $x_2 \in \mathcal{R}_2$, ..., $x_m \in \mathcal{R}_m$ we set

$$\overline{x}_m = (x_1, \ldots, x_m) \in \widehat{\mathcal{R}},$$
$$\overline{x}_m^i = (x_1, \ldots, x_{i-1}, x_{i+1}, \ldots, x_m) \in \widehat{\mathcal{R}}^i,$$
$$\overline{x}_m^{ij} = \overline{x}_m^{ji} = (x_1, \ldots, x_{i-1}, x_{i+1}, \ldots, x_{j-1}, x_{j+1} \ldots, x_m) \in \widehat{\mathcal{R}}^{ij}.$$

Occasionally we shall also write $\overline{x}_r$ for $(x_1, \ldots, x_r) \in \prod_{i=1}^{r} \mathcal{R}_i$, where r can be different from m, and make use of $\overline{x}_r^i$ and $\overline{x}_r^{ij}$ which are defined as above. So, for example, $\overline{x}_r^r = \overline{x}_{r-1}$.

In the present chapter we will study FI's on sets $\mathcal{R}_i$. In Chapter 4 we shall be operating in a more general framework than that just outlined above. Namely, each $\mathcal{R}_i$ will be the image of an arbitrary set $\mathcal{S}_i$, i.e., we will be dealing with surjective maps $\alpha_i : \mathcal{S}_i \to \mathcal{R}_i$. But we shall leave until Chapter 4 the discussion of this more general situation.

In very broad terms FI theory concerns itself with equations involving sums of products of functions. In writing down such equations there are two ways to proceed: either (a) write down an equation literally involving just functions, or (b) write down equations in values taken on by the functions. To clarify this point with a simple illustration, if $f, g : \mathcal{R}_1 \times \mathcal{R}_2 \rightarrow \mathcal{Q}$ and we want to assert that the product (pointwise, not the composite) of f and g is 0, we can either write (a) $fg = 0$ or (b) $f(x_1, x_2)g(x_1, x_2) = 0$ for all $x_1 \in \mathcal{R}_1$, $x_2 \in \mathcal{R}_2$. Thus the *function* approach (a) has the advantage of conciseness and of not surfeiting the reader with extra parentheses etc., whereas the *function-value* approach (b) spells things out in detail and does not leave so much to the reader's imagination. In the present chapter we will consistently use the function-value approach. However, since this approach, though accurate and rigorous, can be somewhat clumsy and long-winded, we shall sometimes take the liberty of abbreviating the notation. If $F : \widehat{\mathcal{R}^i} \rightarrow \mathcal{Q}$ and $G : \widehat{\mathcal{R}^{ij}} \rightarrow \mathcal{Q}$ we can, of course, write a function-value as

$$F(x_1, \ldots, x_{i-1}, x_{i+1}, \ldots, x_m), \quad \text{etc.} \tag{3.1}$$

But, using the shorthand for m-tuples given above, we can write

$$F(\overline{x}_m^i), \quad G(\overline{x}_m^{ij}) \tag{3.2}$$

for function-values. Finally, in the interest of extreme brevity, we may further shorten (3.2) to

$$F^i, \quad G^{ij}. \tag{3.3}$$

We stress that care must be taken in using this ultra-condensed notation. The presence of the superscript i signifies that F^i denotes the value of F evaluated on the $(m-1)$-tuple formed by omitting the i-th component x_i of the element $\overline{x}_m$. Since nothing in the notation F^i gives any indication of what element F is being applied to, it must be tacitly but firmly understood from the particular context exactly what element $\overline{x}_m$ is involved. Of course, if the element $\overline{x}_m$ is further altered, e.g., by substituting other entries in some of its components, then one can no longer use this brief notation but instead must go back to using the original from (3.1).

For example, the most basic FI's we shall be dealing with have summands of the form

$$E_i(x_1, x_2, \ldots, x_{i-1}, x_{i+1}, \ldots, x_m)x_i$$

where $E_i : \widehat{\mathcal{R}^i} \rightarrow \mathcal{Q}$. This expression can be made more concise by using notation introduced in (3.2):

$$E_i(\overline{x}_m^i)x_i,$$

and can be further rewritten using (3.3) as

$$E_i^i x_i.$$

Thus, the fundamental FI

$$\sum_{i \in \mathcal{I}} E_i(\overline{x}_m^i) x_i + \sum_{j \in \mathcal{J}} x_j F_j(\overline{x}_m^j) = 0$$

can be rewritten as

$$\sum_{i \in \mathcal{I}} E_i^i x_i + \sum_{j \in \mathcal{J}} x_j F_j^j = 0.$$

The appearance of E adorned with the same subscript and superscript i may look a bit strange and even redundant. However, in view of the preceding remarks, there is a good reason for the necessity of both to appear. The subscript i is needed to tell which of functions $E_1, E_2, \ldots, E_m$ is involved and the superscript i is needed to say that the value of E_i at $\overline{x}_m^i$ is being called for. Of course, there are more complicated situations where some but not all of the superscripts also appear as superscripts. For instance, the (also important) identity

$$E_{iu}(\overline{x}_m^i) = \sum_{\substack{j \in \mathcal{J}, \\ j \neq i}} \sum_{v=0}^{b} t^v x_j p_{iujv}(\overline{x}_m^{ij}) + \sum_{v=0}^{b} \lambda_{iuv}(\overline{x}_m^i) t^v$$

in the abbreviated notation reads as

$$E_{iu}^i = \sum_{\substack{j \in \mathcal{J}, \\ j \neq i}} \sum_{v=0}^{b} t^v x_j p_{iujv}^{ij} + \sum_{v=0}^{b} \lambda_{iuv}^i t^v.$$

When performing the t-substitution operation (see the preceding chapter!) we shall use some additional abbreviations. For $H : \widehat{\mathcal{R}} \to \mathcal{Q}$ we shall write

$$H(x_k t) \quad \text{for} \quad H(x_1, \ldots, x_{k-1}, x_k t, x_{k+1}, \ldots, x_m)$$

and for $F : \widehat{\mathcal{R}}^i \to \mathcal{Q}$ we shall write

$$F^i(x_k t) \quad \text{for}$$
$$F(x_1, \ldots, x_{i-1}, x_{i+1}, \ldots, x_{k-1}, x_k t, x_{k+1}, \ldots, x_m) \quad \text{if } i < k,$$
$$\text{and} \quad F(x_1, \ldots, x_{k-1}, x_k t, x_{k+1}, \ldots, x_{i-1}, x_{i+1}, \ldots, x_m) \quad \text{if } i > k.$$

Similarly we define $F^i(t x_k)$.

In the above identities, and in fact throughout the book, $\mathcal{I}$ and $\mathcal{J}$ are subsets of $\{1, 2, \ldots, m\}$. The case when one of them is empty is not excluded; here we shall use the following convention: *The sum over the empty set of indices is zero.* Further conventions are necessary to cover the cases where $m = 1$ or $m = 2$. If $m = 1$, then a map $E : \widehat{\mathcal{R}}^i \to \mathcal{Q}$ (here of course i is necessarily equal to 1) should be understood as a constant, i.e., it can be identified by an element in $\mathcal{Q}$. Similarly, if $m = 2$ and we consider $F : \widehat{\mathcal{R}}^{ij} \to \mathcal{Q}$ (here of course $\{i, j\} = \{1, 2\}$), then F is just an element in $\mathcal{Q}$.

3.2 *d*-Free Sets

As above, let $m \in \mathbb{N}$, let $\mathcal{I}, \mathcal{J} \subseteq \{1, 2, \ldots, m\}$, and let $\mathcal{R}_1, \mathcal{R}_2, \ldots, \mathcal{R}_m$ be nonempty subsets of $\mathcal{Q}$. Further, let $E_i : \widehat{\mathcal{R}}^i \to \mathcal{Q}$, $i \in \mathcal{I}$, and $F_j : \widehat{\mathcal{R}}^j \to \mathcal{Q}$, $j \in \mathcal{J}$, be arbitrary maps. In this section we shall consider only *basic* functional identities (FI's)

$$\sum_{i \in \mathcal{I}} E_i(\overline{x}_m^i) x_i + \sum_{j \in \mathcal{J}} x_j F_j(\overline{x}_m^j) = 0 \quad \text{for all } \overline{x}_m \in \widehat{\mathcal{R}}, \tag{3.4}$$

$$\sum_{i \in \mathcal{I}} E_i(\overline{x}_m^i) x_i + \sum_{j \in \mathcal{J}} x_j F_j(\overline{x}_m^j) \in \mathcal{C} \quad \text{for all } \overline{x}_m \in \widehat{\mathcal{R}}. \tag{3.5}$$

These FI's are practically the same as those treated in Part I when introducing the concept of a strongly *d*-free ring. The only difference is that our maps now have different domains and ranges.

We proceed as in Part I. A natural possibility when (3.4) (and hence also (3.5)) is fulfilled is when there exist maps

$$p_{ij} : \widehat{\mathcal{R}}^{ij} \to \mathcal{Q}, \quad i \in \mathcal{I}, \ j \in \mathcal{J}, \ i \neq j,$$
$$\lambda_k : \widehat{\mathcal{R}}^k \to \mathcal{C}, \quad k \in \mathcal{I} \cup \mathcal{J},$$

such that

$$E_i(\overline{x}_m^i) = \sum_{\substack{j \in \mathcal{J}, \\ j \neq i}} x_j p_{ij}(\overline{x}_m^{ij}) + \lambda_i(\overline{x}_m^i), \quad i \in \mathcal{I},$$

$$F_j(\overline{x}_m^j) = - \sum_{\substack{i \in \mathcal{I}, \\ i \neq j}} p_{ij}(\overline{x}_m^{ij}) x_i - \lambda_j(\overline{x}_m^j), \quad j \in \mathcal{J}, \tag{3.6}$$

$$\lambda_k = 0 \quad \text{if} \quad k \notin \mathcal{I} \cap \mathcal{J}.$$

Indeed, one can readily check that (3.6) is a solution of both (3.4) and (3.5). We shall refer to (3.6) as a *standard* solution of (3.4) and (3.5).

If $\mathcal{J} = \emptyset$ (resp. $\mathcal{I} = \emptyset$), then according to our convention that the sum over $\emptyset$ is 0, (3.4) can be rewritten as

$$\sum_{i \in \mathcal{I}} E_i(\overline{x}_m^i) x_i = 0 \quad \text{for all } \overline{x}_m \in \widehat{\mathcal{R}}, \tag{3.7}$$

$$\sum_{j \in \mathcal{J}} x_j F_j(\overline{x}_m^j) = 0 \quad \text{for all } \overline{x}_m \in \widehat{\mathcal{R}}. \tag{3.8}$$

Similarly, special cases of (3.5) are

$$\sum_{i \in \mathcal{I}} E_i(\overline{x}_m^i) x_i \in \mathcal{C} \quad \text{for all } \overline{x}_m \in \widehat{\mathcal{R}}, \tag{3.9}$$

$$\sum_{j \in \mathcal{J}} x_j F_j(\overline{x}_m^j) \in \mathcal{C} \quad \text{for all } \overline{x}_m \in \widehat{\mathcal{R}}. \tag{3.10}$$

It follows from the definition that the standard solution of (3.7) and (3.9) is $E_i = 0$ for each i. Indeed,

$$\sum_{\substack{j \in \mathcal{J}, \\ j \neq i}} x_j p_{ij}(\bar{x}_m^{ij}) = 0$$

because $\mathcal{J} = \emptyset$, whereas $\lambda_i = 0$ because $i \notin \mathcal{I} \cap \mathcal{J}$. Similarly, the standard solution of (3.8) and (3.10) is $F_j = 0$ for each j.

Definition 3.1. $\widehat{\mathcal{R}}$ is said to be a *d-free subset* of $\mathcal{Q}^m$, where $d \in \mathbb{N}$, if for all $\mathcal{I}, \mathcal{J} \subseteq \{1, 2, \ldots, m\}$ the following two conditions are satisfied:

(a) If $\max\{|\mathcal{I}|, |\mathcal{J}|\} \leq d$, then (3.4) implies (3.6).

(b) If $\max\{|\mathcal{I}|, |\mathcal{J}|\} \leq d - 1$, then (3.5) implies (3.6).

If each $\mathcal{R}_k = \mathcal{R}$, then $\mathcal{R}$ is called a *d-free subset* of $\mathcal{Q}$ provided that $\widehat{\mathcal{R}} = \mathcal{R}^m$ is a d-free subset of $\mathcal{Q}^m$ for every $m \in \mathbb{N}$.

Let us make several observations concerning this definition.

1. The reader must not get the impression that the same functions E_i and F_j must simultaneously satisfy (3.4) and (3.5). Indeed, (a) and (b) are separate statements each with their own set of tacitly understood quantifiers.

2. The reader should recall from the discussion in Part I that neither of (a) and (b) implies the other.

3. It is worth pointing out that if $\widehat{\mathcal{R}}$ is d-free and $m \leq d - 1$, then the conclusions of (a) and (b) both hold.

4. If $\widehat{\mathcal{R}}$ is d-free and (3.5) is satisfied with $\max\{|\mathcal{I}|, |\mathcal{J}|\} \leq d - 1$, then the left-hand side of (3.5) is equal to 0, i.e. (3.4) holds. So in Definition 3.1 we could in fact replace (b) by the condition: If $\max\{|\mathcal{I}|, |\mathcal{J}|\} \leq d - 1$, then (3.5) implies (3.4).

5. If $m = 1$ or $m = 2$, then one has to take into account the conventions mentioned at the end of the preceding section. For example, if (3.5) is satisfied with $m = 1$ and $\mathcal{I} = \mathcal{J} = \{1\}$, then E_1 and F_1 are just constants and we have that

$$E_1 x_1 + x_1 F_1 \in \mathcal{C} \quad \text{for all } x_1 \in \mathcal{R}.$$

If $\mathcal{R}$ is 2-free, then $F_1 = -E_1 \in \mathcal{C}$.

6. In general the d-freeness of a given set depends on the choice of $\mathcal{Q}$; that is, it is possible that, say, $\mathcal{R}$ is a d-free subset of $\mathcal{Q}$ but is not a d-free subset of another ring $\mathcal{Q}'$ that contains $\mathcal{R}$ as a subset (regardless of whether or not one of $\mathcal{Q}$ and $\mathcal{Q}'$ is contained in another one). Anyway, since in this chapter the ring $\mathcal{Q}$ is fixed, we shall often simply write that $\widehat{\mathcal{R}}$ *is d-free* (resp. $\mathcal{R}$ *is d-free*), having in mind that it is a d-free subset of $\mathcal{Q}^m$ (resp. $\mathcal{Q}$).

7. The preceding remark indirectly points out an important difference between the concepts of d-freeness and strong d-freeness. The concept of a strong d-free (unital!) ring $\mathcal{A}$ is really a much "stronger" one since we require that our

FI's must have only standard solutions in *any* unital $(\mathcal{A}, \mathcal{A})$-bimodule $\mathcal{M}$. Clearly, if a ring is strongly d-free, then it is in particular a d-free subset of itself. Moreover, a strongly d-free ring is a d-free subset of every ring $\mathcal{Q} \supseteq \mathcal{A}$ having the same unity as $\mathcal{A}$.

8. Because of applications we are primarily interested in the case when each $\mathcal{R}_k = \mathcal{R}$ (as it was always the case in Part I). Now the reader may wonder whether the fact that the $\mathcal{R}_i$'s are allowed to be distinct may simply be an example of excessive generality. One might first reply that the added generality does not cause any essential change in length of or complexity of arguments, and moreover one can hope that this more general context might turn out to be useful (perhaps in the theory of graded algebras). But a stronger motivation will present itself very shortly: in what is the main goal of this section we shall see that the proof of Corollary 3.5 (whose statement concerns a fixed set $\mathcal{R}$) requires the use of the general definition of d-freeness where some of the components are distinct.

The next lemma is similar to Lemma 2.8. Its proof is simple and almost the same as that of Lemma 2.8. Therefore we omit it.

Lemma 3.2. *Let $\widehat{\mathcal{R}}$ be a d-free subset of $\mathcal{Q}^m$. Then:*

(i) *If $\emptyset \neq \mathcal{K} \subseteq \{1, 2, \ldots, m\}$, then $\prod_{k \in \mathcal{K}} \mathcal{R}_k$ is a d-free subset of $\mathcal{Q}^{|\mathcal{K}|}$.*

(ii) *If $|\mathcal{I}| \leq d$, then (3.7) implies that each $E_i = 0$.*

(iii) *If $|\mathcal{J}| \leq d$, then (3.8) implies that each $F_j = 0$.*

(iv) *If $|\mathcal{I}| \leq d - 1$, then (3.9) implies that each $E_i = 0$.*

(v) *If $|\mathcal{J}| \leq d - 1$, then (3.10) implies that each $F_j = 0$.*

(vi) *If $\max\{|\mathcal{I}|, |\mathcal{J}|\} \leq d$, then the p_{ij}'s and λ_i's (from (3.6)) are unique.*

(vii) *Suppose that each $\mathcal{R}_i$ is an additive subgroup of $\mathcal{Q}$. If $\max\{|\mathcal{I}|, |\mathcal{J}|\} \leq d$ and all E_i's and F_j's are $(m-1)$-additive maps, then all p_{ij}'s and λ_i's (from (3.6)) are $(m-2)$-additive and $(m-1)$-additive, respectively.*

Let us add to (vii) that in case $\mathcal{Q}$ is an algebra over a commutative ring $\mathcal{F}$ and the $\mathcal{R}_i$'s are $\mathcal{F}$-submodules, then we can replace multiadditivity by multilinearity; that is, under the same assumption the $(m-1)$-linearity of E_i's and F_j's implies the $(m-2)$-linearity and $(m-1)$-linearity of p_{ij}'s and λ_i's, respectively.

Lemma 3.3. *Let $\mathcal{R}$ be a d-free subset of $\mathcal{Q}$. Then:*

(i) *If $q \in \mathcal{Q}$ is such that $q\mathcal{R} = 0$ (or $\mathcal{R}q = 0$), then $q = 0$.*

(ii) *If $d \geq 2$ and $q \in \mathcal{Q}$ is such that $q\mathcal{R} \subseteq \mathcal{C}$ (or $\mathcal{R}q \subseteq \mathcal{C}$), then $q = 0$.*

(iii) *If $q \in \mathcal{Q}$ is such that $[q, \mathcal{R}] = 0$, then $q \in \mathcal{C}$.*

(iv) *If $d \geq 2$ and $q \in \mathcal{Q}$ is such that $[q, \mathcal{R}] \subseteq \mathcal{C}$, then $q \in \mathcal{C}$.*

(v) *If $\lambda \in \mathcal{C}^*$, then $\lambda\mathcal{R}$ is a d-free subset of $\mathcal{Q}$.*

(vi) *If $0 \neq e \in \mathcal{C}$ is an idempotent, then $e\mathcal{R}$ is a d-free subset of $e\mathcal{Q}$.*

(vii) *If $d \geq 3$, $\mathcal{R}$ is an additive subgroup of $\mathcal{Q}$, and $\mathcal{C}$ is a field with $\operatorname{char}(\mathcal{C}) \neq 2$, then $\deg(\mathcal{R}) \geq 3$.*

Proof. The first four statements are trivial. For example, if $q\mathcal{R} = 0$, then $qx = 0$ is an FI on $\mathcal{R}$ and so $q = 0$ because $\mathcal{R}$ is d-free.

To prove (v), note that the map $x \mapsto \lambda x$, $x \in \mathcal{R}$, gives rise to a bijective correspondence between FI's and their solutions on $\mathcal{R}$ and $\lambda\mathcal{R}$.

Next, to prove (vi), consider for example (3.5) for the set $e\mathcal{R}$, that is

$$\sum_{i \in \mathcal{I}} E_i(ex_1, \ldots, ex_{i-1}, ex_{i+1}, \ldots, ex_m)ex_i$$

$$+ \sum_{j \in \mathcal{J}} ex_j F_j(ex_1, \ldots, ex_{j-1}, ex_{j+1}, \ldots, ex_m) \in \mathcal{C}$$

for all $\bar{x}_m \in \mathcal{R}^m$ with E_i, F_j mapping into $e\mathcal{Q}$. But we can consider this as an FI on $\mathcal{R}$ with maps

$$E_i(ex_1, \ldots, ex_{i-1}, ex_{i+1}, \ldots, ex_m)e, \ eF_j(ex_1, \ldots, ex_{j-1}, ex_{j+1}, \ldots, ex_m).$$

Using our assumption together with $E_i e = E_i$ and $F_j = eF_j$ the desired conclusion follows at once.

It remains to prove (vii). Suppose, on the contrary, that $\deg(\mathcal{R}) \leq 2$. If $x \in \mathcal{R} \setminus \mathcal{C}$, then $\deg(x) = 2$, and so there exists a unique $\tau(x) \in \mathcal{C}$ such that $x^2 - \tau(x)x \in \mathcal{C}$. For $x \in \mathcal{R} \cap \mathcal{C}$ we set $\tau(x) = 2x$. It is easy to see that

$$\tau(\lambda x + \mu) = \lambda\tau(x) + 2\mu \tag{3.11}$$

for all $x \in \mathcal{R}$ and $\lambda, \mu \in \mathcal{C}$ such that $\lambda x + \mu \in \mathcal{R}$. We may regard τ as a map from $\mathcal{R}$ into $\mathcal{C}$. We claim that τ is additive. So pick $u, v \in \mathcal{R}$ and let us show that $\tau(u + v) = \tau(u) + \tau(v)$. Suppose first that $u, v, 1$ are linearly dependent over $\mathcal{C}$. Without loss of generality we may assume that $v = \alpha u + \beta$ for some $\alpha, \beta \in \mathcal{C}$. Using (3.11) we have

$$\tau(u + v) = \tau((1 + \alpha)u + \beta) = (1 + \alpha)\tau(u) + 2\beta = \tau(u) + \tau(v),$$

as desired. So suppose that $u, v, 1$ are linearly independent over $\mathcal{C}$. We have

$$\left(2\tau(u) - \tau(u + v) - \tau(u - v)\right)u + \left(2\tau(v) - \tau(u + v) + \tau(u - v)\right)v$$

$$= \left((u + v)^2 - \tau(u + v)(u + v)\right) + \left((u - v)^2 - \tau(u - v)(u - v)\right)$$

$$- 2\left(u^2 - \tau(u)u\right) - 2\left(v^2 - \tau(v)v\right) \in \mathcal{C},$$

since $x^2 - \tau(x)x \in \mathcal{C}$ for every $x \in \mathcal{R}$. Consequently, $2\tau(u) - \tau(u+v) - \tau(u-v) = 0$ and $2\tau(v) - \tau(u + v) + \tau(u - v) = 0$. Since $\operatorname{char}(\mathcal{C}) \neq 2$ this readily implies

$\tau(u+v) = \tau(u) + \tau(v)$. Thus, τ is indeed an additive map. Hence $E : \mathcal{R} \to \mathcal{Q}$ defined by $E(x) = x - \tau(x)$ is also an additive map, and it satisfies $E(x)x \in \mathcal{C}$ for all $x \in \mathcal{R}$. Linearizing we get $E(x)y + E(y)x \in \mathcal{C}$ for all $x, y \in \mathcal{R}$. Since $\mathcal{R}$ is a 3-free subset of $\mathcal{Q}$ it follows that $E = 0$, meaning that $\mathcal{R} \subseteq \mathcal{C}$. But a subset of $\mathcal{C}$ of course cannot be 3-free. This contradiction shows that $\deg(\mathcal{R}) \geq 3$. □

The assertion (iii) tells us that a d-free subset $\mathcal{R}$ must have trivial centralizer in $\mathcal{Q}$, i.e., $\mathcal{C}(\mathcal{R}) = \mathcal{C}$. So, in some sense, a d-free subset must necessarily be a "big piece" of $\mathcal{Q}$.

We continue with a very useful result which in some situations considerably simplifies establishing d-freeness of sets.

Theorem 3.4. *Let $\mathcal{P}_i \subseteq \mathcal{R}_i \subseteq \mathcal{Q}$, $i = 1, 2, \ldots, m$, be nonempty sets. If $\widehat{\mathcal{P}}$ is a d-free subset of $\mathcal{Q}^m$, then $\widehat{\mathcal{R}}$ is d-free as well.*

Proof. It suffices to prove the theorem for the case where $\mathcal{P}_i = \mathcal{R}_i$ holds for each i except one. Namely, if we know that the theorem holds true in such a case, then repeated applications (i.e., m times) of this particular case yield the theorem in its full generality. Moreover, by interchanging the indices if necessary, we may assume that $\mathcal{P}_i = \mathcal{R}_i$ for $i = 1, 2, \ldots, m - 1$.

We will prove (b), leaving the similar proof of (a) to the reader. Thus we suppose (3.5) is satisfied on $\widehat{\mathcal{R}}$. We consider separately two cases: the case where $m \notin \mathcal{I} \cup \mathcal{J}$ and the case where $m \in \mathcal{I} \cup \mathcal{J}$. The second case is more difficult, but it will be reduced to the first one.

Suppose first that $m \notin \mathcal{I} \cup \mathcal{J}$. This case is easy, we just have to consider our FI by treating x_m as fixed and not as a variable. This is the idea, let us explain it also formally. For every $x_m \in \mathcal{R}_m$ one defines maps $E_{ix_m}(\overline{x}^i_{m-1}) = E_i(\overline{x}^i_m)$ and $F_{jx_m}(\overline{x}^j_{m-1}) = F_j(\overline{x}^j_m)$. Then we have (now written in the abbreviated notation)

$$\sum_{i \in \mathcal{I}} E^i_{ix_m} x_i + \sum_{j \in \mathcal{J}} x_j F^j_{jx_m} \in \mathcal{C}$$

in $\prod_{i=1}^{m-1} \mathcal{P}_k$. Since $\prod_{i=1}^{m-1} \mathcal{P}_k$ is d-free by Lemma 3.2 (i) one can readily complete the proof in this case.

Now consider the second case where $m \in \mathcal{I} \cup \mathcal{J}$. We may assume that $m \in \mathcal{I}$. Since (3.5) is in particular satisfied on $\widehat{\mathcal{P}}$ it follows, since $\widehat{\mathcal{P}}$ is d-free, that all E_i's and F_j's are of standard form on $\widehat{\mathcal{P}}$. Since E_m and (in case also $m \in \mathcal{J}$) F_m are defined on $\prod_{i=1}^{m-1} \mathcal{P}_k = \prod_{i=1}^{m-1} \mathcal{R}_k$ this means that

$$E^m_m = \sum_{\substack{j \in \mathcal{J}, \\ j \neq m}} x_j p^{mj}_{mj} + \lambda^m_m,$$

and (in case also $m \in \mathcal{J}$)

$$F^m_m = -\sum_{\substack{i \in \mathcal{I}, \\ i \neq m}} p^{im}_{im} x_i - \lambda^m_m$$

Definition 3.8. $\widehat{\mathcal{R}}$ is said to be a $(t; d)$-*free subset of* $\mathcal{Q}^m$, where $t \in \mathcal{Q}$ and $d \in \mathbb{N}$, if for all $\mathcal{I}, \mathcal{J} \subseteq \{1, 2, \ldots, m\}$, and all integers $a, b \geq 0$ the following two conditions are satisfied:

(a) If $\max\{|\mathcal{I}| + a, |\mathcal{J}| + b\} \leq d$, then (3.18) implies (3.20).

(b) If $\max\{|\mathcal{I}| + a, |\mathcal{J}| + b\} \leq d - 1$, then (3.19) implies (3.20).

If each $\mathcal{R}_k = \mathcal{R}$, then $\mathcal{R}$ is called a $(t; d)$-*free subset of* $\mathcal{Q}$ provided that $\widehat{\mathcal{R}} = \mathcal{R}^m$ is a $(t; d)$-free subset of $\mathcal{Q}^m$ for every $m \in \mathbb{N}$.

Now, this definition might be a bit surprising if comparing it with the definition of a strongly $(t; d)$-free ring. Now we do require two conditions, (a) and (b), while in Definition 2.12 only one, a version of (a), was sufficient since an appropriate version of (b) follows from it (Corollary 2.16). Glancing through the proof of this one can notice that arguments also work in the present context with one exception: the t-substitution operation used in the proof of Lemma 2.15 may not be applicable simply because tx_i may not lie in $\mathcal{R}_i$ for every $x_i \in \mathcal{R}_i$. So we have to add the additional assumption that $t\mathcal{R}_i \subseteq \mathcal{R}_i$ for every i; we shall write this in short as $t\widehat{\mathcal{R}} \subseteq \widehat{\mathcal{R}}$. For clarity we shall state below all these in a systematic manner.

As usual, we begin by writing special cases of (3.18) where one of the sets $\mathcal{I}$ and $\mathcal{J}$ is $\emptyset$,

$$\sum_{i \in \mathcal{I}} \sum_{u=0}^{a} E_{iu}(\bar{x}_m^i) x_i t^u = 0 \text{ for all } \bar{x}_m \in \widehat{\mathcal{R}}, \tag{3.21}$$

$$\sum_{j \in \mathcal{J}} \sum_{v=0}^{b} t^v x_j F_{jv}(\bar{x}_m^j) = 0 \text{ for all } \bar{x}_m \in \widehat{\mathcal{R}}, \tag{3.22}$$

and also the following two somewhat more general relations:

$$\sum_{i \in \mathcal{I}} \sum_{u=0}^{a} E_{iu}(\bar{x}_m^i) x_i t^u \in \mathcal{C}(t) \text{ for all } \bar{x}_m \in \widehat{\mathcal{R}}, \tag{3.23}$$

$$\sum_{j \in \mathcal{J}} \sum_{v=0}^{b} t^v x_j F_{jv}(\bar{x}_m^j) \in \mathcal{C}(t) \text{ for all } \bar{x}_m \in \widehat{\mathcal{R}} \tag{3.24}$$

(recall that $\mathcal{C}(t)$ denotes the centralizer of t in $\mathcal{Q}$).

The proofs of the following four statements are practically the same as those of Lemmas 2.13, 2.14, 2.15 and Corollary 2.16, so we omit them.

Lemma 3.9. *Let $\widehat{\mathcal{R}}$ be a $(t; d)$-free subset of $\mathcal{Q}^m$. Then:*

(i) *If $\emptyset \neq \mathcal{K} \subseteq \{1, 2, \ldots, m\}$, then $\prod_{k \in \mathcal{K}} \mathcal{R}_k$ is a $(t; d)$-free subset of $\mathcal{Q}^{|\mathcal{K}|}$.*

(ii) *If $|\mathcal{I}| + a \leq d$, then (3.21) implies that each $E_{iu} = 0$.*

(iii) *If $|\mathcal{J}| + b \leq d$, then (3.22) implies that each $F_{jv} = 0$.*

(iv) *If $|\mathcal{I}| + a \leq d - 1$, then (3.23) implies that each $E_{iu} = 0$.*

(v) *If $|\mathcal{J}| + b \leq d - 1$, then (3.24) implies that each $F_{jv} = 0$.*

(vi) *If $\max\{|\mathcal{I}| + a, |\mathcal{J}| + b\} \leq d$, then the p_{iujv}'s and λ_{iuv}'s (from (3.20)) are unique.*

(vii) *Suppose that each $\mathcal{R}_i$ is an additive subgroup of Q. If $\max\{|\mathcal{I}|+a, |\mathcal{J}|+b\} \leq d$ and all E_{iu}'s and F_{jv}'s are $(m-1)$-additive, then all p_{iujv}'s and λ_{iuv}'s (from (3.20)) are $(m-2)$-additive and $(m-1)$-additive, respectively.*

Lemma 3.10. *If there exists a $(t; d)$-free subset of Q^m, then $\deg(t) \geq d$.*

Lemma 3.11. *Suppose there exist maps $\mu_w : \widehat{\mathcal{R}} \to C$, $0 \leq w \leq c$, such that*

$$\sum_{i \in \mathcal{I}} \sum_{u=0}^{a} E_{iu}(\overline{x}_m^i) x_i t^u + \sum_{j \in \mathcal{J}} \sum_{v=0}^{b} t^v x_j F_{jv}(\overline{x}_m^j) = \sum_{w=0}^{c} \mu_w(\overline{x}_m) t^w$$

for all $\overline{x}_m \in \widehat{\mathcal{R}}$. If $\widehat{\mathcal{R}}$ is a $(t; d)$-free subset of Q^m, $t\widehat{\mathcal{R}} \subseteq \widehat{\mathcal{R}}$ and $\max\{a + |\mathcal{I}|, c + |\mathcal{J}|\} \leq d - 1$, then each $\mu_w = 0$.

Corollary 3.12. *If $t\widehat{\mathcal{R}} \subseteq \widehat{\mathcal{R}}$, then the condition (b) (from Definition 3.8) follows from the condition (a).*

A statement similar to Corollary 2.17 is in the present context a trivial consequence of the definition (just take $a = b = 0$):

Lemma 3.13. *If $\widehat{\mathcal{R}}$ is a $(t; d)$-free subset of Q^m for some $t \in Q$, then $\widehat{\mathcal{R}}$ is d-free.*

With a similar proof as that of Theorem 3.4 one can get the $(t; d)$-free analogue of Theorem 3.4.

Theorem 3.14. *Let $\mathcal{P}_i \subseteq \mathcal{R}_i \subseteq Q$, $i = 1, 2, \ldots, m$, be nonempty sets. If $\widehat{\mathcal{P}}$ is a $(t; d)$-free subset of Q^m, then $\widehat{\mathcal{R}}$ is $(t; d)$-free as well.*

As will become evident as the book unfolds, knowledge that a certain set is d-free gives one in certain situations a powerful weapon when trying to establish various results about that set. The problem to prove that a set is d-free, however, may be a difficult task. It has been already indicated in Part I that a possible, and in fact quite common way to accomplish this task is to prove that the set in question is actually $(t; d)$-free for some carefully chosen t. But sometimes a more indirect method of showing that d-freeness of $\widehat{\mathcal{R}}$ can be derived from $(t; d')$-freeness of a closely related set $\widehat{\mathcal{P}}$ must be used. The ultimate goal of this section is to prove Theorem 3.16 that illustrates this method. It appears to be difficult to give a quick proof of Theorem 3.16, jumping directly from $\widehat{\mathcal{P}}$ to $\widehat{\mathcal{R}}$. Rather, following in a similar vein the method of proof of Theorem 3.4, the proof will be accomplished in an incremental fashion of m steps, at each step replacing one more $\mathcal{P}_i$ by an $\mathcal{R}_i$.

First, let us fix some notation and some conditions which will be referred to in Lemma 3.15 and in the proof of Theorem 3.16. Let $t \in Q$ and let $\widehat{\mathcal{P}}$ be a $(t; d+1)$-free subset of Q^m. Further, let $\epsilon_1, \ldots, \epsilon_m \in C^*$ and let $\mathcal{R}_i = \{tx_i + \epsilon_i x_i t \mid x_i \in \mathcal{P}_i\}$, $i = 1, 2, \ldots, m$. By $\mathcal{I}$, $\mathcal{J}$, and $\mathcal{K}$ we denote subsets of $\{1, 2, \ldots, m\}$. As we shall see, $\mathcal{K}$ will play a specific role. We set $\mathcal{T}_\mathcal{K} = \widehat{\mathcal{U}}$ where $\mathcal{U}_i = \mathcal{R}_i$ if $i \in \mathcal{K}$ and $\mathcal{U}_i = \mathcal{P}_i$ if $i \notin \mathcal{K}$. Accordingly, $\mathcal{T}_\mathcal{K}^i$ denotes $\widehat{\mathcal{U}}^i$ and $\mathcal{T}_\mathcal{K}^{ij}$ denotes $\widehat{\mathcal{U}}^{ij}$. So, for example, if $m = 3$ and $\mathcal{K} = \{1, 3\}$, then $\mathcal{T}_\mathcal{K} = \mathcal{R}_1 \times \mathcal{P}_2 \times \mathcal{R}_3$, $\mathcal{T}_\mathcal{K}^1 = \mathcal{P}_2 \times \mathcal{R}_3$, and $\mathcal{T}_\mathcal{K}^{12} = \mathcal{R}_3$. Of course, if $\mathcal{K} = \emptyset$, then $\mathcal{T}_\mathcal{K} = \widehat{\mathcal{P}}$, and if $\mathcal{K} = \{1, 2, \ldots, m\}$, then $\mathcal{T}_\mathcal{K} = \widehat{\mathcal{R}}$.

Lemma 3.15. *Under the above conditions let $H_{jv} : \mathcal{T}_\mathcal{K} \to Q$, $j \in \mathcal{J}$, $v = 0, 1$ be maps such that $H_{j1} = 0$ for all $j \in \mathcal{J} \cap \mathcal{K}$. Assume that $|\mathcal{J}| \leq d - 1$ and*

$$\sum_{j \in \mathcal{J}} \sum_{v=0}^{1} t^v x_j H_{jv}^j \in \sum_{i=0}^{\infty} Ct^i \quad \text{for all } \overline{x}_m \in \mathcal{T}_\mathcal{K}. \tag{3.25}$$

Then each $H_{jv} = 0$.

Proof. Assume first that $\mathcal{J} \cap \mathcal{K} = \emptyset$. This case is in fact trivial. Namely, we just have to consider (3.25) in appropriate fashion: the x_j's with $j \in \mathcal{J}$ must be treated as variables while the x_l's with $l \notin \mathcal{J}$ must be treated as fixed (though arbitrary). Since $\prod_{j \in \mathcal{J}} \mathcal{P}_j$ is a $(t; d+1)$-free subset of $Q^{|\mathcal{J}|}$ by Lemma 3.9 (i), and $|\mathcal{J}| + 1 \leq (d+1) - 1$, it follows from Lemma 3.9 (v) that each $H_{jv} = 0$.

We proceed by induction on $|\mathcal{K}|$. If $|\mathcal{K}| = 0$, then $\mathcal{J} \cap \mathcal{K} = \emptyset$, so this case is a subcase of the one just treated. Now consider the inductive case $|\mathcal{K}| \geq 1$. We may assume that $\mathcal{J} \cap \mathcal{K} \neq \emptyset$, so without loss of generality $m \in \mathcal{J} \cap \mathcal{K}$. Note that by assumption $H_{m1} = 0$. Set $\overline{\mathcal{K}} = \mathcal{K} \setminus \{m\}$. Instead of $x_m \in \mathcal{R}_m$ we can, admittedly with abuse of notation, write $tx_m + \epsilon_m x_m t$ with $x_m \in \mathcal{P}_m$. Then (3.25) can be rewritten as

$$\sum_{j \in \mathcal{J}} \sum_{v=0}^{1} t^v x_j \overline{H}_{jv}^j \in \sum_{i=0}^{\infty} Ct^i \quad \text{for all } \overline{x}_m \in \mathcal{T}_{\overline{\mathcal{K}}},$$

where

$$\overline{H}_{jv}(\overline{x}_m^j) = H_{jv}(\overline{x}_{m-1}^j, tx_m + \epsilon_m x_m t), \quad j \in \mathcal{J}, \ j \neq m, \ v = 0, 1,$$
$$\overline{H}_{m0}(\overline{x}_{m-1}) = \epsilon_m t H_{m0}(\overline{x}_{m-1}),$$
$$\overline{H}_{m1}(\overline{x}_{m-1}) = H_{m0}(\overline{x}_{m-1}).$$

Clearly $\overline{H}_{j1} = 0$ for all $j \in \mathcal{J} \cap \overline{\mathcal{K}}$, and so by induction we conclude that all $\overline{H}_{jv}$'s are 0. A glance at the above relationships then shows that the original H_{jv}'s must all be 0 (recall also that an arbitrary element in $\mathcal{R}_m$ is of the form $tx_m + \epsilon_m x_m t$ with $x_m \in \mathcal{P}_m$). $\quad\square$

The symmetric analogue of Lemma 3.15, in which summands $H_{iu} x_i t^u$ are involved, is clearly proved in the same way. We shall not bother to state this as a separate lemma.

The proof of the next theorem depends upon a technical assertion (denoted by $(\star)$, see below). In FI theory it often happens that a result having a clear and striking statement is derived from another result that might appear somewhat artificial and lengthy. These latter results are hardly interesting in their own right, just the method of the proof forces us to deal with them. Therefore we have decided not to state them as, say, lemmas, but rather incorporate them in the proofs of "nice" results.

Theorem 3.16. *Let $t \in \mathcal{Q}$ with $\deg(t) \geq 3$, let $\widehat{\mathcal{P}}$ be a $(t; d+1)$-free subset of $\mathcal{Q}^m$, and let $\epsilon_1, \ldots, \epsilon_m \in \mathcal{C}^*$. If sets $\mathcal{R}_i$ are such that $tx_i + \epsilon_i x_i t \in \mathcal{R}_i$ for all $x_i \in \mathcal{P}_i$, $i = 1, \ldots, m$, then $\widehat{\mathcal{R}}$ is a d-free subset of $\mathcal{Q}^m$.*

Proof. First of all we note that in view of Theorem 3.4 there is no loss of generality in assuming that $\mathcal{R}_i = \{tx_i + \epsilon_i x_i t \mid x_i \in \mathcal{P}_i\}$.

Our goal is to prove the following assertion:

$(\star)$ Let $E_{iu} : \mathcal{T}_{\mathcal{K}}^i \to \mathcal{Q}$, $F_{jv} : \mathcal{T}_{\mathcal{K}}^j \to \mathcal{Q}$, $u, v = 0, 1$, be maps such that $E_{i1} = 0$ for all $i \in \mathcal{I} \cap \mathcal{K}$ and $F_{j1} = 0$ for all $j \in \mathcal{J} \cap \mathcal{K}$. Suppose that either

(i) $\max\{|\mathcal{I}|, |\mathcal{J}|\} \leq d$ and

$$\sum_{i \in \mathcal{I}} \sum_{u=0}^{1} E_{iu}^i x_i t^u + \sum_{j \in \mathcal{J}} \sum_{v=0}^{1} t^v x_j F_{jv}^j = 0 \quad \text{for all } \bar{x}_m \in \mathcal{T}_{\mathcal{K}},$$

or

(ii) $\max\{|\mathcal{I}|, |\mathcal{J}|\} \leq d - 1$ and

$$\sum_{i \in \mathcal{I}} \sum_{u=0}^{1} E_{iu}^i x_i t^u + \sum_{j \in \mathcal{J}} \sum_{v=0}^{1} t^v x_j F_{jv}^j \in \mathcal{C} \quad \text{for all } \bar{x}_m \in \mathcal{T}_{\mathcal{K}}.$$

Then there exist maps $p_{iujv} : \mathcal{T}_{\mathcal{K}}^{ij} \to \mathcal{Q}$, $\lambda_{kuv} : \mathcal{T}_{\mathcal{K}}^k \to \mathcal{C}$, $i \in \mathcal{I}$, $j \in \mathcal{J}$, $i \neq j$, $u, v = 0, 1$, $k \in \mathcal{I} \cup \mathcal{J}$, such that

$$E_{iu}^i = \sum_{\substack{j \in \mathcal{J}, \ v=0 \\ j \neq i}}^{1} t^v x_j p_{iujv}^{ij} + \sum_{v=0}^{1} \lambda_{iuv}^i t^v, \ i \in \mathcal{I}, \ u = 0, 1,$$

$$F_{jv}^j = -\sum_{\substack{i \in \mathcal{I}, \ u=0 \\ i \neq j}}^{1} p_{iujv}^{ij} x_i t^u - \sum_{u=0}^{1} \lambda_{juv}^j t^u, \ j \in \mathcal{J}, \ v = 0, 1,$$

$\lambda_{kuv} = 0$ if $k \notin \mathcal{I} \cap \mathcal{J}$, or $k \in \mathcal{I} \cap \mathcal{J} \cap \mathcal{K}$ and either $u = 1$ or $v = 1$,

$p_{iujv} = 0$ if $i \in \mathcal{I} \cap \mathcal{K}$ and $u = 1$, or $j \in \mathcal{J} \cap \mathcal{K}$ and $v = 1$.

We shall consider only the case when (i) holds true. The case when (ii) is satisfied can be considered similarly.

We shall prove $(\star)$ by induction on $|\mathcal{I}|+|\mathcal{J}|$. If $|\mathcal{I}|+|\mathcal{J}| = 0$, then $\mathcal{I} = \emptyset = \mathcal{J}$ and there is nothing to prove. In the inductive case $|\mathcal{I}| + |\mathcal{J}| > 0$, first assume that $(\mathcal{I} \cup \mathcal{J}) \cap \mathcal{K} = \emptyset$. This is the easier case which can be handled, similarly as in the proof of the previous lemma, by treating x_l, $l \notin \mathcal{I} \cup \mathcal{J}$ as fixed. Then the desired conclusion follows immediately from $\max\{|\mathcal{I}|+1, |\mathcal{J}|+1\} \le d+1$ and the fact that $\prod_{k \in \mathcal{I} \cup \mathcal{J}} \mathcal{P}_k$ is a $(t; d+1)$-free subset of $\mathcal{Q}^{|\mathcal{I} \cup \mathcal{J}|}$ (Lemma 3.9 (i)).

Within the main induction process on $|\mathcal{I}|+|\mathcal{J}|$ we now proceed by induction on $|\mathcal{K}|$. The case $|\mathcal{K}| = 0$ follows from the preceding paragraph. In the inductive case $|\mathcal{K}| \ge 1$, we may, again in view of the preceding paragraph, assume without loss of generality that $m \in \mathcal{I} \cap \mathcal{K}$. Hence $E_{m1} = 0$ and, in case $m \in \mathcal{J}$, also $F_{m1} = 0$. Setting $\overline{\mathcal{K}} = \mathcal{K} \setminus \{m\}$ and replacing $x_m \in \mathcal{R}_m$ by $t x_m + \epsilon_m x_m t$, $x_m \in \mathcal{P}_m$, we get

$$\sum_{i \in \mathcal{I}} \sum_{u=0}^{1} \overline{E}_{iu}^{i} x_i t^u + \sum_{j \in \mathcal{J}} \sum_{v=0}^{1} t^v x_j \overline{F}_{jv}^{j} = 0 \tag{3.26}$$

for all $\overline{x}_m \in \mathcal{T}_{\overline{\mathcal{K}}}$, where

$$\overline{E}_{iu}(\overline{x}_m^i) = E_{iu}(\overline{x}_{m-1}^i, t x_m + \epsilon_m x_m t), \ i \in \mathcal{I}, \ i \ne m, \ u = 0,1,$$
$$\overline{E}_{m0}(\overline{x}_{m-1}) = E_{m0}(\overline{x}_{m-1})t,$$
$$\overline{E}_{m1}(\overline{x}_{m-1}) = \epsilon_m E_{m0}(\overline{x}_{m-1}),$$
$$\overline{F}_{jv}(\overline{x}_m^j) = F_{jv}(\overline{x}_{m-1}^j, t x_m + \epsilon_m x_m t), \ j \in \mathcal{J}, \ j \ne m, \ v = 0,1,$$

and where, in case $m \in \mathcal{J}$, we have in addition

$$\overline{F}_{m0}(\overline{x}_{m-1}) = \epsilon_m t F_{m0}(\overline{x}_{m-1}),$$
$$\overline{F}_{m1}(\overline{x}_{m-1}) = F_{m0}(\overline{x}_{m-1}).$$

We now separate the argument into two cases, choosing (at the risk of some redundancy) to work through each case separately rather than attempting to combine the cases in a single argument.

Case 1: $m \notin \mathcal{J}$. We note that $\overline{E}_{i1} = 0$ for $i \in \mathcal{I} \cap \overline{\mathcal{K}}$ and $\overline{F}_{j1} = 0$ for $j \in \mathcal{J} \cap \overline{\mathcal{K}}$. Since $|\overline{\mathcal{K}}| = |\mathcal{K}| - 1$, applying the induction assumption to (3.26), we obtain in particular that

$$\overline{E}_{m1}^m = \sum_{j \in \mathcal{J}} \sum_{v=0}^{1} t^v x_j \overline{p}_{m1jv}^{mj}$$

for some $\overline{p}_{m1jv} : \mathcal{T}_{\overline{\mathcal{K}}}^{mj} \to \mathcal{Q}$ with $\overline{p}_{m1j1} = 0$ if $j \in \mathcal{J} \cap \overline{\mathcal{K}}$. Setting $p_{m0jv} = \epsilon_m^{-1} \overline{p}_{m1jv}$ it follows that

$$E_{m0}^m = \sum_{j \in \mathcal{J}} \sum_{v=0}^{1} t^v x_j p_{m0jv}^{mj}$$

with $p_{m0j1} = 0$ if $j \in \mathcal{J} \cap \overline{\mathcal{K}} (= \mathcal{J} \cap \mathcal{K})$. Substituting this into the initial FI from (i) and setting $\overline{\mathcal{I}} = \mathcal{I} \setminus \{m\}$ we get

$$\sum_{i \in \overline{\mathcal{I}}} \sum_{u=0}^{1} E_{iu}^i x_i t^u + \sum_{j \in \mathcal{J}} \sum_{v=0}^{1} t^v x_j \widehat{F}_{jv}^j = 0$$

for all $\overline{x}_m \in \mathcal{T}_{\mathcal{K}}$, where

$$\widehat{F}_{jv}^j = F_{jv}^j + p_{m0jv}^{mj} x_m.$$

Since $|\overline{\mathcal{I}}| + |\mathcal{J}| < |\mathcal{I}| + |\mathcal{J}|$ and since the condition $\widehat{F}_{j1} = 0$ for all $j \in \mathcal{J} \cap \mathcal{K}$ is fulfilled, the result now follows in a straightforward way from the induction hypothesis.

Case 2: $m \in \mathcal{J}$. For future reference we set $\overline{\mathcal{I}} = \mathcal{I} \setminus \{m\}$ and $\overline{\mathcal{J}} = \mathcal{J} \setminus \{m\}$. We note that $\overline{E}_{i1} = 0$ for $i \in \mathcal{I} \cap \overline{\mathcal{K}}$ and $\overline{F}_{j1} = 0$ for $j \in \mathcal{J} \cap \overline{\mathcal{K}}$. Since $|\overline{\mathcal{K}}| = |\mathcal{K}| - 1$, we can then apply the induction assumption to (3.26) to obtain maps $\overline{p}_{iujv} : \mathcal{T}_{\overline{\mathcal{K}}}^{ij} \to \mathcal{Q}$, $i \neq j$, and $\overline{\lambda}_{kuv} : \mathcal{T}_{\overline{\mathcal{K}}}^{k} \to \mathcal{C}$, such that in particular

$$\overline{E}_{mu}^m = \sum_{j \in \overline{\mathcal{J}}} \sum_{v=0}^{1} t^v x_j \overline{p}_{mujv}^{mj} + \sum_{v=0}^{1} \overline{\lambda}_{muv}^m t^v, \quad u = 0, 1, \tag{3.27}$$

$$\overline{F}_{mv}^m = -\sum_{i \in \overline{\mathcal{I}}} \sum_{u=0}^{1} \overline{p}_{iumv}^{im} x_i t^u - \sum_{u=0}^{1} \overline{\lambda}_{muv}^m t^u, \quad v = 0, 1, \tag{3.28}$$

$$\overline{p}_{iujv} = 0 \quad \text{if } i \in \mathcal{I} \cap \overline{\mathcal{K}} \text{ and } u = 1, \text{ or } j \in \mathcal{J} \cap \overline{\mathcal{K}} \text{ and } v = 1. \tag{3.29}$$

Note that

$$\overline{E}_{m0}(\overline{x}_{m-1}) = E_{m0}(\overline{x}_{m-1})t = \epsilon_m^{-1} \overline{E}_{m1}(\overline{x}_{m-1})t$$

and so we obtain from (3.27) that

$$\sum_{j \in \overline{\mathcal{J}}} \sum_{v=0}^{1} t^v x_j H_{jv}^j = \epsilon_m^{-1} \sum_{v=0}^{1} \overline{\lambda}_{m1v}^m t^{v+1} - \sum_{v=0}^{1} \overline{\lambda}_{m0v}^m t^v$$

for all $\overline{x}_{m-1} \in \mathcal{T}_{\overline{\mathcal{K}}}^m$, where $H_{jv}^j = \overline{p}_{m0jv}^{mj} - \epsilon_m^{-1} \overline{p}_{m1jv}^{mj} t$. Since $|\overline{\mathcal{J}}| \leq d-1$ and $H_{j1} = 0$ if $j \in \mathcal{J} \cap \overline{\mathcal{K}}$, Lemma 3.15 implies that the left-hand side of the above identity is equal to 0. Therefore the right-hand side must be 0 too; as $\deg(t) \geq 3$, this yields $\epsilon_m^{-1} \overline{\lambda}_{m10} = \overline{\lambda}_{m01}$ and $\overline{\lambda}_{m00} = 0 = \overline{\lambda}_{m11}$. Set

$$\lambda_{m00} = \epsilon_m^{-1} \overline{\lambda}_{m10} = \overline{\lambda}_{m01},$$

$$p_{m0jv} = \epsilon_m^{-1} \overline{p}_{m1jv}, \quad j \in \mathcal{J}, \ v = 0, 1,$$

$$p_{ium0} = \overline{p}_{ium1}, \quad i \in \mathcal{I}, \ u = 0, 1.$$

be arbitrary functions. The basic FI treated in this section is

$$\sum_{i\in\mathcal{I}}\sum_{u=0}^{a}E_{iu}^{i}x_it^u + \sum_{j\in\mathcal{J}}\sum_{v=0}^{b}t^vx_jF_{jv}^j$$

$$+ \sum_{k\in\mathcal{K}}\sum_{w=0}^{a'}G_{kw}^kx_k^*t^w + \sum_{l\in\mathcal{L}}\sum_{z=0}^{b'}t^zx_l^*H_{lz}^l = 0 \qquad (3.32)$$

for all $\bar{x}_m \in \mathcal{A}^m$. So we have four summands instead of the two familiar ones. As one may expect, formulas in this section shall therefore be necessarily somewhat lengthy and complicated. Anyway, the main ideas are the same as in the previous sections. A *standard solution* of (3.32) is defined as

$$E_{iu}^i = \sum_{\substack{j\in\mathcal{J},\ v=0 \\ j\neq i}}\sum^{b}t^vx_jp_{iujv}^{ij} + \sum_{\substack{l\in\mathcal{L},\ z=0 \\ l\neq i}}\sum^{b'}t^zx_l^*q_{iulz}^{il} + \sum_{v=0}^{b}\lambda_{iuv}^it^v,$$

$$F_{jv}^j = -\sum_{\substack{i\in\mathcal{I},\ u=0 \\ i\neq j}}\sum^{a}p_{iujv}^{ij}x_it^u - \sum_{\substack{k\in\mathcal{K},\ w=0 \\ k\neq j}}\sum^{a'}r_{kwjv}^{kj}x_k^*t^w - \sum_{u=0}^{a}\lambda_{juv}^jt^u,$$

$$G_{kw}^k = \sum_{\substack{j\in\mathcal{J},\ v=0 \\ j\neq k}}\sum^{b}t^vx_jr_{kwjv}^{kj} + \sum_{\substack{l\in\mathcal{L},\ z=0 \\ l\neq k}}\sum^{b'}t^zx_l^*s_{kwlz}^{kl} + \sum_{z=0}^{b'}\mu_{kwz}^kt^z, \qquad (3.33)$$

$$H_{lz}^l = -\sum_{\substack{i\in\mathcal{I},\ u=0 \\ i\neq l}}\sum^{a}q_{iulz}^{il}x_it^u - \sum_{\substack{k\in\mathcal{K},\ w=0 \\ k\neq l}}\sum^{a'}s_{kwlz}^{kl}x_k^*t^w - \sum_{w=0}^{a'}\mu_{lwz}^lt^w,$$

$$\lambda_{kuv} = 0 \quad \text{if } k \notin \mathcal{I}\cap\mathcal{J} \quad\text{and}\quad \mu_{rwz} = 0 \quad \text{if } r \notin \mathcal{K}\cap\mathcal{L}$$

for all $\bar{x}_m \in \mathcal{A}^m$ and all $i \in \mathcal{I}$, $0 \le u \le a$ etc., where

$$p_{iujv}, q_{iulz}, r_{kwjv}, s_{kwlz} : \mathcal{A}^{m-2} \to \mathcal{Q}$$

and

$$\lambda_{iuv}, \mu_{kwz} : \mathcal{A}^{m-1} \to \mathcal{C}.$$

Definition 3.19. $\mathcal{A}$ is said to be a $(*; t; d)$-*free subring* of $\mathcal{Q}$, where $t \in \mathcal{S}\cup\mathcal{K}$ and $d \in \mathbb{N}$, if for all $m \in \mathbb{N}$, all $\mathcal{I}, \mathcal{J} \subseteq \{1, 2, \ldots, m\}$, and all integers $a, b, a', b' \ge 0$ the following condition is satisfied: If $\max\{|\mathcal{I}| + |\mathcal{K}| + a'', |\mathcal{J}| + |\mathcal{L}| + b''\} \le d$, where $a'' = \max\{a, a'\}$ and $b'' = \max\{b, b'\}$, then (3.32) implies (3.33).

The restriction that $\mathcal{A}$ is not just any subset but a subring is not entirely necessary in this definition. We could easily define $(*; t; d)$-free subsets of $\mathcal{Q}$ (or even of $\mathcal{Q}^m$) on which $*$ is defined. But besides giving a dry definition there is not much we could say about such sets (if they are not also subrings). One should consider the concepts such as $(*; t; d)$-freeness and $(t; d)$-freeness primarily as auxiliary ones, helpful when establishing what is of crucial importance for us: the d-freeness of certain sets. We shall therefore confine ourselves only to the situation which we find really important and useful for applications. Just a glance at our main results (Theorems 3.25 and 3.28) hopefully gives some evidence about the utility of the notion just introduced.

We shall not bother stating various comments and remarks concerning Definition 3.19 as we did at similar situations above. Let us just mention in some of the following lemmas and proofs thereof special cases of (3.32) will appear, and the reader should be prepared to divine from (3.33) exactly how a standard solution reads. For example, if no "G" terms appear in (3.32), then in the corresponding standard solutions all r_{kwjv}'s and s_{kwlz}'s are equal to 0 and all μ_{kwz}'s are 0.

In case $\mathcal{K} = \mathcal{L} = \emptyset$, (3.32) coincides with (3.18) and the condition required in Definition 3.19 coincides with the condition (a) from Definition 3.8. Since $\mathcal{A}$ is a subring and $t \in \mathcal{A}$, the condition (b) from Definition 3.8 is automatically fulfilled by Corollary 3.12. Thus we have

Lemma 3.20. *If $\mathcal{A}$ is a $(*; t; d)$-free subring of $\mathcal{Q}$, then $\mathcal{A}$ is $(t; d)$-free.*

Our next goal now is to prove Theorem 3.25 which shows that the converse of Lemma 3.20 "almost" holds. As usual when proving general statements about FI's, this will be done progressively, analyzing various special cases of (3.32). More precisely, we shall consider a sequence of special cases where some of the index sets $\mathcal{I}, \mathcal{J}, \mathcal{K}, \mathcal{L}$ are empty; the proofs, which are inductive in nature, will then be able to fall back on a previously proved case to handle the initial part of the induction.

We remark that in the next lemma, as well as in some other lemmas that follow, obvious alternate forms of the lemmas also hold; the context at hand will dictate which form is applicable.

Lemma 3.21. *Let $\mathcal{A}$ be a $(t; d)$-free subring of $\mathcal{Q}$. If $\mathcal{I} = \mathcal{L} = \emptyset$ and $\max\{|\mathcal{J}| + b + 1, |\mathcal{K}| + a'\} \le d$, then (3.32) has only standard solutions.*

Proof. Since $\mathcal{I} = \mathcal{L} = \emptyset$, (3.32) reduces to

$$\sum_{j \in \mathcal{J}} \sum_{v=0}^{b} t^v x_j F_{jv}^j + \sum_{k \in \mathcal{K}} \sum_{w=0}^{a'} G_{kw}^k x_k^* t^w = 0.$$

Setting $F_{j,b+1} = 0$ for every $j \in \mathcal{J}$ we thus have

$$D(\bar{x}_m) = \sum_{j \in \mathcal{J}} \sum_{v=0}^{b+1} t^v x_j F_{jv}^j + \sum_{k \in \mathcal{K}} \sum_{w=0}^{a'} G_{kw}^k x_k^* t^w = 0. \tag{3.34}$$

Note that in particular $F_{j,b+1} = 0$ if $j \in \mathcal{J} \cap \mathcal{K}$ – the reason for emphasizing this special case of a more general fact is that we shall consider it as a part of our induction assumption. As a matter of fact, we shall prove the following assertion:

($\star$) If maps F_{jv} and G_{kw} satisfy (3.34) and $F_{j,b+1} = 0$ if $j \in \mathcal{J} \cap \mathcal{K}$, then they are of the form

$$F^j_{jv} = - \sum_{\substack{k \in \mathcal{K}, \\ k \neq j}} \sum_{w=0}^{a'} r^{kj}_{kwjv} x^*_k t^w \tag{3.35}$$

and

$$G^k_{kw} = \sum_{\substack{j \in \mathcal{J}, \\ j \neq k}} \sum_{v=0}^{b+1} t^v x_j r^{kj}_{kwjv}. \tag{3.36}$$

for some r_{kwjv}'s, provided that $\max\{|\mathcal{J}| + b + 1, |\mathcal{K}| + a'\} \leq d$.

The conditions in ($\star$) may appear a bit artificial, but the method of the proof forces us to deal with them. The only reason for us to prove ($\star$) is that it implies the lemma. Indeed, one can easily check this, in particular by taking into account that $F_{j,b+1} = 0$ forces $r_{kwj,b+1} = 0$ for all $k \in \mathcal{K}$, $0 \leq w \leq a'$, $j \in \mathcal{J}$ (see Lemma 3.9 (ii)).

So let us prove ($\star$). First of all we note that without loss of generality we may assume that $\mathcal{J} \cap \mathcal{K} \neq \emptyset$. Namely, if $\mathcal{J} \cap \mathcal{K} = \emptyset$, then we can regard (3.34) as a special case of (3.18) and hence simply apply the definition of $(t;d)$-freeness to obtain the desired conclusion. There is an apparent obstacle since the second summand in (3.34) involves x^*_k instead of x_k. However, there is no harm in replacing x^*_k by x_k (to be formally correct we would introduce new variables $y_k = x^*_k$), since this does not effect the first summand.

So assume that, say, $1 \in \mathcal{J} \cap \mathcal{K}$. Using the t-substitution operation, this time by computing $D(tx_1) - \epsilon D(\overline{x}_m)t$, we get

$$\sum_{j \in \mathcal{J}} \sum_{v=0}^{b+1} t^v x_j \widehat{F}^j_{jv} + \sum_{\substack{k \in \mathcal{K}, \\ k \neq 1}} \sum_{w=0}^{a'+1} \widehat{G}^k_{kw} x^*_k t^w = 0 \tag{3.37}$$

where

$$\widehat{F}^1_{10} = -\epsilon F^1_{10} t,$$
$$\widehat{F}^1_{1v} = F^1_{1,v-1} - \epsilon F^1_{1v} t, \quad 1 \leq v \leq b,$$
$$\widehat{F}^1_{1,b+1} = F^1_{1b},$$
$$\widehat{F}^j_{jv} = F^j_{jv}(tx_1) - \epsilon F^j_{jv} t, \quad j \neq 1, 0 \leq v \leq b+1,$$
$$\widehat{G}^k_{k0} = G^k_{k0}(tx_1), \quad k \neq 1,$$
$$\widehat{G}^k_{kw} = G^k_{kw}(tx_1) - \epsilon G^k_{k,w-1}, \quad k \neq 1, 0 \leq w \leq a',$$
$$\widehat{G}^k_{k,a'+1} = -\epsilon G^k_{ka'}, \quad k \neq 1.$$

We shall prove ($\star$) by induction on $|\mathcal{K}|$. In case $|\mathcal{K}| = 1$, i.e., $\mathcal{K} = \{1\}$, (3.37) reduces to

$$\sum_{j \in \mathcal{J}} \sum_{v=0}^{b+1} t^v x_j \widehat{F}^j_{jv} = 0$$

which yields that each $\widehat{F}_{jv} = 0$ (Lemma 3.9 (iii)). In view of the definition of $\widehat{F}_{1v}$ this clearly implies that each $F_{1v} = 0$ as well. But then (3.34) can be rewritten as

$$\sum_{j \in \overline{\mathcal{J}}} \sum_{v=0}^{b+1} t^v x_j F^j_{jv} + \sum_{w=0}^{a'} G^1_{1w} x_1^* t^w = 0$$

where $\overline{\mathcal{J}} = \mathcal{J} \setminus \{1\}$. Since $\overline{\mathcal{J}} \cap \{1\} = \emptyset$, this situation is, as noted above, trivial. So this case is settled.

Now assume $|\mathcal{K}| \geq 2$ and set $\overline{\mathcal{K}} = \mathcal{K} \setminus \{1\}$. Observe that $\widehat{F}_{j,b+1} = 0$ if $j \in \mathcal{J} \cap \overline{\mathcal{K}}$ and $|\overline{\mathcal{K}}| + a' + 1 = |\mathcal{K}| + a'$. Therefore we may apply induction to (3.37). In particular it follows that

$$G^k_{kw}(tx_1) - \epsilon G^k_{k,w-1} = \sum_{\substack{j \in \mathcal{J}, \\ j \neq k}} \sum_{v=0}^{b+1} t^v x_j \widehat{r}^{kj}_{kwjv}, \quad 1 \leq w \leq a',$$

$$-\epsilon G^k_{ka'} = \sum_{\substack{j \in \mathcal{J}, \\ j \neq k}} \sum_{v=0}^{b+1} t^v x_j \widehat{r}^{kj}_{k,a'+1,jv}$$

for every $k \in \overline{\mathcal{K}}$ and some $\widehat{r}_{kwjv} : \mathcal{A}^{m-2} \to \mathcal{Q}$. Rearranging terms in these recursive formulas we infer that the G_{kw}'s are of the form (3.36) for every $k \in \overline{\mathcal{K}}$ and some r_{kwjv}'s. Now, $|\mathcal{K}| \geq 2$ and so, for example, $2 \in \mathcal{K}$. Therefore we may repeat the preceding process with respect to x_2 (i.e., computing $D(tx_2) - \epsilon D(\overline{x}_m)t$, etc.; if $2 \notin \mathcal{J}$ the argument is even simpler). This in particular shows that (3.36) holds also for $k = 1$. Using (3.36) in (3.34) we obtain

$$\sum_{j \in \mathcal{J}} \sum_{v=0}^{b+1} t^v x_j \left\{ F^j_{jv} + \sum_{\substack{k \in \mathcal{K}, \\ k \neq j}} \sum_{w=0}^{a'} r^{kj}_{kwjv} x_k^* t^w \right\} = 0.$$

Lemma 3.9 (iii) now tells us that F_{jv}'s are of the form (3.35), which completes the proof of ($\star$). □

Lemma 3.22. *Let $\mathcal{A}$ be a $(t; d)$-free subring of $\mathcal{Q}$. If $\mathcal{L} = \emptyset$ and $\max\{|\mathcal{I}| + a, |\mathcal{J}| + b + 1, |\mathcal{I}| + |\mathcal{K}| + a'\} \leq d$, then (3.32) has only standard solutions.*

Proof. We proceed by induction on $|\mathcal{I}|$. Lemma 3.21 tells us that the result is true if $|\mathcal{I}| = 0$. So assume $|\mathcal{I}| > 0$. The proof that follows is a rather straightforward

Proof. We have to prove that (3.32) has only standard solutions provided that $\max\{|\mathcal{I}|+|\mathcal{K}|+a'', |\mathcal{J}|+|\mathcal{L}|+b''\} \le d$ where $a'' = \max\{a, a'\}$ and $b'' = \max\{b, b'\}$.

It suffices to treat the case where $a = a'$ and $b = b'$. Indeed, suppose that the theorem is true in this special case, and assume that, say, $b < b'$. Then we set $F_{jv} = 0$ for all $b+1 \le v \le b'$ and $j \in \mathcal{J}$, so that we can rewrite (3.32) with b' playing the role of b. By our assumption we can then express all F'_{jv}'s (including the "redundant" ones for $b+1 \le v \le b'$) through their standard forms. Applying Lemma 3.23 (with $d+1$ playing the role of d) it follows that all p_{iujv}, r_{kwjv} are zero for $b+1 \le v \le b'$; moreover, Lemma 3.10 then implies that also each $\lambda_{juv} = 0$, $b+1 \le v \le b'$. Therefore we can simply omit writing these maps when expressing E_{iu}'s and G_{kw}'s by their standard forms. Similarly we discuss the cases when $b > b'$, $a < a'$ and $a > a'$; in each case we conclude that (3.33) holds.

So assume that $a = a' = a''$ and $b = b' = b''$. We proceed by induction on $|\mathcal{I}|+|\mathcal{L}|$. If $|\mathcal{I}|+|\mathcal{L}| = 0$, i.e., $\mathcal{I} = \mathcal{L} = \emptyset$, then using $\max\{|\mathcal{J}|+b+1, |\mathcal{K}|+a\} \le d+1$ we see that the result follows from Lemma 3.21 (with $d+1$ playing the role of d). In the inductive case $|\mathcal{I}| + |\mathcal{L}| > 0$ we may assume that $1 \in \mathcal{I} \cup \mathcal{L}$. Substituting $x_1 x_{m+1}^*$ for x_1 to (3.32) and using $(x_1 x_{m+1}^*)^* = x_{m+1} x_1^*$ we see that

$$\sum_{\substack{i \in \mathcal{I}, \\ i \ne 1}}^{a} \sum_{u=0} E_{iu}^i (x_1 x_{m+1}^*) x_i t^u + \sum_{v=0}^{b} t^v x_1 \{x_{m+1}^* F_{1v}^1\} + \sum_{\substack{j \in \mathcal{J}, \\ j \ne 1}}^{b} \sum_{v=0} t^v x_j F_{jv}^j (x_1 x_{m+1}^*)$$

$$+ \sum_{z=0}^{b} t^z x_{m+1} \{x_1^* H_{1z}^1\} + \sum_{w=0}^{a} \{G_{1w}^1 x_{m+1}\} x_1^* t^w + \sum_{\substack{k \in \mathcal{K}, \\ k \ne 1}}^{a} \sum_{w=0} G_{kw}^k (x_1 x_{m+1}^*) x_k^* t^w$$

$$+ \sum_{u=0}^{a} \{E_{1u}^1 x_1\} x_{m+1}^* t^u + \sum_{\substack{l \in \mathcal{L}, \\ l \ne 1}}^{b} \sum_{z=0} t^z x_l^* H_{lz}^l (x_1 x_{m+1}^*) = 0,$$

where it is understood that each $E_{1u} = 0$ (resp. $F_{1v} = 0$, $G_{1w} = 0$, $H_{1z} = 0$) provided that $1 \notin \mathcal{I}$ (resp. $1 \notin \mathcal{J}$, $1 \notin \mathcal{K}$, $1 \notin \mathcal{L}$). This can be rewritten as

$$\sum_{i \in \mathcal{I}'}^{a} \sum_{u=0} \overline{E}_{iu}^i x_i t^u + \sum_{j \in \mathcal{J}'}^{b} \sum_{v=0} t^v x_j \overline{F}_{jv}^j$$

$$+ \sum_{k \in \mathcal{K}'}^{a} \sum_{w=0} \overline{G}_{kw}^k x_k^* t^w + \sum_{l \in \mathcal{L}'}^{b} \sum_{z=0} t^z x_l^* \overline{H}_{lz}^l = 0, \qquad (3.39)$$

where

$$\mathcal{I}' = \mathcal{I} \setminus \{1\},$$

$$\mathcal{J}' = \begin{cases} \mathcal{J} \cup \{m+1\} & \text{if } 1 \in \mathcal{L}, \\ \mathcal{J} & \text{if } 1 \notin \mathcal{L}, \end{cases}$$

$$\mathcal{K}' = \begin{cases} \mathcal{K} \cup \{m+1\} & \text{if } 1 \in \mathcal{I}, \\ \mathcal{K} & \text{if } 1 \notin \mathcal{I}, \end{cases}$$

$$\mathcal{L}' = \mathcal{L} \setminus \{1\},$$

and

$$\overline{E}_{iu}(\overline{x}_{m+1}^i) = E_{iu}^i(x_1 x_{m+1}^*), \quad i \in \mathcal{I}', \ 0 \le u \le a,$$
$$\overline{F}_{1v}(\overline{x}_{m+1}^1) = x_{m+1}^* F_{1v}(\overline{x}_m^1), \quad 0 \le v \le b,$$
$$\overline{F}_{jv}(\overline{x}_{m+1}^j) = F_{jv}^j(x_1 x_{m+1}^*), \quad j \in \mathcal{J}', \ j \ne 1, m+1, \ 0 \le v \le b,$$
$$\overline{F}_{m+1,v}(\overline{x}_{m+1}^{m+1}) = x_1^* H_{1v}(\overline{x}_m^1), \quad 0 \le v \le b,$$
$$\overline{G}_{1w}(\overline{x}_{m+1}^1) = G_{1w}(\overline{x}_m^1) x_{m+1}, \quad 0 \le w \le a,$$
$$\overline{G}_{kw}(\overline{x}_{m+1}^k) = G_{kw}^k(x_1 x_{m+1}^*), \quad k \in \mathcal{K}', \ k \ne 1, m+1, \ 0 \le w \le a,$$
$$\overline{G}_{m+1,w}(\overline{x}_{m+1}^{m+1}) = E_{1w}(\overline{x}_m^1) x_1, \quad 0 \le w \le a,$$
$$\overline{H}_{lz}(\overline{x}_{m+1}^l) = H_{lz}^l(x_1 x_{m+1}^*), \quad l \in \mathcal{L}', \ 0 \le z \le b,$$

Observing that $|\mathcal{I}'| + |\mathcal{K}'| = |\mathcal{I}| + |\mathcal{K}|$, $|\mathcal{J}'| + |\mathcal{L}'| = |\mathcal{J}| + |\mathcal{L}|$ and $|\mathcal{I}'| + |\mathcal{L}'| < |\mathcal{I}| + |\mathcal{L}|$ we see that we are in a position to apply the induction assumption to (3.39). Therefore all the maps, in particular the $\overline{F}_{m+1,v}$'s and $\overline{G}_{m+1,w}$'s, have standard solutions. Accordingly, we have

$$x_1^* H_{1v}^1 = -\sum_{i \in \mathcal{I}'} \sum_{u=0}^a \overline{p}_{iu,m+1,v}^{i,m+1} x_i t^u - \sum_{k \in \mathcal{K}'} \sum_{w=0}^a \overline{r}_{kw,m+1,v}^{k,m+1} x_k^* t^w,$$

$$E_{1w}^1 x_1 = \sum_{j \in \mathcal{J}'} \sum_{v=0}^b t^v x_j \overline{r}_{m+1,wjv}^{m+1,j} + \sum_{l \in \mathcal{L}'} \sum_{z=0}^b t^z x_l^* \overline{s}_{m+1,wlz}^{m+1,l}$$

for all $0 \le v \le b$, $0 \le w \le a$ (here all $\lambda_{m+1,uv}$'s and $\mu_{m+1,wz}$'s are equal to zero because $m+1 \notin \mathcal{I}'$ and $m+1 \notin \mathcal{L}'$). We can interpret these two identities so that Lemma 3.22 is applicable (with $d+1$ playing the role of d). To be precise, to get the right setting for applying this lemma we have to replace x_1 by x_1^* in the first identity, while in the second identity we just refer to the left-right symmetry. Thus, since $|\mathcal{I}'| \le |\mathcal{I}|$ we may apply Lemma 3.22 to the first identity to obtain

$$H_{1z}^1 = -\sum_{i \in \mathcal{I}'} \sum_{u=0}^a q_{iu1z}^{i1} x_i t^u - \sum_{\substack{k \in \mathcal{K}, \\ k \ne 1}} \sum_{w=0}^a s_{kw1z}^{k1} x_k^* t^w - \sum_{w=0}^a \mu_{1wz}^1 t^w \qquad (3.40)$$

(here v is replaced by z). Note that each $\mu_{1wz} = 0$ if $1 \notin \mathcal{K}$ and all terms in (3.40) are equal to zero if $1 \notin \mathcal{L}$. Similarly, using $|\mathcal{L}'| \le |\mathcal{L}|$ we infer from the second identity that

$$E_{1u}^1 = \sum_{\substack{j \in \mathcal{J}, \\ j \ne 1}} \sum_{v=0}^b t^v x_j p_{1ujv}^{1j} + \sum_{l \in \mathcal{L}'} \sum_{z=0}^b t^z x_l^* q_{1ulz}^{1l} + \sum_{v=0}^b \lambda_{1uv}^1 t^v \qquad (3.41)$$

(here w is replaced by u). As above we note that all the terms in (3.41) are zero if $1 \notin \mathcal{I}$ and each $\lambda_{1uv} = 0$ if $1 \notin \mathcal{J}$. Substituting (3.40) and (3.41) in (3.32) we get

$$\sum_{i \in \mathcal{I}'} \sum_{u=0}^{a} \{E_{iu}^i - \sum_{z=0}^{b} t^z x_1^* q_{iu1z}^{i1}\} x_i t^u$$

$$+ \sum_{v=0}^{b} t^v x_1 \{F_{1v}^1 + \sum_{u=0}^{a} \lambda_{1uv}^1 t^u\} + \sum_{\substack{j \in \mathcal{J}, \\ j \neq 1}} \sum_{v=0}^{b} t^v x_j \{F_{jv}^j + \sum_{u=0}^{a} p_{1ujv}^{1j} x_1 t^u\}$$

$$+ \sum_{w=0}^{a} \{G_{1w}^1 - \sum_{z=0}^{b} \mu_{1wz}^1(\overline{x}_m) t^z\} x_1^* t^w + \sum_{\substack{k \in \mathcal{K}, \\ k \neq 1}} \sum_{w=0}^{a} \{G_{kw}^k - \sum_{z=0}^{b} t^z x_1^* s_{kw1z}^{k1}\} x_k^* t^w$$

$$+ \sum_{l \in \mathcal{L}'} \sum_{z=0}^{b} t^z x_l^* \{H_{lz}^l + \sum_{u=0}^{a} q_{1ulz}^{1l} x_1 t^u\} = 0.$$

We now have $|\mathcal{I}'| + |\mathcal{L}'| < |\mathcal{I}| + |\mathcal{L}|$ and using the induction assumption one can easily complete the proof.   □

Theorem 3.25 is certainly interesting in its own right. In particular it tells us that in $(t; d+1)$-free subrings one has also control of FI's of the form

$$\sum_{i \in \mathcal{I}} E_i^i x_i + \sum_{j \in \mathcal{J}} x_j F_{jv}^j + \sum_{k \in \mathcal{K}} G_{kw}^k x_k^* + \sum_{l \in \mathcal{L}} x_l^* H_{lz}^l = 0$$

provided that $\max\{|\mathcal{I}| + |\mathcal{K}|, |\mathcal{J}| + |\mathcal{L}|\} \leq d$ (consider the case $a = b = a' = b' = 0$ in Definition 3.19). However, the main reason for which we find Theorem 3.25 important is that we shall use it as a crucial tool in establishing $(t; d)$-freeness (and consequently d-freeness) of $\mathcal{S}$ and $\mathcal{K}$ (Theorem 3.28). This is the main and the final goal of this section, and to achieve it we again first need some auxiliary results.

Lemma 3.26. *Let $\mathcal{A}$ be a $(t; d)$-free subring of $\mathcal{Q}$. Suppose there exist $\lambda_r : \mathcal{A}^m \to \mathcal{C}$, $0 \leq r \leq c$, such that*

$$\sum_{i \in \mathcal{I}} \sum_{u=0}^{a} E_{iu}^i x_i t^u + \sum_{j \in \mathcal{J}} \sum_{v=0}^{b} t^v x_j F_{jv}^j + \sum_{k \in \mathcal{K}} \sum_{w=0}^{a'} G_{kw}^k x_k^* t^w$$

$$+ \sum_{l \in \mathcal{L}} \sum_{z=0}^{b'} t^z x_l^* H_{lz}^l = \sum_{r=0}^{c} \lambda_r(\overline{x}_m) t^r$$

for all $\overline{x}_m \in \mathcal{A}^m$. If $\max\{|\mathcal{I}| + |\mathcal{K}| + a + 1, |\mathcal{K}| + a' + 1, |\mathcal{J}| + |\mathcal{L}| + c\} \leq d - 1$, then each $\lambda_r = 0$.

Proof. We use a similar approach as in the proof of Lemma 3.21. Set $E_{i,a+1} = G_{k,a'+1} = 0$ for $i \in \mathcal{I}$ and $k \in \mathcal{K}$, and consider

$$D(\overline{x}_m) = \sum_{i \in \mathcal{I}} \sum_{u=0}^{a+1} E^i_{iu} x_i t^u + \sum_{j \in \mathcal{J}} \sum_{v=0}^{b} t^v x_j F^j_{jv}$$

$$+ \sum_{k \in \mathcal{K}} \sum_{w=0}^{a'+1} G^k_{kw} x^*_k t^w + \sum_{l \in \mathcal{L}} \sum_{z=0}^{b'} t^z x^*_l H^l_{lz} = \sum_{r=0}^{c} \lambda_r(\overline{x}_m) t^r. \quad (3.42)$$

We emphasize that in particular $E_{i,a+1} = 0$ if $i \in \mathcal{I} \cap \mathcal{L}$ and $G_{k,a'+1} = 0$ if $k \in \mathcal{K} \cap \mathcal{J}$.

We will consider a slightly more general situation than the one appearing in the statement of the lemma. Our goal is to show that the following is true:

($\star$) If maps $E_{iu}, F_{jv}, G_{kw}, H_{lz}, \lambda_r$ are such that $E_{i,a+1} = 0$ if $i \in \mathcal{I} \cap \mathcal{L}$ and $G_{k,a'+1} = 0$ if $k \in \mathcal{K} \cap \mathcal{J}$, then (3.42) together with $\max\{|\mathcal{I}| + |\mathcal{K}| + a + 1, |\mathcal{K}| + a' + 1, |\mathcal{J}| + |\mathcal{L}| + c\} \leq d - 1$ implies that each $\lambda_r = 0$.

We prove ($\star$) by induction on $|\mathcal{J}| + |\mathcal{L}|$. Suppose $|\mathcal{J}| + |\mathcal{L}| = 0$. Then Lemma 3.23 tells us that each $E_{iu} = 0$ and each $G_{kw} = 0$. Hence $D(\overline{x}_m) = 0$ and accordingly $\sum_{r=0}^{c} \lambda_r(\overline{x}_m) t^r = 0$. Since $\deg(t) \geq d$ by Lemma 3.10 and $d > c$ by assumption, we conclude that each $\lambda_r = 0$.

We may now assume that $|\mathcal{J}| + |\mathcal{L}| > 0$. Of course it is enough to show that $\lambda_c = 0$. We consider two cases: when $|\mathcal{J}| \neq 0$ and when $|\mathcal{J}| = 0$.

Suppose first that $|\mathcal{J}| \neq 0$, say $1 \in \mathcal{J}$. Computing $D(tx_1) - tD(\overline{x}_m)$ we get

$$\sum_{i \in \mathcal{I}} \sum_{u=0}^{a+1} \widehat{E}^i_{iu} x_i t^u + \sum_{\substack{j \in \mathcal{J}, \\ j \neq 1}} \sum_{v=0}^{b+1} t^v x_j \widehat{F}^j_{jv} + \sum_{k \in \mathcal{K}} \sum_{w=0}^{a'+1} \widehat{G}^k_{kw} x^*_k t^w$$

$$+ \sum_{l \in \mathcal{L}} \sum_{z=0}^{b'+1} t^z x^*_l \widehat{H}^l_{lz} = \sum_{r=0}^{c+1} \widehat{\lambda}_r(\overline{x}_m) t^r$$

for appropriate $\widehat{E}_{iu}, \widehat{F}_{jv}, \widehat{G}_{kw}, \widehat{H}_{lz}, \widehat{\lambda}_w$ (here we already made use of the assumption that $G_{1,a'+1} = 0$ if $1 \in \mathcal{K}$, since otherwise the term $\widehat{G}_{1,a'+2}$ would also appear). One can easily check that $\widehat{\lambda}_{c+1} = -\lambda_c$, $\widehat{E}_{i,a+1} = 0$ if $i \in \mathcal{I} \cap \mathcal{L}$, and $\widehat{G}_{k,a'+1} = 0$ if $k \in \mathcal{K} \cap (\mathcal{J} \setminus \{1\})$ (incidentally, $\widehat{G}_{1,a'+1}$ may not be zero and that is why we cannot prove the lemma directly). Clearly $|\mathcal{J} \setminus \{1\}| + |\mathcal{L}| + c + 1 = |\mathcal{J}| + |\mathcal{L}| + c$ and hence we can apply the induction assumption to obtain $\lambda_c = -\widehat{\lambda}_{c+1} = 0$.

If $|\mathcal{J}| = 0$, then we may assume that $1 \in \mathcal{L}$. This case is slightly easier since the second summand of (3.42) does not appear. We now compute $D(x_1 t) - \epsilon t D(\overline{x}_m)$; this will result in being able to replace $\mathcal{L}$ by $\mathcal{L} \setminus \{1\}$. Since the details of the proof are entirely similar to those of the preceding case, we leave it for the reader to complete the proof.

Thus ($\star$) is proved. Note that the lemma follows immediately. $\square$

We now have enough information to tackle the question on the $(t; d)$-freeness of $\mathcal{S}$ and $\mathcal{K}$. For convenience we recall the FI's appearing in the definition of $(t; d)$-freeness of a subset $\mathcal{R}$. These are

$$\sum_{i \in \mathcal{I}} \sum_{u=0}^{a} E^i_{iu} x_i t^u + \sum_{j \in \mathcal{J}} \sum_{v=0}^{b} t^v x_j F^j_{jv} = 0 \quad \text{for all } \overline{x}_m \in \mathcal{R}^m, \tag{3.43}$$

and

$$\sum_{i \in \mathcal{I}} \sum_{u=0}^{a} E^i_{iu} x_i t^u + \sum_{j \in \mathcal{J}} \sum_{v=0}^{b} t^v x_j F^j_{jv} \in \mathcal{C} \quad \text{for all } \overline{x}_m \in \mathcal{R}^m. \tag{3.44}$$

Their standard solution is defined by

$$E^i_{iu} = \sum_{\substack{j \in \mathcal{J}, \\ j \neq i}} \sum_{v=0}^{b} t^v x_j p^{ij}_{iujv} + \sum_{v=0}^{b} \lambda^i_{iuv} t^v,$$

$$F^j_{jv} = -\sum_{\substack{i \in \mathcal{I}, \\ i \neq j}} \sum_{u=0}^{a} p^{ij}_{iujv} x_i t^u - \sum_{u=0}^{a} \lambda^j_{juv} t^u, \tag{3.45}$$

$$\lambda_{kuv} = 0 \quad \text{if } k \notin \mathcal{I} \cap \mathcal{J},$$

where $p_{ij} : \mathcal{R}^{m-2} \to \mathcal{Q}$ and $\lambda_k : \mathcal{R}^{m-1} \to \mathcal{C}$.

In the next lemma we will come very close to the main result, Theorem 3.28; in fact, the lemma partially even supersedes the theorem. It considers (3.43) and (3.44) on the sets $\{x + x^* \mid x \in \mathcal{A}\}$ and $\{x - x^* \mid x \in \mathcal{A}\}$ which are clearly subsets of $\mathcal{S}$ and $\mathcal{K}$, respectively. Quite often (for example, in the case when $\mathcal{A}$ is an algebra over a field of characteristic not 2) these sets are in fact equal to $\mathcal{S}$ and $\mathcal{K}$, respectively.

Lemma 3.27. *Let $\mathcal{A}$ be a $(t; d)$-free subring of $\mathcal{Q}$, let $\epsilon = 1$ or $\epsilon = -1$, and let $\mathcal{R} = \{x + \epsilon x^* \mid x \in \mathcal{A}\}$. Then:*

(i) *If $\max\{2|\mathcal{I}| + a, 2|\mathcal{J}| + b\} \leq d - 1$, then (3.43) has only standard solutions.*

(ii) *If $\max\{2|\mathcal{I}| + a, 2|\mathcal{J}| + b\} \leq d - 2$, then (3.44) has only standard solutions.*

Proof. For every $x \in \mathcal{A}$ we shall write

$$\widehat{x} = x + \epsilon x^*.$$

So $\mathcal{R} = \{\widehat{x} \mid x \in \mathcal{A}\}$. We define new maps

$$\widehat{E}_{iu}(\overline{x}^i_m) = E_{iu}(\widehat{x}_1, \dots, \widehat{x}_{i-1}, \widehat{x}_{i+1}, \dots, \widehat{x}_m), \quad i \in \mathcal{I},$$

$$\widehat{F}_{jv}(\overline{x}^j_m) = F_{jv}(\widehat{x}_1, \dots, \widehat{x}_{j-1}, \widehat{x}_{j+1}, \dots, \widehat{x}_m), \quad j \in \mathcal{J},$$

$$\widehat{G}_{iu}(\overline{x}^i_m) = \epsilon E_{iu}(\widehat{x}_1, \dots, \widehat{x}_{i-1}, \widehat{x}_{i+1}, \dots, \widehat{x}_m), \quad i \in \mathcal{I},$$

$$\widehat{H}_{jv}(\overline{x}^j_m) = \epsilon F_{jv}(\widehat{x}_1, \dots, \widehat{x}_{j-1}, \widehat{x}_{j+1}, \dots, \widehat{x}_m), \quad j \in \mathcal{J}$$

for all $\bar{x}_m \in \mathcal{A}^m$. Further, set

$$D(\bar{x}_m) = \sum_{i \in \mathcal{I}} \sum_{u=0}^{a} \widehat{E}^i_{iu} x_i t^u + \sum_{j \in \mathcal{J}} \sum_{v=0}^{b} t^v x_j \widehat{F}^j_{jv}$$

$$+ \sum_{i \in \mathcal{I}} \sum_{u=0}^{a} \widehat{G}^i_{iu} x_i^* t^u + \sum_{j \in \mathcal{J}} \sum_{v=0}^{b} t^v x_j^* \widehat{H}^j_{jv}.$$

It is clear that (3.43) implies $D(\bar{x}_m) = 0$, and similarly, (3.44) implies $D(\bar{x}_m) \in \mathcal{C}$. Assuming the condition required in (ii), i.e., $\max\{2|\mathcal{I}|+a, 2|\mathcal{J}|+b\} \leq d-2$, it follows that $\max\{2|\mathcal{I}|+a+1, 2|\mathcal{J}|\} \leq d-1$. This is exactly what is needed to apply Lemma 3.26 (for $c = 0$), showing that $D(\bar{x}_m) \in \mathcal{C}$ forces $D(\bar{x}_m) = 0$. Accordingly, in order to prove the lemma we may assume that $D(\bar{x}_m) = 0$ and $\max\{2|\mathcal{I}| + a, 2|\mathcal{J}| + b\} \leq d - 1$; our goal is to derive (3.45) from these two assumptions.

Theorem 3.25 tells us that $\mathcal{A}$ is $(*; t; d-1)$-free. Therefore $D(\bar{x}_m) = 0$ implies that

$$\widehat{E}^i_{iu} = \sum_{\substack{j \in \mathcal{J}, \ v=0 \\ j \neq i}}^{b} t^v x_j \widehat{p}^{ij}_{iujv} + \sum_{\substack{j \in \mathcal{J}, \ v=0 \\ j \neq i}}^{b} t^v x_j^* \widehat{q}^{ij}_{iujv} + \sum_{v=0}^{b} \widehat{\lambda}^i_{iuv} t^v,$$

$$\widehat{F}^j_{jv} = -\sum_{\substack{i \in \mathcal{I}, \ u=0 \\ i \neq j}}^{a} \widehat{p}^{ij}_{iujv} x_i t^u - \sum_{\substack{i \in \mathcal{I}, \ u=0 \\ i \neq j}}^{a} \widehat{r}^{ij}_{iujv} x_i^* t^u - \sum_{u=0}^{a} \widehat{\lambda}^j_{juv} t^u,$$

$$\widehat{G}^i_{iu} = \sum_{\substack{j \in \mathcal{J}, \ v=0 \\ j \neq i}}^{b} t^v x_j \widehat{r}^{ij}_{iujv} + \sum_{\substack{j \in \mathcal{J}, \ v=0 \\ j \neq i}}^{b} t^v x_j^* \widehat{s}^{ij}_{iujv} + \sum_{v=0}^{b} \widehat{\mu}^i_{iuv} t^v,$$

$$\widehat{H}^j_{jv} = -\sum_{\substack{i \in \mathcal{I}, \ u=0 \\ i \neq j}}^{a} \widehat{q}^{ij}_{iujv} x_i t^u - \sum_{\substack{i \in \mathcal{I}, \ u=0 \\ i \neq j}}^{a} \widehat{s}^{ij}_{iujv} x_i^* t^u - \sum_{u=0}^{a} \widehat{\mu}^j_{juv} t^u,$$

$$\widehat{\lambda}_{kuv} = \widehat{\mu}_{kuv} = 0 \quad \text{if } k \notin \mathcal{I} \cap \mathcal{J}$$

for some $\widehat{p}^{ij}_{iujv} : \mathcal{A}^{m-2} \to \mathcal{Q}$, $\widehat{\lambda}_{kuv} : \mathcal{A}^{m-1} \to \mathcal{C}$, etc. However, $\widehat{F}_{jv} = \epsilon \widehat{H}_{jv}$ and so after comparing the right-hand sides of the second and the fourth identities we may apply Lemma 3.23 to obtain

$$\widehat{p}_{iujv} = \epsilon \widehat{q}_{iujv}, \quad \widehat{r}_{iujv} = \epsilon \widehat{s}_{iujv}, \quad i \in \mathcal{I}, \ j \in \mathcal{J},$$

and apply Lemma 3.10 to obtain $\widehat{\lambda}_{juv} = \epsilon \widehat{\mu}_{juv}$, $j \in \mathcal{J}$; the latter clearly yields that in fact

$$\widehat{\lambda}_{kuv} = \epsilon \widehat{\mu}_{kuv}, \quad k \in \mathcal{I} \cup \mathcal{J}.$$

Similarly, $\widehat{E}_{iu} = \epsilon \widehat{G}_{iu}$ and hence

$$\widehat{p}_{iujv} = \epsilon \widehat{r}_{iujv}, \quad \widehat{q}_{iujv} = \epsilon \widehat{s}_{iujv}, \quad i \in \mathcal{I}, \ j \in \mathcal{J}.$$

Consequently, we have

$$\widehat{p}_{iujv} = \widehat{s}_{iujv} = \epsilon \widehat{q}_{iujv} = \epsilon \widehat{r}_{iujv}, \quad i \in \mathcal{I}, \; j \in \mathcal{J}.$$

Hence it follows that

$$E_{iu}(\widehat{x}_1, \dots, \widehat{x}_{i-1}, \widehat{x}_{i+1}, \dots, \widehat{x}_m)$$

$$= \sum_{\substack{j \in \mathcal{J}, \; v=0 \\ j \neq i}}^{b} t^v \widehat{x}_j \widehat{p}_{iujv}(\overline{x}_m^{ij}) + \sum_{v=0}^{b} \widehat{\lambda}_{iuv}(\overline{x}_m^i) t^v, \tag{3.46}$$

and

$$F_{jv}(\widehat{x}_1, \dots, \widehat{x}_{j-1}, \widehat{x}_{j+1}, \dots, \widehat{x}_m)$$

$$= -\sum_{\substack{i \in \mathcal{I}, \; u=0 \\ i \neq j}}^{a} \widehat{p}_{iujv}(\overline{x}_m^{ij}) \widehat{x}_i t^u - \sum_{u=0}^{a} \widehat{\lambda}_{juv}(\overline{x}_m^j) t^u. \tag{3.47}$$

Let us denote by t_i arbitrary elements in $\mathcal{A}$ satisfying $\widehat{t}_i = 0$ (that is, the t_i's are skew elements if $\epsilon = 1$, and are symmetric elements if $\epsilon = -1$). Replace each x_i in (3.47) by $y_i = x_i + t_i$. The left-hand side clearly remains unchanged, and so the same must be true for the right-hand side. This means that

$$\sum_{\substack{i \in \mathcal{I}, \; u=0 \\ i \neq j}}^{a} \widehat{p}_{iujv}(\overline{x}_m^{ij}) \widehat{x}_i t^u + \sum_{u=0}^{a} \widehat{\lambda}_{juv}(\overline{x}_m^j) t^u$$

$$= \sum_{\substack{i \in \mathcal{I}, \; u=0 \\ i \neq j}}^{a} \widehat{p}_{iujv}(\overline{y}_m^{ij}) \widehat{x}_i t^u + \sum_{u=0}^{a} \widehat{\lambda}_{juv}(\overline{y}_m^j) t^u$$

for all x_i's and all t_i's. Now rewrite this identity as

$$\sum_{\substack{i \in \mathcal{I}, \; u=0 \\ i \neq j}}^{a} \left\{ \widehat{p}_{iujv}(\overline{x}_m^{ij}) - \widehat{p}_{iujv}(\overline{y}_m^{ij}) \right\} (x_i + \epsilon x_i^*) t^u = \sum_{u=0}^{a} \left\{ \widehat{\lambda}_{juv}(\overline{y}_m^j) - \widehat{\lambda}_{juv}(\overline{x}_m^j) \right\} t^u.$$

Considering x_i's as variables and t_i's as fixed we see that Lemma 3.23 can be applied. Hence it follows that $\widehat{p}_{iujv}(\overline{x}_m^{ij}) = \widehat{p}_{iujv}(\overline{y}_m^{ij})$ for all i, j, u, v, $i \neq j$. Consequently, $\widehat{\lambda}_{juv}(\overline{x}_m^j) = \widehat{\lambda}_{juv}(\overline{y}_m^j)$ for all j, u, v by Lemma 3.10. Since $\widehat{\lambda}_{kuv} = 0$ if $k \notin \mathcal{I} \cap \mathcal{J}$, the latter does not hold only for $j \in \mathcal{J}$ but also for those from $\mathcal{I}$.

The conclusion from the preceding paragraph makes it possible for us to define maps $p_{iujv} : \mathcal{R}^{m-2} \to \mathcal{Q}$ and $\lambda_{kuv} : \mathcal{R}^{m-1} \to \mathcal{C}$ by

$$p_{iujv}\left((\widehat{x}_1, \widehat{x}_2, \dots, \widehat{x}_m)^{ij} \right) = \widehat{p}_{iujv}\left((x_1, x_2, \dots, x_m)^{ij} \right),$$

$$\lambda_{kuv}\left((\widehat{x}_1, \widehat{x}_2, \dots, \widehat{x}_m)^k \right) = \widehat{\lambda}_{kuv}\left((x_1, x_2, \dots, x_m)^k \right).$$

Indeed, if $x_i, y_i \in \mathcal{A}$ are such that $\widehat{x}_i = \widehat{y}_i$, then $t_i = y_i - x_i$ satisfies $\widehat{t}_i = 0$ and therefore

$$p_{iujv}\left((\widehat{x}_1, \widehat{x}_2, \ldots, \widehat{x}_m)^{ij}\right) = \widehat{p}_{iujv}\left((x_1, x_2, \ldots, x_m)^{ij}\right)$$

$$= \widehat{p}_{iujv}\left((x_1 + t_1, x_2 + t_2, \ldots, x_m + t_m)^{ij}\right) = p_{iujv}\left((\widehat{y}_1, \widehat{y}_2, \ldots, \widehat{y}_m)^{ij}\right),$$

showing that p_{uijv} is well-defined. Similarly we see that each $\widehat{\lambda}_{kuv}$ is well-defined. Now it is clear from (3.46) and (3.47) that (3.45) holds.   □

We are now in a position to state the main result of this section. Let us first recall that throughout this section we are tacitly assuming that $\mathcal{A} \subseteq \mathcal{Q}$ is a ring with involution and $t \in \mathcal{S} \cup \mathcal{K}$. In this setting we have

Theorem 3.28. *If $\mathcal{A}$ is a $(t; 2d + 1)$-free subring of $\mathcal{Q}$, then $\mathcal{S}$ and $\mathcal{K}$ are $(t; d)$-free subsets of $\mathcal{Q}$. In particular, $\mathcal{S}$ and $\mathcal{K}$ are d-free.*

Proof. The sets $\{x + x^* \mid x \in \mathcal{A}\}$ and $\{x - x^* \mid x \in \mathcal{A}\}$ are clearly subsets of $\mathcal{S}$ and $\mathcal{K}$, respectively. According to Theorem 3.14 it suffices to show that these two sets are $(t; d)$-free. That is, we have to show that (3.43) (respectively, (3.44)) has only standard solutions on these sets provided that $\max\{|\mathcal{I}| + a, |\mathcal{J}| + b\} \leq d$ (respectively, $\max\{|\mathcal{I}| + a, |\mathcal{J}| + b\} \leq d - 1$). Note that Lemma 3.27 shows that this is true indeed even under the slightly milder assumption that $\max\{2|\mathcal{I}| + a, 2|\mathcal{J}| + b\} \leq 2d$ (respectively, $\max\{2|\mathcal{I}| + a, 2|\mathcal{J}| + b\} \leq 2d - 1$).   □

Literature and Comments. The concepts of d-free and $(t; d)$-free sets were introduced by Beidar and Chebotar in [29], and the main results from Sections 3.2 and 3.4 are taken from this paper. The two constructions from Section 3.3 appear in [24] as auxiliary results needed in the study of Lie maps on prime rings. In Section 3.5 we have basically followed the papers [22, 38]. These papers do not introduce the notion of $(\ast; t; d)$-freeness and they consider the case of prime rings only. However, the arguments essentially still work in the more general context treated in Section 3.5.

Chapter 4

Functional Identities on d-Free Sets

Chapter 3 was primarily devoted to constructing new d-free sets from given d-free or $(t; d)$-free sets. Now we turn our attention to the study of FI's on d-free sets. Of course, by the very definition one can say everything that is possible about the basic FI's through which d-free sets were introduced. But what we intend to show is that one can analyze also some other FI's on d-free sets, some of them considerably more complicated than the basic ones.

4.1 Introducing the General Setting

In Section 3.1 we set forth the framework needed in order to define the notion of d-free sets and variations thereof, and we laid out the notational rules for describing the very simplest kinds of FI's. The functions involved essentially stemmed from those of the form $E_i(\overline{x}_m^i)x_i$ and $x_j F_j(\overline{x}_m^j)$. In Chapter 3 the theory we discussed only involved these simple functions and certain extensions thereof, e.g., $E_{iu}(\overline{x}_m^i)x_i t^u$ (Section 3.4) and $G_i(\overline{x}_m^i)x_i^*$ (Section 3.5).

In the present chapter we shall extend the theory in two directions. First (already alluded to in Section 3.1) we shall need a more general framework, one in which arbitrary sets $\mathcal{S}_i$ together with fixed maps $\alpha_i : \mathcal{S}_i \to \mathcal{Q}$, are required. The notion of d-freeness for such "pairs" $(\mathcal{S}_i; \alpha_i)$ is given in Section 4.2; the main result of this section shows that this extended notion of d-freeness fortunately coincides with the original notion of d-freeness for the sets $\mathcal{R}_i = \mathcal{S}_i^{\alpha_i}$ (we will always write the α_i's as exponents). An example involving Lie homomorphisms, which is presented later in this section, provides a good illustration for the need of this extended framework. Secondly, we will greatly expand the complexity of functions appearing in the theory, thus it is appropriate at this point to develop the requisite notation. Furthermore, we shall describe this notation from the *function* approach as well as from the *function-value* approach, since the *function* notation is called for in Section 4.3.

The setting where the general theory takes place is the following: $\mathcal{Q}$ is a unital ring with center $\mathcal{C}$, m is a fixed positive integer, $\mathcal{S}_1, \mathcal{S}_2, \ldots, \mathcal{S}_m$ are fixed arbitrary nonempty sets, and $\alpha_i : \mathcal{S}_i \to \mathcal{Q}$, $i = 1, 2, \ldots, m$, are fixed maps. We set

$$\mathcal{R}_i = \mathcal{S}_i^{\alpha_i}.$$

We shall write $\widehat{\mathcal{S}}$ for $\mathcal{S}_1 \times \mathcal{S}_2 \times \ldots \times \mathcal{S}_m$, $\widehat{\mathcal{R}}$ for $\mathcal{R}_1 \times \mathcal{R}_2 \times \ldots \times \mathcal{R}_m$, and in the same fashion as in Chapter 3 we define $\widehat{\mathcal{S}}^i$, $\widehat{\mathcal{S}}^{ij}$, etc. In the present chapter the m-tuple $\bar{x}_m = (x_1, x_2, \ldots, x_m)$ will usually be an element from $\widehat{\mathcal{S}}$. The meaning of $\bar{x}_m^i$ and $\bar{x}_m^{ij}$ will be the same as in Chapter 3.

We will often (in fact, usually) be interested in the case when $\mathcal{S} = \mathcal{S}_1 = \mathcal{S}_2 = \ldots = \mathcal{S}_m$ and $\alpha = \alpha_1 = \alpha_2 = \ldots = \alpha_m$; here we will write $\mathcal{S}^m$ in place of $\widehat{\mathcal{S}}$. However, since there may be potential applications to certain areas such as graded algebras, we will set forth the initial basic theory as far as possible using the more general notation (this will not involve appreciably more difficulty in the ensuing arguments).

In the general FI theory the basic "building blocks" are two types of functions. One arises from "multilinear" monomials $X_{i_1} X_{i_2} \ldots X_{i_p}$ (i.e., the i_k's are distinct), and the other arises from arbitrary functions

$$B : \mathcal{S}_{j_1} \times \mathcal{S}_{j_2} \times \ldots \times \mathcal{S}_{j_n} \to \mathcal{Q}, \quad 0 \leq n < m$$

(or, in the case of a single $\mathcal{S}$, $B : \mathcal{S}^n \to \mathcal{Q}$); if $n = 0$, then it should be understood that B is an element in $\mathcal{Q}$.

If we follow the *function* notation path, then it is a good idea that all functions can be interpreted as having the same common domain, namely $\widehat{\mathcal{S}}$ (or $\mathcal{S}^m$). With this in mind, given a sequence $(i_1, i_2, \ldots, i_p)$ of distinct elements from the set $\{1, 2, \ldots, m\}$ we define a *monomial function* $M : \widehat{\mathcal{S}} \to \mathcal{Q}$ to be given by

$$(x_1, x_2, \ldots, x_m) \mapsto x_{i_1}^{\alpha_{i_1}} x_{i_2}^{\alpha_{i_2}} \ldots x_{i_p}^{\alpha_{i_p}}.$$

We let $\mathrm{dom}(M)$ denote the set $\{i_1, i_2, \ldots, i_p\}$. Here M is said to have *degree* p, $\deg(M) = p$. In particular we have $X_i : \widehat{\mathcal{S}} \to \mathcal{Q}$ given by

$$(x_1, x_2, \ldots, x_m) \mapsto x_i^{\alpha_i}$$

and it is obvious that the monomial function M just defined above is indeed the product $X_{i_1} X_{i_2} \ldots X_{i_p}$ of the individual functions X_{i_k}. The function $M : \bar{x}_m \mapsto 1$ will be considered as a monomial function with $\mathrm{dom}(M) = \emptyset$ and $\deg(M) = 0$. For every monomial function $M = X_{i_1} X_{i_2} \ldots X_{i_p}$ we define

$$\widehat{\mathcal{S}}^M = \mathcal{S}_{j_1} \times \mathcal{S}_{j_2} \times \ldots \times \mathcal{S}_{j_{m-p}}.$$

where $\{i_1, \ldots, i_p\} \cup \{j_1, \ldots, j_{m-p}\}$ is a partition of $\{1, 2, \ldots, m\}$ and $j_1 < j_2 < \ldots < j_{m-p}$.

Given a function $B : \mathcal{S}_{j_1} \times \mathcal{S}_{j_2} \times \ldots \times \mathcal{S}_{j_n} \to \mathcal{Q}, 0 \leq n < m, j_1 < j_2 < \ldots < j_n$, we replace B by the function (which by a slight abuse of notation we continue to denote by B) from $\widehat{\mathcal{S}}$ into $\mathcal{Q}$ given by

$$(x_1, x_2, \ldots, x_m) \mapsto B(x_{j_1}, x_{j_2}, \ldots, x_{j_n}).$$

Here one could refer to all the x_k's where $k \notin \{j_1, j_2, \ldots, j_n\}$ as the "silent variables" in B. The *function-value* approach is self-evident: simply apply monomial functions M and arbitrary functions B to a typical element $\overline{x}_m \in \widehat{\mathcal{S}}$ as indicated in our discussion of the *function* approach. Thus we have respective function values $M(\overline{x}_m) = x_{i_1}^{\alpha_{i_1}} x_{i_2}^{\alpha_{i_2}} \ldots x_{i_p}^{\alpha_{i_p}}$ and $B(\overline{x}_m) = B(x_{j_1}, x_{j_2}, \ldots, x_{j_n})$. In case each $\mathcal{S}_i$ is the same $\mathcal{S}$, given a function $B : \mathcal{S}^n \to \mathcal{Q}$, there are of course various ways of producing values of B in $\mathcal{Q}$, namely for any choice of sequence $(j_1, j_2, \ldots, j_n)$ we have the value $B(x_{j_1}, x_{j_2}, \ldots, x_{j_n})$. As we shall presently see the context of the situation will always make it clear which selection of $(j_1, j_2, \ldots, j_n)$ to make.

We are now in a position to describe precisely the types of functions that are studied in FI theory. We refer as always to the general setting indicated near the beginning of this section.

We will first describe these functions using the *function* notation. We begin by fixing n such that $0 \leq n < m$. We next let $\{i_1, \ldots, i_p\} \cup \{k_1, \ldots, k_r\} \cup \{j_1, \ldots, j_n\}$ be a partition of $\{1, \ldots, m\}$ with $p + r = m - n$ and $j_1 < j_2 < \ldots < j_n$. Note that, whereas n is fixed, p and r are free to vary within the imposed restriction. Let M be the monomial function $X_{i_1} X_{i_2} \ldots X_{i_p}$ and N the monomial function $X_{k_1} X_{k_2} \ldots X_{k_r}$. Note that MN is again a monomial function and $\widehat{\mathcal{S}}^{MN} = \mathcal{S}_{j_1} \times \mathcal{S}_{j_2} \times \ldots \times \mathcal{S}_{j_n}$. For each such pair M, N let there be given a function

$$B_{M,N} : \widehat{\mathcal{S}}^{MN} \to \mathcal{Q}$$

(recall that $B_{M,N}$ is just an element in $\mathcal{Q}$ in case $n = 0$) where, as indicated earlier, by a slight abuse of notation we also denote its extension to $\widehat{\mathcal{S}}$ by $B_{M,N}$:

$$(x_1, \ldots, x_m) \mapsto B_{M,N}(x_{j_1}, \ldots, x_{j_n}).$$

Note that the subscript "M, N" fulfills a dual purpose: it serves to index the function and at the same time to indicate the "silent" variables. Then the basic type of function we wish to consider is one of the form

$$\sum_{M,N} M B_{M,N} N. \tag{4.1}$$

It is understood that this summation is subject to the basic restrictions, i.e., M and N are monomial functions such that $\mathrm{dom}(M) \cap \mathrm{dom}(N) = \emptyset$ and $\deg(M) + \deg(N) = m - n$, and in addition is subject to any restrictions that the particular problem at hand might impose. We shall call such a function a *core* function. For each summand in (4.1) the corresponding function $B_{M,N}$ will be called the *middle function*; M will be referred to as the *left* monomial (function), and N the *right* monomial (function). The middle function always has arity n (fixed). We shall say that $B_{U,V}$ is a *leftmost middle function* if $B_{M,N} = 0$ whenever $\deg(M) < \deg(U)$ (i.e., U has the minimal degree among left monomials that "really" appear in (4.1)). Similarly we define a *rightmost middle function*.

The basic functions involved in the definition of *d*-free sets are, of course, examples of core functions. In this situation we have $\mathcal{S}_k = \mathcal{R}_k$ and $\alpha_k = \mathrm{id}_{\mathcal{R}_k}$. These were presented in *function-value* notation in the identity

$$\sum_{i\in\mathcal{I}} E_i(\overline{x}_m^i)x_i + \sum_{j\in\mathcal{J}} x_j F_j(\overline{x}_m^j) = 0. \tag{4.2}$$

Here $n = m - 1$, with $p = 0, r = 1$ in the first summation and $p = 1, r = 0$ in the second summation. In *function* notation (4.2) is, strictly speaking, written as

$$\sum_{i\in\mathcal{I}} E_{1,X_i} X_i + \sum_{j\in\mathcal{J}} X_j F_{X_j,1} = 0. \tag{4.3}$$

In practice, one may take the further liberty of rewriting (4.3) as

$$\sum_{i\in\mathcal{I}} E_i X_i + \sum_{j\in\mathcal{J}} X_j F_j = 0.$$

Here, of course, the E_i's are leftmost middle functions, and the F_j's are rightmost middle functions.

To gain some practice with the various notations described above, we will follow through with a "real life" example involving Lie homomorphisms. Let $\mathcal{S}$ and $\mathcal{Q}$ be rings and let α be a Lie homomorphism from $\mathcal{S}$ into $\mathcal{Q}$. Then $[(x^2)^\alpha, x^\alpha] = [x^2, x]^\alpha = 0$ for all $x \in \mathcal{S}$. The usual linearization process results in

$$[(x_1x_2 + x_2x_1)^\alpha, x_3^\alpha] + [(x_1x_3 + x_3x_1)^\alpha, x_2^\alpha] + [(x_2x_3 + x_3x_2)^\alpha, x_1^\alpha] = 0$$

for all $x_1, x_2, x_3 \in \mathcal{S}$. Define $B : \mathcal{S}^2 \to \mathcal{Q}$ according to $B(x, y) = (xy + yx)^\alpha$ for all $x, y \in \mathcal{S}$. For $i = 1, 2, 3$ define $B_{1,X_i} = B = B_{X_i,1}$. In complete *function-value* notation the last identity reads

$$\sum_{i=1}^{3} B_{1,X_i}(\overline{x}_3)X_i(\overline{x}_3) - \sum_{i=1}^{3} X_i(\overline{x}_3)B_{X_i,1}(\overline{x}_3) = 0 \tag{4.4}$$

for all $\overline{x}_3 \in \mathcal{S}^3$. Notice here that in this example all the functions B_{1,X_i} and $B_{X_i,1}$ are equal to the same function B but in (4.4) they take on different values. Of course we have just written (4.4) to show how a particular problem fits into the general *function-value* notation. In practice, using appropriate definitions,

$X_i(\overline{x}_3) = x_i^\alpha$, and the fact that each function is the same B we would, writing the summation out in full, replace (4.4) by

$$B(x_2, x_3)x_1^\alpha + B(x_1, x_3)x_2^\alpha + B(x_1, x_2)x_3^\alpha$$
$$- x_1^\alpha B(x_2, x_3) - x_2^\alpha B(x_1, x_3) - x_3^\alpha B(x_1, x_2) = 0 \quad (4.5)$$

for all $x_1, x_2, x_3 \in \mathcal{S}$.

Redoing this example in *function* notation we now have X_i, $i = 1, 2, 3$, denoting the monomial functions. We define $B_{1,X_i} : \mathcal{S}^3 \to \mathcal{Q}$ by $\overline{x}_3 \mapsto B(\overline{x}_3^i)$, etc. and write (4.5) as

$$\sum_{i=1}^{3} B_{1,X_i} X_i - \sum_{i=1}^{3} X_i B_{X_i,1} = 0.$$

Another type of functions that is of special importance for us are the so-called *quasi-polynomials*, defined as follows. Let $L = X_{i_1} \ldots X_{i_p}$ be an arbitrary monomial of degree p, $0 \le p \le m$, and let there be given a function

$$\lambda_L : \mathcal{S}_{j_1} \times \mathcal{S}_{j_2} \times \ldots \times \mathcal{S}_{j_n} \to \mathcal{C}$$

where $\{i_1, \ldots, i_p\} \cup \{j_1, \ldots, j_n\}$ is a partition of $\{1, 2, \ldots, m\}$, with $j_1 < j_2 < \ldots < j_n$ (if $p = m$, then λ_L is an element in $\mathcal{C}$). Following the already established convention we shall continue to denote by λ_L the function from $\widehat{\mathcal{S}}$ to $\mathcal{C}$ given by

$$\overline{x}_m \mapsto \lambda_L(x_{j_1}, \ldots, x_{j_n})$$

(if $p = m$, then λ_L is a constant function).

Now, in *function* notation, a function of the form

$$P = \sum_L \lambda_L L \quad (4.6)$$

will be called a *quasi-polynomial*. The λ_L's will be referred to as the *coefficients* of P, and λ_1 will be called the *central coefficient*. If at least one coefficient λ_L is nonzero, then we shall say that the *degree* of P is m. If every $\lambda_L = 0$, then we define the degree of P to be $-\infty$. A quasi-polynomial of degree 0 is just a nonzero element in $\mathcal{C}$ (which we identify by a constant function).

Incidentally, let us point out that a quasi-polynomial of degree m consists of summands which all also have degree m; the degree is basically just the number of variables involved.

So, in function-value notation a quasi-polynomial of degree 1 is a function

$$P(x_1) = \lambda x_1^{\alpha_1} + \mu(x_1)$$

where $\lambda \in \mathcal{C}$ and $\mu : \mathcal{S}_1 \to \mathcal{C}$ is a central coefficient (and of course at least one of λ and μ is not 0). In function notation we would write this as

$$P = \lambda_{X_1} X_1 + \lambda_1 \quad \text{where } \lambda_{X_1} = \lambda \text{ and } \lambda_1 = \mu.$$

A quasi-polynomial of degree 2 can be, in function-value notation, written as

$$P(x_1, x_2) = \lambda_1 x_1^{\alpha_1} x_2^{\alpha_2} + \lambda_2 x_2^{\alpha_2} x_1^{\alpha_1} + \mu_1(x_1) x_2^{\alpha_2} + \mu_2(x_2) x_1^{\alpha_1} + \nu(x_1, x_2)$$

with

$$\lambda_1, \lambda_2 \in \mathcal{C}, \ \mu_1 : \mathcal{S}_1 \to \mathcal{C}, \ \mu_2 : \mathcal{S}_2 \to \mathcal{C}, \ \nu : \mathcal{S}_1 \times \mathcal{S}_2 \to \mathcal{C}.$$

Further, a quasi-polynomial of degree 3 consists of summands such as

$$\lambda_1 x_1^{\alpha_1} x_2^{\alpha_2} x_3^{\alpha_3}, \ \mu_1(x_1) x_2^{\alpha_2} x_3^{\alpha_3}, \ \nu_1(x_1, x_2) x_3^{\alpha_3}, \ \text{etc.}$$

Quasi-polynomials may be regarded as the "good guys" in FI theory: often one is able to show that a function $B_{M,N}$ appearing in (4.1) must turn out to be a quasi-polynomial. In the example considered in this section, i.e., the one arising from a Lie homomorphism and then leading to the FI (4.5), one can expect that under suitable assumptions the map B should be a quasi-polynomial of degree 2 with (referring to the above notation) $\lambda_1 = \lambda_2$ and $\mu_1 = \mu_2$.

The most general type of an FI we shall consider in this book is one in which one equates a core function with a quasi-polynomial:

$$\sum_{M,N} M B_{M,N} N = \sum_L \lambda_L L.$$

Here it is understood that the quasi-polynomial is regarded as a known quantity while the middle functions of the core function are to be regarded as unknown. This will be the topic of Section 4.3.

4.2 d-Free Pairs

We begin with a simple example illustrating some of our purposes in the present as well as in the next sections. Let $\mathcal{R}$ be an additive subgroup of $\mathcal{Q}$ and let $F : \mathcal{R} \to \mathcal{Q}$ be an additive commuting map, i.e., F satisfies

$$[F(x), x] = 0 \quad \text{for all } x \in \mathcal{R} \tag{4.7}$$

(cf. Example 1.5). A standard example of such a map is given by

$$F(x) = \lambda x + \mu(x), \ \lambda \in \mathcal{C}, \ \mu : \mathcal{R} \to \mathcal{C}, \tag{4.8}$$

where μ is of course an additive map. Which assumptions on $\mathcal{R}$ should one require in order to conclude that F is of a standard form (4.8)? In Example 1.5 we gave an answer to this question for the special case where $\mathcal{R}$ is a ring satisfying some conditions. Now we would of course like to obtain an answer connected to the present context. The FI (4.7) is not exactly of the form enabling us to use the d-freeness condition, but it is not far from such a one. Namely, all we have to do is to linearize (4.7) to get

$$F(x_2)x_1 + F(x_1)x_2 - x_1 F(x_2) - x_2 F(x_1) = 0 \quad \text{for all } x_1, x_2 \in \mathcal{R}.$$

Assuming that $\mathcal{R}$ is 2-free it follows that there are $p_{12}, p_{21} \in \mathcal{Q}$ and $\lambda_1, \lambda_2 : \mathcal{R} \to \mathcal{C}$ such that

$$F(x_2) = x_2 p_{12} + \lambda_1(x_2), \qquad F(x_1) = x_1 p_{21} + \lambda_2(x_1),$$
$$-F(x_2) = -p_{21} x_2 - \lambda_1(x_2), \qquad -F(x_1) = -p_{12} x_1 - \lambda_2(x_1)$$

for all $x_1, x_2 \in \mathcal{R}$. Comparing, for example, the first and the last formula we get $[x, p_{12}] = \lambda_2(x) - \lambda_1(x) \in \mathcal{C}$ for all $x \in \mathcal{R}$, which yields $p_{12} \in \mathcal{C}$ and $\lambda_2 = \lambda_1$ (see Lemma 3.3 (iv)). Similarly we find out that $p_{21} = p_{12}$. Now setting $\lambda = p_{12} = p_{21} \in \mathcal{C}$ and $\mu = \lambda_1 = \lambda_2$ we see that F is indeed of a standard form (4.8).

Thus (4.7) implies (4.8) if $\mathcal{R}$ is 2-free. Now we extend the FI (4.7) as follows. Let $\mathcal{S}$ be an additive group (not necessarily a subset of $\mathcal{Q}$), let $\alpha : \mathcal{S} \to \mathcal{Q}$ be an additive map, and let $F : \mathcal{S} \to \mathcal{Q}$ be an additive map satisfying

$$[F(x), x^\alpha] = 0 \quad \text{for all } x \in \mathcal{S}. \tag{4.9}$$

We define a standard solution of (4.9) in the obvious fashion, that is,

$$F(x) = \lambda x^\alpha + \mu(x), \quad \lambda \in \mathcal{C}, \quad \mu : \mathcal{S} \to \mathcal{C}; \tag{4.10}$$

note that this is the same as saying that F is a quasi-polynomial of degree 1. The question that now naturally appears is: Does (4.9) imply (4.10) if $\mathcal{S}^\alpha$ is a 2-free subset of $\mathcal{Q}$?

It should be mentioned that similar questions can naturally appear. For example, a Lie homomorphism α from a ring $\mathcal{S}$ to a ring $\mathcal{R}$ gives rise to the FI $[B(x, x), x^\alpha] = 0$ where B is a biadditive map (cf. the preceding section). In order to describe the form of α one must first describe the form of B.

Let us first remark that the answer to our question is clearly "yes" in case α is injective. Namely, in this case (4.9) yields $[(F\alpha^{-1})(y), y] = 0$ for all $y \in \mathcal{S}^\alpha$, which basically reduces our problem to the one treated above (with $F\alpha^{-1}$ playing the role of F). If α is not injective, then this does not seem to be entirely obvious since d-freeness assumption can not be used directly. Anyhow, the answer will follow easily from Theorem 4.3 below. This theorem considers a much more general situation which we shall now describe.

We shall now work in the framework introduced in Section 4.1. In particular, $\mathcal{S}_1, \mathcal{S}_2, \ldots, \mathcal{S}_m$ will be arbitrary nonempty sets, $\alpha_k : \mathcal{S}_k \to \mathcal{Q}$, $1 \leq k \leq m$ will be arbitrary functions, and we define $\mathcal{R}_k = \mathcal{S}_k^{\alpha_k} \subseteq \mathcal{Q}$. Further, we set

$$\widehat{\alpha} = (\alpha_1, \alpha_2, \ldots, \alpha_m).$$

We shall consider $\widehat{\alpha}$ as a map from $\widehat{\mathcal{S}}$ onto $\widehat{\mathcal{R}}$ given by

$$\widehat{\alpha}(x_1, x_2, \ldots, x_m) = (x_1^{\alpha_1}, x_2^{\alpha_2}, \ldots, x_m^{\alpha_m}).$$

The definitions of $\widehat{\alpha}^i : \widehat{\mathcal{S}}^i \to \widehat{\mathcal{R}}^i$ and $\widehat{\alpha}^{ij} : \widehat{\mathcal{S}}^{ij} \to \widehat{\mathcal{R}}^{ij}$ are self-explanatory.

Let $\mathcal{I}, \mathcal{J}$ have the usual meaning, and let $E_i : \widehat{\mathcal{S}}^i \to \mathcal{Q}$, $i \in \mathcal{I}$, and $F_j : \widehat{\mathcal{S}}^j \to \mathcal{Q}$, $j \in \mathcal{J}$. Consider the FI's

$$\sum_{i \in \mathcal{I}} E_i(\overline{x}_m^i) x_i^{\alpha_i} + \sum_{j \in \mathcal{J}} x_j^{\alpha_j} F_j(\overline{x}_m^j) = 0 \quad \text{for all } \overline{x}_m \in \widehat{\mathcal{S}}, \tag{4.11}$$

$$\sum_{i \in \mathcal{I}} E_i(\overline{x}_m^i) x_i^{\alpha_i} + \sum_{j \in \mathcal{J}} x_j^{\alpha_j} F_j(\overline{x}_m^j) \in \mathcal{C} \quad \text{for all } \overline{x}_m \in \widehat{\mathcal{S}}. \tag{4.12}$$

Of course, if each $\mathcal{S}_k = \mathcal{R}_k$ and each $\alpha_k = \mathrm{id}_{\mathcal{R}_k}$, then these are just the familiar FI's appearing in the definition of a d-free subset. It is now quite clear how to proceed. We define a *standard solution* of (4.11) and (4.12) by

$$E_i(\overline{x}_m^i) = \sum_{\substack{j \in \mathcal{J}, \\ j \neq i}} x_j^{\alpha_j} p_{ij}(\overline{x}_m^{ij}) + \lambda_i(\overline{x}_m^i), \quad i \in \mathcal{I},$$

$$F_j(\overline{x}_m^j) = -\sum_{\substack{i \in \mathcal{I}, \\ i \neq j}} p_{ij}(\overline{x}_m^{ij}) x_i^{\alpha_i} - \lambda_j(\overline{x}_m^j), \quad j \in \mathcal{J}, \tag{4.13}$$

$$\lambda_k = 0 \quad \text{if} \quad k \notin \mathcal{I} \cap \mathcal{J},$$

where

$$p_{ij} : \widehat{\mathcal{S}}^{ij} \to \mathcal{Q}, \ i \in \mathcal{I}, \ j \in \mathcal{J}, \ i \neq j,$$

$$\lambda_k : \widehat{\mathcal{S}}^k \to \mathcal{C}, \ k \in \mathcal{I} \cup \mathcal{J}.$$

Definition 4.1. A *pair* $(\widehat{\mathcal{S}}; \widehat{\alpha})$ is said to be *d-free*, where $d \in \mathbb{N}$, if for all $\mathcal{I}, \mathcal{J} \subseteq \{1, 2, \ldots, m\}$ the following two conditions are satisfied:

(a) If $\max\{|\mathcal{I}|, |\mathcal{J}|\} \leq d$, then (4.11) implies (4.13).

(b) If $\max\{|\mathcal{I}|, |\mathcal{J}|\} \leq d - 1$, then (4.12) implies (4.13).

We could easily establish a full analogue of Lemma 3.2. Let us, however, record only the appropriate version of its assertion (vii). The proof is standard and straightforward, so we omit it.

Lemma 4.2. *Let $(\widehat{\mathcal{S}}; \widehat{\alpha})$ be a d-free pair. Suppose that each $\mathcal{S}_i$ is an additive group and that all E_i's and F_j's are $(m-1)$-additive maps. If $\max\{|\mathcal{I}|, |\mathcal{J}|\} \leq d$, then all p_{ij}'s and λ_i's (from (4.13)) are $(m-2)$-additive and $(m-1)$-additive, respectively.*

Definition 4.1 is obviously a generalization of the definition of a d-free subset of $\mathcal{Q}^m$ (i.e., if each $\mathcal{S}_k = \mathcal{R}_k$ and each $\alpha_k = \mathrm{id}_{\mathcal{R}_k}$, then these two definitions coincide). The reader might feel uneasy expecting that this higher level of generality will cause further complications in arguments and notation. But then the next theorem shall come as a relief. Its message is clear: the only properties of the α_i's that are relevant for us are their ranges, and the study of d-free pairs can be immediately transferred to that of d-free sets.

Theorem 4.3. *The pair $(\widehat{\mathcal{S}}; \widehat{\alpha})$ is d-free if and only if the set $\widehat{\mathcal{R}}$ is d-free.*

Proof. Note that there exist maps $\beta_k : \mathcal{R}_k \to \mathcal{S}_k$ such that $\alpha_k \beta_k = \mathrm{id}_{\mathcal{R}_k}$. Thus $\widehat{\beta} = (\beta_1, \beta_2, \ldots, \beta_m)$ is an injection of $\widehat{\mathcal{R}}$ into $\widehat{\mathcal{S}}$ such that $\widehat{\alpha}\widehat{\beta} = \mathrm{id}_{\widehat{\mathcal{R}}}$.

Assume first that $(\widehat{\mathcal{S}}; \widehat{\alpha})$ is d-free and let us show that $\widehat{\mathcal{R}}$ satisfies condition (a) of Definition 3.1. Thus we begin with the basic identity

$$\sum_{i \in \mathcal{I}} E_i^i y_i + \sum_{j \in \mathcal{J}} y_j F_j^j = 0 \quad \text{for all } \overline{y}_m \in \widehat{\mathcal{R}},$$

with $\max\{|\mathcal{I}|, |\mathcal{J}|\} \le d$. We define $\widetilde{E}_i : \widehat{\mathcal{S}}^i \to \mathcal{Q}$ to be the composite function $E_i \widehat{\alpha}^i$. Similarly $\widetilde{F}_j = F_j \widehat{\alpha}^j$. Clearly

$$\sum_{i \in \mathcal{I}} \widetilde{E}_i^i x_i^{\alpha_i} + \sum_{j \in \mathcal{J}} x_j^{\alpha_j} \widetilde{F}_j^j = 0 \quad \text{for all } \overline{x}_m \in \widehat{\mathcal{S}}.$$

Since $(\widehat{\mathcal{S}}; \widehat{\alpha})$ is d-free, there exist $\widetilde{p}_{ij} : \widehat{\mathcal{S}}^{ij} \to \mathcal{Q}$ and $\widetilde{\lambda}_k : \widehat{\mathcal{S}}^k \to \mathcal{C}$ (subject to the usual conditions) such that

$$\widetilde{E}^i = \sum_{\substack{j \in \mathcal{J}, \\ j \ne i}} x_j^{\alpha_j} \widetilde{p}_{ij}^{ij} + \widetilde{\lambda}^i, \quad i \in \mathcal{I},$$

$$\widetilde{F}^j = -\sum_{\substack{i \in \mathcal{I}, \\ i \ne j}} \widetilde{p}_{ij}^{ij} x_i^{\alpha_i} - \widetilde{\lambda}^j, \quad j \in \mathcal{J},$$

$$\widetilde{\lambda}_k = 0 \quad \text{if } k \notin \mathcal{I} \cap \mathcal{J}.$$

We define $p_{ij} = \widetilde{p}_{ij} \widehat{\beta}^{ij}$ and $\lambda_k = \widetilde{\lambda}_k \widehat{\beta}^k$. Using the fact that $\widehat{\alpha}^k \widehat{\beta}^k = \mathrm{id}_{\widehat{\mathcal{R}}^k}$, it is then a routine matter to verify that the p_{ij}'s and λ_k's provide a standard solution for the E_i's and F_j's. Condition (b) can be checked similarly.

We now assume that $\widehat{\mathcal{R}}$ is d-free. As above we will just show that condition (a) of Definition 4.1 holds and leave the similar verification of condition (b) to the reader. Therefore our job is to show that the following FI on $\widehat{\mathcal{S}}$:

$$D(\overline{x}_m) = \sum_{i \in \mathcal{I}} E_i^i x_i^{\alpha_i} + \sum_{j \in \mathcal{J}} x_j^{\alpha_j} F_j^j = 0$$

has only a standard solution.

The proof is by induction on $|\mathcal{I} \cup \mathcal{J}|$. If $|\mathcal{I} \cup \mathcal{J}| = 1$, then 1-freeness of $\widehat{\mathcal{R}}$ forces E_1 to map $\mathcal{S}_2 \times \ldots \times \mathcal{S}_m$ into $\mathcal{C}$ (with $F_1 = -E_1$) and we are done.

In the inductive situation we will first treat the case where $\mathcal{I} \cup \mathcal{J} = \{1, 2, \ldots, m\}$. We set $\epsilon_k = \beta_k \alpha_k$, making the important observation that $\alpha_k \epsilon_k = \alpha_k$.

We first claim that each E_i has the desired form when acting on $\mathcal{S}_1^{\epsilon_1} \times \ldots \times \mathcal{S}_{i-1}^{\epsilon_{i-1}} \times \mathcal{S}_{i+1}^{\epsilon_{i+1}} \times \ldots \times \mathcal{S}_m^{\epsilon_m}$ (similarly for each F_j). Indeed, defining $\overline{E}_i = E_i \widehat{\beta}^i$,

$\overline{F}_j = F_j\widehat{\beta}^j$ and using the fact that for every $y_i \in \mathcal{R}_i$ we have $y_i = x_i^{\alpha_i}$ where $x_i = y_i^{\beta_i}$, it is clear that the FI

$$\sum_{i \in \mathcal{I}} \overline{E}_i(\overline{y}_m^i)y_i + \sum_{j \in \mathcal{J}} y_j\overline{F}_j(\overline{y}_m^j) = 0 \quad \text{for all } \overline{y}_m \in \widehat{\mathcal{R}}$$

holds. Since $\widehat{\mathcal{R}}$ is d-free there exist $\overline{p}_{ij} : \widehat{\mathcal{R}}^{ij} \to \mathcal{Q}$, $\overline{\lambda}_k : \widehat{\mathcal{R}}^k \to \mathcal{C}$ such that

$$\overline{E}_i(\overline{y}_m^i) = \sum_{\substack{j \in \mathcal{J} \\ j \neq i}} y_j\overline{p}_{ij}(\overline{y}_m^{ij}) + \overline{\lambda}_i(\overline{y}_m^i), \quad i \in \mathcal{I},$$

$$\overline{F}_j(\overline{y}_m^j) = -\sum_{\substack{i \in \mathcal{I} \\ i \neq j}} \overline{p}_{ij}(\overline{y}_m^{ij})y_i - \overline{\lambda}_j(\overline{y}_m^j), \quad j \in \mathcal{J},$$

$$\overline{\lambda}_k = 0 \quad \text{if } k \notin \mathcal{I} \cap \mathcal{J}.$$

Now set $p_{ij} = \overline{p}_{ij}\widehat{\alpha}^{ij}$ and $\lambda_k = \overline{\lambda}_k\widehat{\alpha}^k$. For $x_i \in \mathcal{S}_i$ let $u_i = x_i^{\epsilon_i}$ and $y_i = u_i^{\alpha_i}$. Then it follows readily that

$$E_i(\overline{u}_m^i) = \sum_{\substack{j \in \mathcal{J} \\ j \neq i}} u_j^{\alpha_j}p_{ij}(\overline{u}_m^{ij}) + \lambda_i(\overline{u}_m^i), \quad i \in \mathcal{I},$$

$$F_j(\overline{u}_m^j) = -\sum_{\substack{i \in \mathcal{I} \\ i \neq j}} p_{ij}(\overline{u}_m^{ij})u_i^{\alpha_i} - \lambda_j(\overline{u}_m^j), \quad j \in \mathcal{J},$$

$$\lambda_k = 0 \quad \text{if } k \notin \mathcal{I} \cap \mathcal{J},$$

and our claim is established. This would complete the proof provided that each ϵ_k was $\mathrm{id}_{\mathcal{S}_k}$, i.e., each α_k was injective. But of course our point is that the theorem is true without this assumption, so there is more to the proof.

For each $s = 1, 2, \ldots, m$ we set $\mathcal{I}_s = \mathcal{I} \setminus \{s\}$, $\mathcal{J}_s = \mathcal{J} \setminus \{s\}$. Further, given $\overline{x}_m \in \widehat{\mathcal{S}}$, for each $s = 0, 1, 2, \ldots, m$ we set

$$\widetilde{x}_s = (x_1, \ldots, x_s, x_{s+1}^{\epsilon_{s+1}}, \ldots, x_m^{\epsilon_m}).$$

and note that $E_s(\widetilde{x}_s^s) = E_s(\widetilde{x}_{s-1}^s)$ (similarly for F_s). This observation plus the fact that $x_k^{\alpha_k} = (x_k^{\epsilon_k})^{\alpha_k}$ for all k enable us to conclude that

$$0 = D(\widetilde{x}_s) - D(\widetilde{x}_{s-1}) = \sum_{i \in \mathcal{I}_s} G_{si}(\overline{x}_m^i)x_i^{\alpha_i} + \sum_{j \in \mathcal{J}_s} x_j^{\alpha_j}H_{sj}(\overline{x}_m^j)$$

where

$$G_{si}(\overline{x}_m^i) = E_i(\widetilde{x}_s^i) - E_i(\widetilde{x}_{s-1}^i) \quad \text{and} \quad H_{sj}(\overline{x}_m^j) = F_j(\widetilde{x}_s^j) - F_j(\widetilde{x}_{s-1}^j).$$

By the induction assumption each G_{si} and H_{sj} can be written in standard form. For $\overline{x}_m \in \widehat{\mathcal{S}}$ we note that

$$E_i(\overline{x}_m^i) = \sum_{s=1}^{m} G_{si}(\overline{x}_m^i) + E_i(\overline{u}_m^i),$$

where $u_k = x_k^{\epsilon_k}$. A similar result holds for each F_j. In view of the preceding claim, the proof has now been completed for the special case that $\mathcal{I} \cup \mathcal{J} = \{1, 2, \ldots, m\}$.

In the general case we may assume without loss of generality that $\mathcal{I} \cup \mathcal{J} = \{1, 2, \ldots, r\}$, where $1 \le r < m$. Let $z \in \mathcal{S}_{r+1} \times \ldots \times \mathcal{S}_m$ and in $\mathcal{S}_1 \times \ldots \times \mathcal{S}_r$ define $E_{zi}(\overline{x}_r^i) = E_i(\overline{x}_r^i, z)$ for $i \in \mathcal{I}$ and $F_{zj}(\overline{x}_r^j) = F_j(\overline{x}_r^j, z)$ for $j \in \mathcal{J}$. Clearly

$$\sum_{i \in \mathcal{I}} E_{zi}^i x_i^{\alpha_i} + \sum_{j \in \mathcal{J}} x_j^{\alpha_j} F_{zj}^j = 0$$

holds in $\mathcal{S}_1 \times \ldots \times \mathcal{S}_r$. Since $\mathcal{R}_1 \times \ldots \times \mathcal{R}_r$ is d-free by Lemma 3.2 (i) the special case above applies here and so there exist p_{zij} and λ_{zk} such that

$$E_{zi}^i = \sum_{\substack{j \in \mathcal{J} \\ j \ne i}} x_j^{\alpha_j} p_{zij}^{ij} + \lambda_{zi}^i, \quad i \in \mathcal{I},$$

$$F_{zj}^j = -\sum_{\substack{i \in \mathcal{I} \\ i \ne j}} p_{zij}^{ij} x_i^{\alpha_i} - \lambda_{zj}^j, \quad j \in \mathcal{J},$$

$$\lambda_{zk} = 0 \quad \text{if } k \notin \mathcal{I} \cap \mathcal{J}.$$

The proof is then completed by defining $p_{ij}(\overline{x}_r^{ij}, z) = p_{zij}(\overline{x}_r^{ij})$ and $\lambda_k(\overline{x}_r^k, z) = \lambda_{zk}(\overline{x}_r^k)$ for all $(\overline{x}_r, z) \in \widehat{\mathcal{S}}$. $\qquad \square$

Now it is clear that the answer to our question on (4.9) is "yes" in any case. One just has to follow the arguments concerning the simpler case (4.7) and apply Theorem 4.3 at appropriate places.

4.3   Quasi-polynomials and Core Functions

In retrospect, proving that a map F from the previous section (i.e., the one satisfying $[F(x), x^\alpha] = 0$) must be a quasi-polynomial is fairly easy (of course after having Theorem 4.3 in hand). However, the argument consists of several steps, that is, we have to use the properties of d-free sets iteratively. The reader can imagine that in a more complicated example (say involving more variables) the necessary procedure needed for arriving at the expected conclusion could be extremely tedious. One of our purposes in this section is to find devices which will make it possible for us to avoid such lengthy procedures.

The notions of core functions and quasi-polynomials were defined in Section 4.1, and the reader should consult this section for complete details concerning the framework surrounding the definition of these key notions. We emphasize that this framework is that of "pairs" given in Section 4.2. In particular we point out, in view of Theorem 4.3, any assumption of d-freeness of $\widehat{\mathcal{R}}$ we may make in the ensuing theorems automatically holds for a corresponding pair $(\widehat{\mathcal{S}}; \widehat{\alpha})$.

Let us recall from Section 4.1 that whenever dealing with a core function $\sum_{M,N} M B_{M,N} N$ we are tacitly assuming that M, N are monomial functions with

$\mathrm{dom}(M) \cap \mathrm{dom}(N) = \emptyset$ and $\deg(M) + \deg(N)$ fixed. The primary goal of this section is to show that, given a (known) quasi-polynomial P, if a core function is equal to P, then under appropriate d-freeness conditions the *middle* functions $B_{M,N}$ must in fact be quasi-polynomials. The main result in this regard is Theorem 4.10, which makes the key inductive step toward accomplishing this goal.

We begin with a useful and frequently used result that says that under favorable circumstances if a quasi-polynomial is the zero function, then each of its coefficients must be zero. Before stating it we first remark that every quasi-polynomial $P = \sum_L \lambda_L L$ of degree $m \geq 1$ can be represented as

$$P = \sum_{i=1}^{m} X_i P_i + \lambda_1 \tag{4.14}$$

where each P_i is either 0 or is a quasi-polynomial of degree $m - 1$. This is an immediate consequence of the obvious fact that every L with $\deg(L) \geq 1$ can be written as $L = X_i M$ for suitable i where $\deg(M) = \deg(L) - 1$.

Lemma 4.4. *Let $P = \sum_L \lambda_L L$ be a quasi-polynomial of degree $\leq m$, and suppose that either $\widehat{\mathcal{R}}$ is m-free and $\lambda_1 = 0$ or $\widehat{\mathcal{R}}$ is $(m+1)$-free. Then $P = 0$ if and only if each $\lambda_L = 0$.*

Proof. The proof is by induction on m. If $m = 1$, then we have $\lambda_{X_1} X_1 = -\lambda_1$. By either d-freeness condition we have $\lambda_{X_1} = 0$ (and hence $\lambda_1 = 0$). So let $m > 1$. Writing P as in (4.14) we see that $\sum_{i=1}^{m} X_i P_i = -\lambda_1$. The d-freeness assumptions then say that each $P_i = 0$ (and hence $\lambda_1 = 0$). Now by our induction hypothesis it follows that all coefficients of each P_i are 0, which clearly implies that each $\lambda_L = 0$. $\square$

In the special case where each $\mathcal{S}_i = \mathcal{R}_i = \mathcal{R}$, each $\alpha_i = \mathrm{id}_{\mathcal{R}_i}$, and $\lambda_L = 0$ whenever $\deg(L) < m$, Lemma 4.4 gives

Lemma 4.5. *Let $f(x_1, x_2, \ldots, x_m) \in \mathcal{C}\langle X \rangle$ be a nonzero multilinear polynomial. If $\mathcal{R}$ is an m-free subset of $\mathcal{Q}$, then f is not a polynomial identity on $\mathcal{R}$.*

Let us record another simple consequence of Lemma 4.4.

Lemma 4.6. *Suppose that each $\mathcal{S}_i$ is an additive group and each α_i is an additive function. Let $P = \sum_L \lambda_L L$ be a quasi-polynomial of degree $\leq m$, and suppose that either $\widehat{\mathcal{R}}$ is m-free and $\lambda_1 = 0$ or $\widehat{\mathcal{R}}$ is $(m+1)$-free. If P is a multiadditive function, then all its coefficients λ_L are multiadditive functions.*

Proof. Note that relations such as

$$P(x_1 + x_1', x_2, \ldots, x_m) - P(x_1, x_2, \ldots, x_m) - P(x_1', x_2, \ldots, x_m) = 0$$

when rewritten in terms of λ_L's can be interpreted so that Lemma 4.4 can be used; using this one can easily complete the proof. $\square$

Remark 4.7. When dealing with algebras (instead of with rings), it is more useful to state Lemma 4.6 for multilinear (instead of multiadditive) functions. The
necessary changes in the proof are obvious.

Admittedly we have stated Lemma 4.6 slightly inaccurately. First of all it
should be mentioned that we have treated λ_L as a function defined on $\mathcal{S}_{j_1} \times \mathcal{S}_{j_2} \times
\ldots \times \mathcal{S}_{j_n}$ and not on $\widehat{\mathcal{S}}$ (cf. Section 4.1). So, by multiadditivity of λ_L we actually
mean that λ_L is n-additive. The case when $\deg(L) = m$ still needs an additional
explanation. Namely, in this case λ_L is an element in $\mathcal{C}$ which we identify with
a constant function. Now if we define, for convenience, that a 0-additive function
is the same as a constant function, then the statement of Lemma 4.6 becomes
unambiguous.

In our next result we consider a special case of our main goal. Namely, if
a core function is equal to a quasi-polynomial and if the degree p of each left
monomial is fixed, and hence, since the arity n of the middle functions is also
fixed, the degree r of each right monomial is fixed as well, then, under suitable
conditions, each middle function must be a quasi-polynomial.

Lemma 4.8. *Let n, p, r be nonnegative integers with $n < m$ and $n+p+r = m$, and
for each pair M, N with $\deg(M) = p$, $\deg(N) = r$ let $B_{M,N} : \widehat{\mathcal{S}}^{MN} \to \mathcal{Q}$. Suppose*

$$\sum_{M,N} M B_{M,N} N = P \tag{4.15}$$

*where P is a quasi-polynomial with central term λ_1, and suppose that either $\widehat{\mathcal{R}}$ is
m-free and $\lambda_1 = 0$ or $\widehat{\mathcal{R}}$ is $(m+1)$-free. Then each $B_{M,N}$ is a quasi-polynomial.*

Proof. The proof is by induction on m. For $m = 1$ we have $n = 0$ and may assume
that $p = 1$. Then $B_{X_1,1} = a \in \mathcal{Q}$ and (4.15) reads $X_1 a = \lambda_{X_1} X_1 + \lambda_1$, in other
words

$$X_1(a - \lambda_{X_1}) = \lambda_1$$

for suitable $\lambda_{X_1} \in \mathcal{C}$, $\lambda_1 : \mathcal{S} \to \mathcal{C}$. Therefore by the d-freeness assumptions
$a - \lambda_{X_1} = 0$, i.e., $B_{X,1} \in \mathcal{C}$.

In the inductive case we have $m > 1$ and we may assume that $p > 0$. Since
each monomial function M may be written as $M = X_i K$ for suitable i, where
$\deg(K) = p - 1$ and K does not involve X_i, (4.15) may be written as

$$\sum_{i=1}^{m} X_i \left(\sum_{K,N} K B_{X_i K, N} N \right) = \sum_{i=1}^{m} X_i P_i + \lambda_1$$

where P_i is a quasi-polynomial and $\lambda_1 : \widehat{\mathcal{S}} \to \mathcal{C}$. Again by the d-freeness assumptions we conclude that for each i,

$$\sum_{K,N} K B_{X_i K, N} N = P_i. \tag{4.16}$$

Since (4.16) does not involve X_i the induction hypothesis then asserts that each $B_{M,N} = B_{X_i K, N}$ is a quasi-polynomial. □

The sole purpose of the following lemma is that it is needed in the proof of the main theorem of this section (Theorem 4.10). The reader will recognize the FI in this lemma as being a slight generalization of the basic identity (4.12). We also remark that in this lemma we exceptionally allow that n equals m.

Lemma 4.9. *Let $0 \le n \le m$, and for each M with $\deg(M) = m - n$ let $B_M : \widehat{\mathcal{S}}^M \to \mathcal{Q}$. Suppose*

$$\sum_M M B_M = \sum_{i=1}^m F_i X_i + \lambda, \qquad (4.17)$$

where $F_i : \widehat{\mathcal{S}}^{X_i} \to \mathcal{Q}$ and $\lambda : \widehat{\mathcal{S}} \to \mathcal{C}$. Suppose that either $\widehat{\mathcal{R}}$ is m-free and $\lambda = 0$ or $\widehat{\mathcal{R}}$ is $(m+1)$-free. Then each B_M is of the form

$$B_M = \sum_{i \notin \mathrm{dom}(M)} p_{iM} X_i + \mu_M$$

where $p_{iM} : \widehat{\mathcal{S}}^{MX_i} \to \mathcal{Q}$ and $\mu_M : \widehat{\mathcal{S}}^M \to \mathcal{C}$.

Proof. The proof is by induction on $\deg(M) = m - n$. If $\deg(M) = 0$ we are immediately done. Now rewrite (4.17) as

$$\sum_{j=1}^m X_j \left(\sum_K K B_{X_j K} \right) - \sum_{i=1}^m F_i X_i = \lambda$$

where $\deg(K) = \deg(M) - 1$. The d-freeness assumptions imply that for each j,

$$\sum_K K B_{X_j K} = \sum_{i \neq j} p_{ij} X_i + \mu_j$$

where $p_{ij} : \widehat{\mathcal{S}}^{X_i X_j} :\to \mathcal{Q}$ and $\mu_j : \widehat{\mathcal{S}}^{X_j} \to \mathcal{C}$. By the induction hypothesis each $B_M = B_{X_j K}$ is of the required form. □

We now come to the main theorem of this section. It says in essence that (under suitable conditions) if a core function is equal to a quasi-polynomial, then each (originally unknown) middle function is a sum of terms each of which has some X_i as either a left or right factor (in addition there may be a central term). In general one cannot hope to do better than this, as illustrated by the following simple example. Let $f : \mathcal{S} \to \mathcal{Q}$ be a completely arbitrary function, and consider the identity (for $m = 2$ and $n = 1$):

$$B_{1, X_1} X_1 + X_2 B_{X_2, 1} = 0 \qquad (4.18)$$

Then $B_{1, X_1} = X_2 f$ and $B_{X_2, 1} = -f X_1$ is certainly a solution of (4.18) but clearly nothing further can be said about f.

Theorem 4.10. *Let $0 \leq n < m$, and for each pair M, N with $\deg(MN) = m - n$ let $B_{M,N} : \widehat{\mathcal{S}}^{MN} \to \mathcal{Q}$. Suppose*

$$\sum_{M,N} M B_{M,N} N = P \tag{4.19}$$

where P is a quasi-polynomial with central coefficient λ_1, and suppose that either $\widehat{\mathcal{R}}$ is m-free and $\lambda_1 = 0$ or $\widehat{\mathcal{R}}$ is $(m + 1)$-free. Then for each M, N,

$$B_{M,N} = \sum_{i \notin \mathrm{dom}(MN)} (p_{M,N,X_i} X_i + X_i q_{X_i,M,N}) + \mu_{M,N} \tag{4.20}$$

where $p_{M,N,X_i}, q_{X_i,M,N} : \widehat{\mathcal{S}}^{MNX_i} \to \mathcal{Q}$, $\mu_{M,N} : \widehat{\mathcal{S}}^{MN} \to \mathcal{C}$. Moreover, if $B_{M,N}$ is a rightmost middle function (resp. $B_{M,N}$ is a leftmost middle function), then

$$q_{X_i,M,N} = 0 \quad \text{for all } M \text{ and } i$$
$$(\text{resp.} \quad p_{M,N,X_i} = 0 \quad \text{for all } N \text{ and } i). \tag{4.21}$$

Proof. We first give a brief outline of the proof before embarking on the details. The proof is by induction on m. The case $m = 1$ is quickly taken care of. Using Lemma 4.8 and the induction hypothesis, we are able to assume that v, the minimum degree of those N for which $B_{M,N} \neq 0$ for some M, is equal to 0, and similarly that u, the minimum degree of those M for which $B_{M,N} \neq 0$ for some M, is equal to 0. The d-freeness conditions together with Lemma 4.9 are used to prove the theorem for $B_{M,1}$. This fact, together with Lemma 4.8, enable us to transfer the problem to a framework where m is replaced by $m - 1$, thus allowing us to again apply the induction hypothesis. The theorem for $B_{M,N}$, where $\deg(N) > 1$, is then immediate; a little more care is needed to obtain the desired result for B_{M,X_i}.

For $m = 1$ we have $n = 0$. Then (4.19) reads

$$X_1 a + b X_1 = \lambda_{X_1} X_1 + \lambda_1$$

where $a, b \in \mathcal{Q}$, $\lambda_{X_1} \in \mathcal{C}$, and $\lambda_1 : \mathcal{S}_1 \to \mathcal{C}$. Rearranging terms we have

$$X_1 a + (b - \lambda_{X_1}) X_1 = \lambda_1.$$

By the d-freeness conditions we see that $a, b - \lambda_{X_1} \in \mathcal{C}$, whence $a, b \in \mathcal{C}$.

Now let $m > 1$. Suppose $v > 0$. Since any N in (4.19) can be written as $N = KL$, where $\deg(L) = v$, we may write (4.19) in the form:

$$\sum_{\deg(L)=v} \left[\sum_{M,K} M B_{M,KL} K \right] L$$

Proof. By assumption there is an n-additive map $A : \mathcal{S}^n \to \mathcal{Q}$ such that

$$T(x) = A(x, x, \ldots, x) \quad \text{for all } x \in \mathcal{S}.$$

Define a new n-additive map $B : \mathcal{S}^n \to \mathcal{Q}$ by

$$B(\overline{x}_n) = \sum_{\pi \in S_n} A(x_{\pi(1)}, x_{\pi(2)}, \ldots, x_{\pi(n)}),$$

and note that the complete linearization of (4.28) yields

$$\sum_{i=1}^{n+1} B(\overline{x}_{n+1}^i) x_i^\alpha - \sum_{i=1}^{n+1} x_i^\alpha B(\overline{x}_{n+1}^i) = 0.$$

In particular, B satisfies the conditions of Corollary 4.14 (with $m = n + 1$, $P = 0$ and $c_{M,N} = \pm 1$). As $\mathcal{R}$ is $(n + 1)$-free it follows that B is a quasi-polynomial. Moreover, by Lemma 4.6 its coefficients are multiadditive. Note that this implies that the trace of B can be represented in the form stated in (4.29), i.e.,

$$B(x, x, \ldots, x) = \sum_{i=0}^{n} \lambda_i(x)(x^\alpha)^{n-i},$$

where each λ_i is the trace of an i-additive map. Since $B(x, x, \ldots, x) = n! T(x)$, this proves the corollary. $\qquad\square$

In the statement of Corollary 4.15 we have tacitly used the convention that by a 0-additive function we simply mean a fixed element. So we can rewrite (4.28) as

$$n! T(x) = \lambda_0 (x^\alpha)^n + \lambda_1(x)(x^\alpha)^{n-1} + \lambda_2(x)(x^\alpha)^{n-2} + \ldots + \lambda_{n-1}(x) x^\alpha + \lambda_n(x)$$

where $\lambda_0 \in \mathcal{C}$, λ_1 is an additive map from $\mathcal{S}$ into $\mathcal{C}$, λ_2 is the trace of an biadditive map (i.e., $\lambda_2(x) = \Lambda_2(x, x)$ where $\Lambda_2 : \mathcal{S}^2 \to \mathcal{C}$ is a biadditive map), etc.

In case $\mathcal{S} = \mathcal{R}$ and $\alpha = \mathrm{id}_\mathcal{R}$ we can formulate Corollary 4.15 in terms of commuting maps. Although this is just a particular case of a more general result, we record it because of the special role that commuting maps play in FI theory.

Corollary 4.16. *Let $\mathcal{R}$ be an additive subgroup of $\mathcal{Q}$, and let $T : \mathcal{R} \to \mathcal{Q}$ be the trace of an n-additive map, $n \geq 1$. If T is commuting (i.e., $[T(x), x] = 0$ for all $x \in \mathcal{R}$), $\mathcal{R}$ is $(n + 1)$-free, and $n!$ is invertible in $\mathcal{C}$, then there exist traces of i-additive maps $\lambda_i : \mathcal{R} \to \mathcal{C}$, $i = 0, 1, \ldots, n$ such that $T(x) = \sum_{i=0}^{n} \lambda_i(x) x^{n-i}$ for all $x \in \mathcal{R}$.*

Of lesser importance, but still sometimes useful, are conditions implying that the trace of a multiadditive map must be 0.

Corollary 4.17. *Let S be an additive group, let $\alpha : S \to Q$ be an additive map, let $c_1, c_2 \in C$, and let $T : S \to Q$ be the trace of an n-additive map, $n \geq 1$, such that*

$$c_1 T(x)x^\alpha + c_2 x^\alpha T(x) = 0 \quad \text{for all } x \in S. \tag{4.30}$$

Suppose that $n!$, $c_1 + c_2$, and at least one of c_1, c_2 are invertible in C. If $R = S^\alpha$ is $(n+1)$-free, then $T(x) = 0$ for all $x \in S$.

Proof. The proof is similar to that of Corollary 4.15. We introduce B in the same fashion as above, and then, since one of c_1, c_2 is invertible, using basically the same arguments we arrive at

$$n!T(x) = B(x, x, \dots, x) = \sum_{i=0}^{n} \lambda_i(x)(x^\alpha)^{n-i}$$

for every $x \in S$. Since $n!$ is invertible it follows in particular that $T(x)$ always commutes with x^α. Therefore (4.30), together with the assumption that $c_1 + c_2$ is invertible, gives $T(x)x^\alpha = 0$ for every $x \in S$. The complete linearization of this identity yields

$$\sum_{i=1}^{n+1} B(\bar{x}_{n+1}^i)x_i^\alpha = 0.$$

Since R is $(n+1)$-free, it follows that $B = 0$. In particular, we obtain $n!T(x) = B(x, x, \dots, x) = 0$, and hence $T(x) = 0$ for every $x \in S$. □

Remark 4.18. It is easy to see, by inspecting the proof, that the conclusion of Corollary 4.17 still holds if we replace the condition (4.30) by the condition $c_1 T(x)x^\alpha + c_2 x^\alpha T(x) \in C$ for every $x \in S$ provided that we assume the $(n+2)$-freeness of R (instead of only $(n+1)$-freeness). Similarly, in Corollary 4.15 we may replace (4.28) by $[T(x), x^\alpha] \in C$ if we assume that R is $(n+2)$-free.

In Part III we will often need the "nonlinear version" of Lemma 4.4. The simplest way to explain what we mean by this is through an example.

Example 4.19. Let $\lambda \in C$, let $\mu, \nu : S^2 \to C$ be biadditive maps, and let $\varepsilon : S^4 \to C$ be a 4-additive map. We define $W : S^2 \to Q$ by

$$W(x_1, x_2) = \lambda x_1^\alpha x_2^\alpha x_1^\alpha x_2^\alpha + \mu(x_1, x_1)x_2^\alpha x_2^\alpha + \nu(x_1, x_2)x_1^\alpha x_2^\alpha + \varepsilon(x_1, x_1, x_2, x_2).$$

Of course, W is a not a quasi-polynomial (in the above sense) since both x_1 and x_2 appear twice (instead of only once) in each of the terms on the right-hand side. Assume that

$$W(x_1, x_2) = 0 \quad \text{for all } x_1, x_2 \in S. \tag{4.31}$$

It is then possible to conclude, under appropriate assumptions on $R = S^\alpha$, that $\lambda = \mu(x_1, x_1) = \nu(x_1, x_2) = \varepsilon(x_1, x_1, x_2, x_2) = 0$? Unfortunately, Lemma 4.4 is not directly applicable in this situation. However, we can linearize (4.31) on both variables, i.e., we replace x_1 by $x_1 + x_3$ and x_2 by $x_2 + x_4$, and thereby obtain

$$\lambda x_1^\alpha x_2^\alpha x_3^\alpha x_4^\alpha + \lambda x_3^\alpha x_2^\alpha x_1^\alpha x_4^\alpha + \lambda x_1^\alpha x_4^\alpha x_3^\alpha x_2^\alpha + \lambda x_3^\alpha x_4^\alpha x_1^\alpha x_2^\alpha$$

$$+ \Big(\mu(x_1, x_3) + \mu(x_3, x_1) \Big) x_2^\alpha x_4^\alpha + \Big(\mu(x_1, x_3) + \mu(x_3, x_1) \Big) x_4^\alpha x_2^\alpha$$

$$+ \nu(x_1, x_2) x_3^\alpha x_4^\alpha + \nu(x_3, x_2) x_1^\alpha x_4^\alpha + \nu(x_1, x_4) x_3^\alpha x_2^\alpha + \nu(x_3, x_4) x_1^\alpha x_2^\alpha$$

$$+ \varepsilon(x_1, x_3, x_2, x_4) + \varepsilon(x_3, x_1, x_2, x_4) + \varepsilon(x_1, x_3, x_4, x_2) + \varepsilon(x_3, x_1, x_4, x_2) = 0.$$

Now Lemma 4.4 tells us that

$$\lambda = 0, \quad \mu(x_1, x_3) + \mu(x_3, x_1) = 0, \quad \nu(x_1, x_2) = 0$$
$$\varepsilon(x_1, x_3, x_2, x_4) + \varepsilon(x_3, x_1, x_2, x_4) + \varepsilon(x_1, x_3, x_4, x_2) + \varepsilon(x_3, x_1, x_4, x_2) = 0$$

for all $x_1, x_2, x_3, x_4 \in \mathcal{S}$, provided that $\mathcal{R}$ is a 5-free subset of $\mathcal{Q}$ (if ε was 0, then 4-freeness would be enough). In particular, $2\mu(x_1, x_1) = 4\varepsilon(x_1, x_1, x_2, x_2) = 0$ for all $x_1, x_2 \in \mathcal{S}$. So we can get the desired conclusion that $\lambda = \mu(x_1, x_1) = \nu(x_1, x_2) = \varepsilon(x_1, x_1, x_2, x_2) = 0$, but we have to pay a small price for this: the assumption that $\mathcal{Q}$ is 2-torsion free must be imposed.

It should be obvious to the reader that this example can be greatly generalized. On the other hand, it may also be obvious that a statement covering the general case for which the procedure from this example works requires introducing further notation and tedious arguing. This is one of these situations where everything seems intuitively clear, but a formal explanation is necessarily somewhat lengthy. We shall therefore use a somewhat informal style, and thereby try to avoid a dreary exposition.

We first have to introduce the notion of a general "nonlinear quasi polynomial" $W = W(x_1, \dots, x_n)$ of degree d. Of course, n denotes the number of variables involved, so we must have $d \geq n$ (in Example 4.19 we have $n = 2$ and $d = 4$); in fact, we are only interested in the case when $d > n$ since the $d = n$ case corresponds to the "ordinary" quasi-polynomial. Further, to each $i \in \{1, \dots, n\}$ we attribute a positive integer k_i, which will denote the number of appearances of the variable x_i in *each* of the summands of W (e.g., $k_1 = k_2 = 2$ in Example 4.19). A typical summand of $W(x_1 \dots, x_n)$ is of the form

$$\mu(x_1, \dots, x_1, x_2, \dots, x_2, \dots, x_n, \dots, x_n) x_{j_1}^\alpha x_{j_2}^\alpha \dots x_{j_q}^\alpha, \tag{4.32}$$

where $0 \leq q \leq d$, $\mu : \mathcal{S}^l \to \mathcal{C}$ is a multiadditive map (called a coefficient of W) where $l = d - q$, and each x_i appears exactly k_i times in (4.32). Further, denoting by l_i the number of times x_i appears in $\mu(x_1, \dots, x_{i-1}, x_i, \dots, x_i, x_{i+1}, \dots, x_n)$ we clearly have $l_1 + \dots + l_n = l$, and it should be pointed out that the case when some $l_i = 0$ is not excluded (that is, μ may not depend on some x_i's). Of course, some of the j_i's may be equal, i.e., it is possible that $j_i = j_k$ when $i \neq k$. If $q = d$, then μ should be understood as an element in $\mathcal{C}$, and if $q = 0$, then the expression $x^\alpha_{j_1} x^\alpha_{j_2} \ldots x^\alpha_{j_q}$ should be read as 1 and in this case μ is called the central coefficient of W (just as in the case of usual quasi-polynomials). It should also be understood that different summands of W involve different "nonlinear monomials" $x^\alpha_{j_1} x^\alpha_{j_2} \ldots x^\alpha_{j_q}$.

We remark that $0 \leq l_i \leq k_i$ and that l_i depends on a particular summand of W, while the k_i's are fixed. For instance, W in Example 4.19 consists of summands for which $l_1 = l_2 = 0$, $l_1 = 2$ and $l_2 = 0$, $l_1 = l_2 = 1$, and, in case of the central summand, $l_1 = l_2 = 2$.

Let us add a remark about our notational convention. We have required that in $\mu(x_1, \ldots, x_1, x_2, \ldots, x_2, \ldots, x_n, \ldots, x_n)$ the variables x_i appear in the straightforward order. When facing a concrete nonlinear quasi-polynomial, a different order might appear more natural (say, something like $\mu(x_{i_1}, x_{i_2}, \ldots, x_{i_1}, \ldots, x_{i_p})$). Anyway, in principle we can then replace the given map by the one in which the variables are permuted, and so the simple order that was required can indeed be assumed without loss of generality. Anyway, this requirement is basically needed just because it makes it possible for us to state the next lemma in a more clear way.

Lemma 4.20. *With respect to the preceding notation and conventions, assume that*

$$W(x_1, \ldots, x_n) = 0 \quad \text{for all } x_1 \ldots, x_n \in \mathcal{S}. \tag{4.33}$$

Suppose that either the central coefficient of W is 0 and $\mathcal{R} = \mathcal{S}^\alpha$ is d-free or $\mathcal{R}$ is $(d+1)$-free. Let μ be a coefficient of W from (4.32). Then

$$\sum_{\pi \in S_{l_1}} \sum_{\rho \in S_{l_2}} \cdots \sum_{\sigma \in S_{l_n}} \mu\big(x_{\pi(1)}, \ldots, x_{\pi(l_1)}, y_{\rho(1)}, \ldots, y_{\rho(l_2)}, \ldots, z_{\sigma(1)}, \ldots, z_{\sigma(l_n)}\big) = 0$$

for all $x_i, y_i, \ldots, z_i \in \mathcal{S}$. In particular,

$$l_1! \, l_2! \ldots l_n! \, \mu(x_1, \ldots, x_1, x_2, \ldots, x_2, \ldots, x_n, \ldots, x_n) = 0$$

for all $x_i \in \mathcal{S}$, and $\mu = 0$ if each $l_i \leq 1$.

Proof. Since $k_1, \ldots, k_n$ are fixed, each summand of W can be linearized by the precisely the same formal sequence of steps. Thus the complete linearization P of W is just the sum of the linearizations P_T of each summand T of W. Of course, P is now an ordinary (albeit complicated) quasi-polynomial. It is to be noted that for distinct summands T and T' of W the (multilinear) monomials appearing in P_T have empty intersection with the monomials appearing in $P_{T'}$, and that the coefficients of P_T are linearizations of the coefficients of T. As $P = 0$, Lemma 4.4 tells us that all the coefficients of P_T are zero, which is exactly the conclusion of the lemma. $\qquad\square$

The reader might find this proof a bit unsatisfactory. But a detailed rigorous proof is basically just a notational exercise - the ideas are the same as in Example 4.19. Anyway, in a concrete situation one can always choose to "go on foot" and check the necessary details directly. Typical concrete cases that appear in applications are usually of similar complexity as the one treated in Example 4.19.

Let us restate Lemma 4.20 for the special case when $n = 1$. The formulation then becomes simple and clear.

Corollary 4.21. *Let $\mathcal{S}$ be an additive group, let $\alpha : \mathcal{S} \to \mathcal{Q}$ be an additive map, and let $\mu_l : \mathcal{S}^l \to \mathcal{C}$ be an l-additive map, $l = 0, 1, \ldots, n$. Suppose that*

$$\sum_{l=0}^{n} \mu_l(x, \ldots, x)(x^\alpha)^{n-l} = 0 \quad \text{for all } x \in \mathcal{S}.$$

If either $\mu_n = 0$ and $\mathcal{R} = \mathcal{S}^\alpha$ is n-free or $\mathcal{R}$ is $(n+1)$-free, then $\mu_0 = \mu_1 = 0$ and

$$\sum_{\pi \in S_l} \mu_l(x_{\pi(1)}, \ldots, x_{\pi(l)}) = 0$$

for all $x_1, \ldots, x_l \in \mathcal{S}$, $l = 2, \ldots, n$. In particular, $l!\, \mu_l(x, \ldots, x) = 0$ for all $x \in \mathcal{S}$.

Literature and Comments. Most of the results in this chapter are taken from the papers [29] and [30] by Beidar and Chebotar.

Some further results concerning quasi-polynomials, which are not considered in this book, can be found in [92].

Chapter 5

Functional Identities in (Semi)prime Rings

Up until now we have seen how to construct new d-free sets from given ones (Chapter 3) and have analyzed certain functional identities acting on d-free sets (Chapter 4). But, with the exception of the results from Chapter 2, we have yet to show the existence of important classes of d-free sets. Our main purpose in this chapter is to remedy this situation. Our success in this endeavor has chiefly been in the case of various subsets of a prime ring $\mathcal{A}$ (considered as a subring of its maximal left quotient ring $\mathcal{Q}$). These results will be presented in Section 5.2. They will be obtained as corollaries to the results from Section 5.1 which establish the d-freeness of rings that contain elements satisfying certain technical conditions – specifically, the so-called fractional degree of such elements must be $\geq d$. In Section 5.3 we shall see that the basic result on d-freeness of prime rings can be extended to a more general (and truly more entangled) semiprime setting. Section 5.4 is devoted to commuting traces of multiadditive maps on prime rings; the definitive result is established in the case of biadditive maps. The chapter ends with Section 5.5 which studies generalized functional identities in prime rings.

5.1 The Fractional Degree

Let $\mathcal{A}$ be a subring of a unital ring $\mathcal{Q}$. We are looking for an appropriate "degree" function of an element $t \in \mathcal{A}$ (let us temporarily call it $f(t)$) with the property that if $f(t) \geq d$, then $\mathcal{A}$ is $(t; d)$-free in $\mathcal{Q}$. In Chapter 2 we already found such a function, namely the *strong degree*, $f(t) = s\text{-deg}(t)$. The present section will be devoted to a certain extension of the strong degree, the so-called *fractional degree*, $f(t) = f\text{-}\deg_{\mathcal{Q}}(t)$. The proof of the main result (Theorem 5.6) establishing that $f\text{-}\deg_{\mathcal{Q}}(t) \geq d$ implies $(t; d)$-freeness will be similar, just slightly more involved, than the proof of Theorem 2.19 that gives a similar conclusion with respect to the strong degree.

First we introduce some notation and conventions. By $\mathcal{C}$ we denote the center of $\mathcal{Q}$. Let $\mathcal{S} \neq \emptyset$ and $\mathcal{T}$ be subsets of $\mathcal{Q}$. We set

$$\ell(\mathcal{T}; \mathcal{S}) = \{t \in \mathcal{T} \mid t\mathcal{S} = 0\},$$

i.e., $\ell(\mathcal{T}; \mathcal{S})$ is the *left annihilator* of $\mathcal{S}$ in $\mathcal{T}$. Similarly,

$$r(\mathcal{T}; \mathcal{S}) = \{t \in \mathcal{T} \mid \mathcal{S}t = 0\}$$

is the *right annihilator* of $\mathcal{S}$ in $\mathcal{T}$. Further, $\mathcal{C}(\mathcal{S})$ denotes the centralizer of $\mathcal{S}$ in $\mathcal{Q}$. We assume throughout that

$$\mathcal{C}(\mathcal{A}) = \mathcal{C}, \tag{5.1}$$

that is, only central elements in $\mathcal{Q}$ commute with every element in $\mathcal{A}$.

Next, let

$$\mathrm{Id}_\mathcal{Q}(\mathcal{A}) = \{q \in \mathcal{Q} \mid q\mathcal{A} \cup \mathcal{A}q \subseteq \mathcal{A}\}$$

be the *idealizer* of $\mathcal{A}$ in $\mathcal{Q}$. Clearly $\mathrm{Id}_\mathcal{Q}(\mathcal{A})$ is a unital subring of $\mathcal{Q}$ and it is the largest subring of $\mathcal{Q}$ containing $\mathcal{A}$ as an ideal. Of course if $\mathcal{A}$ contains 1, then $\mathrm{Id}_\mathcal{Q}(\mathcal{A}) = \mathcal{A}$. In fact the involvement of $\mathrm{Id}_\mathcal{Q}(\mathcal{A})$ is not absolutely necessary in everything that follows, i.e., it could be replaced by $\mathcal{A}$ itself; the reader should wait until Lemma 5.25 to see that dealing with $\mathrm{Id}_\mathcal{Q}(\mathcal{A})$ instead of $\mathcal{A}$ can really be useful.

The next definition might strike the reader as a somewhat artificial one. It is indeed a very technical one, but as we shall see in the sequel the need for both conditions (i) and (ii) arises naturally in the study of FI's.

Definition 5.1. An element $a \in \mathrm{Id}_\mathcal{Q}(\mathcal{A})$ is said to be *fractionable in $\mathcal{Q}$* if the following two conditions hold:

(i) If $\varphi : \mathcal{A} \to \mathcal{Q}$ is an additive map such that $\varphi(xay) = ax\varphi(y)$ for all $x, y \in \mathcal{A}$, then there exists $q \in \mathcal{Q}$ such that $\varphi(x) = axq$ for all $x \in \mathcal{A}$;

(ii) $r(\mathcal{Q}; a\mathcal{A}) = \ell(\mathcal{Q}; \mathcal{A}a) = 0$.

Let us state a few simple observations concerning this notion, in particular to give at least some indications why the name "fractionable" was chosen. Suppose that $a \in \mathrm{Id}_\mathcal{Q}(\mathcal{A})$ is invertible with $a^{-1} \in \mathcal{A}$. Then a is fractionable (in $\mathcal{Q}$). Indeed, (ii) is trivially fulfilled, and (i) follows immediately by taking $y = a^{-1}$. Conversely, considering the identity map on $\mathcal{A}$ we see that if $a \in \mathcal{C} \cap \mathrm{Id}_\mathcal{Q}(\mathcal{A})$ is fractionable (in $\mathcal{Q}$), then a is invertible (in $\mathcal{Q}$). Therefore, if $\mathcal{A}$ is a commutative unital domain, then invertible elements in $\mathcal{A}$ coincide with elements that are fractionable in $\mathcal{A}$. On the other hand, every nonzero element a in such a ring $\mathcal{A}$ is fractionable in the field of fractions of $\mathcal{A}$. Indeed, a trivially satisfies (ii), and if φ satisfies the condition of (i), then we have $a\varphi(x) = \varphi(ax) = ax\varphi(1)$ and so $\varphi(x) = ax(a^{-1}\varphi(1))$.

In the next section we shall see that in an important special case the conditions (i) and (ii) are equivalent. In general, however, they are independent. For example, if $\mathcal{A} = \mathcal{Q}$ is a commutative unital domain, then every nonzero element satisfies (ii), but only invertible elements satisfy (i). If $e \neq 0, 1$ is an idempotent in $\mathcal{C}$ and $\mathcal{A} = e\mathcal{Q}$, then $e \in \mathcal{A}$ satisfies (i) but not (ii).

We remark that (ii), more precisely the condition $r(\mathcal{Q}; a\mathcal{A}) = 0$, implies the uniqueness of the element q in (i). Another useful observation is: If $\mathrm{Id}_{\mathcal{Q}}(\mathcal{A})$ contains a fractionable element, then

$$r(\mathcal{Q}; \mathcal{A}) = \ell(\mathcal{Q}; \mathcal{A}) = 0. \tag{5.2}$$

This clearly follows from (ii).

Lemma 5.2. *Let $a, b \in \mathrm{Id}_{\mathcal{Q}}(\mathcal{A})$ be fractionable elements and let $r, s \in \mathcal{Q}$ be such that $axs = rxb$ for all $x \in \mathcal{A}$. Then there exists $\lambda \in \mathcal{C}$ such that $r = \lambda a$ and $s = \lambda b$.*

Proof. Let us first consider the case where $s = r$ and $b = a$, that is $axr = rxa$ for all $x \in \mathcal{A}$. Define $\varphi : \mathcal{A} \to \mathcal{Q}$ by $\varphi(x) = rx$, and note that $\varphi(xay) = rxay = axry = ax\varphi(y)$ for all $x, y \in \mathcal{A}$. Since a is fractionable it follows that there exists $\lambda \in \mathcal{Q}$ such that $\varphi(x) = ax\lambda$ for all $x \in \mathcal{A}$. Thus, $ax\lambda = rx$ for every $x \in \mathcal{A}$, and hence $a(xy)\lambda = r(xy) = (rx)y = (ax\lambda)y$. That is, $a\mathcal{A}[\mathcal{A}, \lambda] = 0$ which yields $[\mathcal{A}, \lambda] = 0$, i.e., $\lambda \in \mathcal{C}$ by (5.1). But then $r = \lambda a$ by (5.2).

The general case can be easily reduced to the one from the preceding paragraph. Indeed, assuming that $axs = rxb$ for all $x \in \mathcal{A}$, it follows, since $xay \in \mathcal{A}$ for all $x, y \in \mathcal{A}$, that $a(xay)s = r(xay)b$. On the other hand, $ax(ays) = axryb$. Comparing both relations we get $(axr - rxa)\mathcal{A}b = 0$. Since b is fractionable it follows that $axr = rxa$ for all $x \in \mathcal{A}$. Consequently, $r = \lambda a$ for some $\lambda \in \mathcal{C}$, and so the initial identity yields $a\mathcal{A}(s - \lambda b) = 0$. Consequently, $s = \lambda b$. □

Recall that $\mathcal{M}(\mathrm{Id}_{\mathcal{Q}}(\mathcal{A}))$ denotes the multiplication ring of $\mathrm{Id}_{\mathcal{Q}}(\mathcal{A})$. As above we define, for convenience, that $t^0 = 1$ for every $t \in \mathcal{Q}$. We are now ready to define what is the central notion of this section.

Definition 5.3. The *fractional degree* of an element $t \in \mathrm{Id}_{\mathcal{Q}}(\mathcal{A})$ is greater than n (in $\mathcal{Q}$) where $n \geq 0$ (notation $f\text{-}\deg_{\mathcal{Q}}(t) > n$), if for every $i = 0, 1, \ldots, n$ there exists $\mathcal{E}_i \in \mathcal{M}(\mathrm{Id}_{\mathcal{Q}}(\mathcal{A}))$ such that

$$\mathcal{E}_i(t^j) = 0 \text{ if } j \neq i, \text{ and } \mathcal{E}_i(t^i) \text{ is fractionable in } \mathcal{Q}.$$

If $f\text{-}\deg_{\mathcal{Q}}(t) > n - 1$ but $f\text{-}\deg_{\mathcal{Q}}(t) \not> n$, then we say that the fractional degree of t is n ($f\text{-}\deg_{\mathcal{Q}}(t) = n$). If $f\text{-}\deg_{\mathcal{Q}}(t) > n$ for every positive integer n, then we write $f\text{-}\deg_{\mathcal{Q}}(t) = \infty$. Finally we set

$$f\text{-}\deg_{\mathcal{Q}}(\mathcal{A}) = \sup\{f\text{-}\deg_{\mathcal{Q}}(t) \mid t \in \mathrm{Id}(\mathcal{A})\}.$$

The similarity to the definition of the strong degree is obvious. In fact the only difference is that in the strong degree case we require that $\mathcal{E}_i(t^i) = 1$, while in the present context we only require that $\mathcal{E}_i(t^i)$ is fractionable. Accordingly, for every unital ring $\mathcal{A}$ and every $t \in \mathcal{A}$ we have

$$s\text{-}\deg(t) \leq f\text{-}\deg_{\mathcal{A}}(t).$$

Our intention now is basically to show that the results on FI's concerning the strong degree can be extended to the more general (and more involved) fractional degree setting. We shall follow the pattern presented in Chapter 2. It should be pointed out, however, that these new results do not completely cover those from Chapter 2. Namely, in the strong degree setting we have considered functions into an arbitrary $(\mathcal{A}, \mathcal{A})$-bimodule $\mathcal{M}$, while now we have to restrict ourselves to the (fixed) ring $\mathcal{Q}$ upon which the fractional degree depends.

For the rest of this section the rings $\mathcal{A} \subseteq \mathcal{Q}$ and the element $t \in \mathrm{Id}_{\mathcal{Q}}(\mathcal{A})$ will be fixed. Therefore we slightly simplify the notation by writing $f\text{-}\deg(.)$ instead of $f\text{-}\deg_{\mathcal{Q}}(.)$, and $\mathrm{Id}(\mathcal{A})$ instead of $\mathrm{Id}_{\mathcal{Q}}(\mathcal{A})$. Recall that $_a M_b$ with $a, b \in \mathrm{Id}(\mathcal{A})$ denotes an element in $\mathcal{M}(\mathrm{Id}(\mathcal{A}))$ defined by $_a M_b(x) = axb$.

Lemma 5.4. *Let* $c_0, c_1, \ldots, c_m, d_0, d_1, \ldots, d_n \in \mathcal{Q}$.

(i) *If* $f\text{-}\deg(t) > m$ *and* $\sum_{i=0}^{m} c_i x t^i = 0$ *for all* $x \in \mathcal{A}$, *then each* $c_i = 0$.

(ii) *If* $f\text{-}\deg(t) > n$ *and* $\sum_{j=0}^{n} t^j x d_j = 0$ *for all* $x \in \mathcal{A}$, *then each* $d_j = 0$.

(iii) *If* $f\text{-}\deg(t) > \max\{n, m\}$ *and* $\sum_{i=0}^{m} c_i x t^i + \sum_{j=0}^{n} t^j x d_j = 0$ *for all* $x \in \mathcal{A}$, *then* $c_i \in \sum_{j=0}^{n} \mathcal{C} t^j$ *and* $d_j \in \sum_{i=0}^{m} \mathcal{C} t^i$ *for all* i, j.

Proof. The proofs of (i) and (ii) are analogous, so we prove only (i). Thus, assume the conditions of (i) are fulfilled, and fix $0 \leq i \leq m$. Then there exists $\mathcal{E} = \sum_{l=1}^{k} a_l M_{b_l} \in \mathcal{M}(\mathrm{Id}(\mathcal{A}))$ such that $\mathcal{E}(t^j) = 0$ if $j \neq i$ and $\mathcal{E}(t^i)$ is fractionable in $\mathcal{Q}$, so in particular $\ell(\mathcal{Q}; \mathcal{A}\mathcal{E}(t^i)) = 0$. Since $x a_l \in \mathcal{A}$ for every $x \in \mathcal{A}$ and $1 \leq l \leq k$ we may substitute $x a_l$ for x in $\sum_{i=0}^{m} c_i x t^i = 0$. Multiplying the identity so obtained from the right by b_l and summing up we arrive at $c_i x \mathcal{E}(t^i) = 0$ for every $x \in \mathcal{A}$. But this forces $c_i = 0$.

Now assume the conditions of (iii) are fulfilled. Note that for every $\mathcal{E} \in \mathcal{M}(\mathrm{Id}(\mathcal{A}))$ we have $\sum_{j=0}^{n} \mathcal{E}(t^j) x d_j + \sum_{i=0}^{m} \mathcal{E}(c_i) x t^i = 0$ for all $x \in \mathcal{A}$. Fix $0 \leq j \leq n$. Since $f\text{-}\deg(t) > n$ we can choose $\mathcal{E}$ so that the last identity reduces to

$$a_j x d_j + \sum_{i=0}^{m} r_i x t^i = 0 \quad \text{for all } x \in \mathcal{A}, \tag{5.3}$$

where $a_j = \mathcal{E}(t^j) \in \mathrm{Id}(\mathcal{A})$ is fractionable and $r_i = \mathcal{E}(c_i) \in \mathcal{Q}$. From (5.3) it follows that $a_j x \mathcal{E}'(d_j) + \sum_{i=0}^{m} r_i x \mathcal{E}'(t^i) = 0$ for every $\mathcal{E}' \in \mathcal{M}(\mathrm{Id}(\mathcal{A}))$. Since $f\text{-}\deg(t) > n$, fixing $0 \leq i \leq m$ we can choose $\mathcal{E}'$ so that the last identity can be written as $a_j x s_j = r_i x b_i$ where $s_j = \mathcal{E}'(d_j)$ and $b_i = -\mathcal{E}'(t^i) \in \mathrm{Id}(\mathcal{A})$ is fractionable. Invoking Lemma 5.2 we see that $r_i = \lambda_i a_j$ for some $\lambda_i \in \mathcal{C}$. But then one can rewrite (5.3) as $a_j x (d_j + \sum_{i=0}^{m} \lambda_i t^i) = 0$. Since a_j is fractionable it follows that $d_j = -\sum_{i=0}^{m} \lambda_i t^i \in \sum_{i=0}^{m} \mathcal{C} t^i$. Similarly we see that $c_i \in \sum_{j=0}^{n} \mathcal{C} t^j$. $\qquad\square$

From (i) we infer that if $f\text{-}\deg(t) > m$, then $\sum_{i=0}^{m} c_i t^i = 0$ with $c_i \in \mathcal{C}$ implies $c_i = 0$. Therefore, c_i's and d_j's (from (iii)) can be expressed as $\mathcal{C}$-linear combinations of powers of t in a unique way.

Now we are ready to tackle FI's. The lemma and (to some extent also) the theorem that follow will be proved in exactly the same way as Lemma 2.18 and Theorem 2.19 above. Nevertheless we shall give the details of the proofs.

As usual, E_{iu}, F_{jv} will denote arbitrary maps from $\mathcal{A}^{m-1}$ into $\mathcal{Q}$, m will be a fixed positive integer, and $\mathcal{I}, \mathcal{J}$ will be subsets of $\{1, 2, \ldots, m\}$. First we consider the FI's

$$\sum_{i \in \mathcal{I}} \sum_{u=0}^{a} E_{iu}^{i} x_{i} t^{u} = 0 \quad \text{for all } \bar{x}_{m} \in \mathcal{A}^{m} \tag{5.4}$$

and

$$\sum_{j \in \mathcal{J}} \sum_{v=0}^{b} t^{v} x_{j} F_{jv}^{j} = 0 \quad \text{for all } \bar{x}_{m} \in \mathcal{A}^{m}. \tag{5.5}$$

Lemma 5.5. *If $f\text{-}\deg(t) \geq |\mathcal{I}| + a$, then (5.4) implies that each $E_{iu} = 0$. Similarly, if $f\text{-}\deg(t) \geq |\mathcal{J}| + b$, then (5.5) implies that each $F_{jv} = 0$.*

Proof. We prove only the first assertion. If $|\mathcal{I}| = 1$, say $\mathcal{I} = \{1\}$, then (5.4) reads as

$$\sum_{u=0}^{a} E_{1u}^{1} x_{1} t^{u} = 0 \quad \text{for every } x_{1} \in \mathcal{A}.$$

Fixing $x_{2}, \ldots, x_{m}$, we may apply Lemma 5.4 (i) to conclude that each $E_{1u} = 0$. Thus we may assume that $|\mathcal{I}| > 1$, say $1, 2 \in \mathcal{I}$. Set

$$H(\bar{x}_{m}) = \sum_{i \in \mathcal{I}} \sum_{u=0}^{a} E_{iu}^{i} x_{i} t^{u} = 0$$

and apply the t-substitution operation. That is,

$$
\begin{aligned}
0 &= H(x_{1}t) - H(\bar{x}_{m})t \\
&= \sum_{\substack{i \in \mathcal{I}, \\ i \neq 1}} E_{i0}^{i}(x_{1}t)x_{i} + \sum_{\substack{i \in \mathcal{I}, \\ i \neq 1}} \sum_{u=1}^{a} \{E_{iu}^{i}(x_{1}t) - E_{i,u-1}^{i}\} x_{i} t^{u} - \sum_{\substack{i \in \mathcal{I}, \\ i \neq 1}} E_{ia}^{i} x_{i} t^{a+1}.
\end{aligned}
$$

The induction assumption yields $E_{ia} = 0$, $E_{iu}^{i}(x_{1}t) - E_{i,u-1}^{i} = 0$ for all $i \neq 1$, $u = 0, 1, \ldots, a$, hence each $E_{iu} = 0$ for $i \neq 1$. Similarly, by replacing the role of x_{1} by x_{2}, we get $E_{1u} = 0$ for all $u = 0, 1, \ldots, a$. $\qquad\square$

Theorem 5.6. *Let $\mathcal{A}$ be a subring of a unital ring $\mathcal{Q}$ such that the centralizer of $\mathcal{A}$ in $\mathcal{Q}$ is equal to the center of $\mathcal{Q}$. Let $t \in \mathrm{Id}_{\mathcal{Q}}(\mathcal{A})$. If $d \in \mathbb{N}$ is such that $f\text{-}\deg_{\mathcal{Q}}(t) \geq d$, then $\mathcal{A}$ is a $(t; d)$-free subset of $\mathcal{Q}$; in particular, $\mathcal{A}$ is d-free.*

Proof. By Corollary 3.12 it suffices to check only that the condition (a) of $(t; d)$-freeness is fulfilled. That is, we want to prove that

$$\sum_{i \in \mathcal{I}} \sum_{u=0}^{a} E_{iu}^{i} x_{i} t^{u} + \sum_{j \in \mathcal{J}} \sum_{v=0}^{b} t^{v} x_{j} F_{jv}^{j} = 0 \tag{5.6}$$

for all $\bar{x}_m \in \mathcal{A}$ with $f\text{-}\deg(t) \geq \max\{|\mathcal{I}| + a, |\mathcal{J}| + b\}$ implies the existence of functions $p_{iujv} : \mathcal{A}^{m-2} \to \mathcal{Q}$ and $\lambda_{kuv} : \mathcal{A}^{m-1} \to \mathcal{C}$ such that

$$E_{iu}^i = \sum_{\substack{j \in \mathcal{J} \\ j \neq i}} \sum_{v=0}^b t^v x_j p_{iujv}^{ij} + \sum_{v=0}^b \lambda_{iuv}^i t^v,$$

$$F_{jv}^j = -\sum_{\substack{i \in \mathcal{I} \\ i \neq j}} \sum_{u=0}^a p_{iujv}^{ij} x_i t^u - \sum_{u=0}^a \lambda_{juv}^j t^u, \qquad (5.7)$$

$$\lambda_{kuv} = 0 \quad \text{if} \quad k \notin \mathcal{I} \cap \mathcal{J}$$

for all $\bar{x}_m \in \mathcal{A}$.

Assuming that all the E_{iu}'s are given according to (5.7) (in particular $\lambda_{iuv} = 0$ if $i \notin \mathcal{J}$ holds) it follows from (5.6) that

$$\sum_{j \in \mathcal{J}} \sum_{v=0}^b t^v x_j \left[F_{jv}^j + \sum_{\substack{i \in \mathcal{I}, \\ i \neq j}} \sum_{u=0}^a p_{iujv}^{ij} x_i t^u + \sum_{u=0}^a \lambda_{juv}^j t^u \right] = 0,$$

and so we infer from Lemma 5.5 that all the F_{jv}'s are given according to (5.7) as well. Similarly, if all the F_{jv}'s are given according to (5.7), then all the E_{iu}'s are also given according to (5.7).

We proceed by induction on $|\mathcal{I}| + |\mathcal{J}|$. Lemma 5.5 covers the cases where $|\mathcal{I}| = 0$ or $|\mathcal{J}| = 0$. Therefore the first case that has to be considered is when $|\mathcal{I}| = 1 = |\mathcal{J}|$. We have to consider separately two subcases: when $\mathcal{I} = \mathcal{J}$ and when $\mathcal{I} \neq \mathcal{J}$.

In the first subcase we may assume that $\mathcal{I} = \{1\} = \mathcal{J}$, so we have

$$\sum_{u=0}^a E_{1u}^1 x_1 t^u + \sum_{v=0}^b t^v x_1 F_{1v}^1 = 0.$$

Fixing $x_2, x_3 \dots, x_m$ we may apply Lemma 5.4 (iii) to conclude that

$$E_{1u}(x_2, \dots, x_m) = \sum_{v=0}^b \lambda_{1uv}(x_2, \dots, x_m) t^v$$

for some $\lambda_{1uv}(x_2, \dots, x_m) \in \mathcal{C}$, $0 \leq u \leq a$, $0 \leq v \leq b$. That is, the E_{1u}'s are given according to (5.7), and hence the F_{1v}'s are given according to (5.7) as well.

In the second subcase we may assume that $\mathcal{I} = \{2\}$ and $\mathcal{J} = \{1\}$. Thus,

$$D(\bar{x}_m) = \sum_{u=0}^a E_{2u}^2 x_2 t^u + \sum_{v=0}^b t^v x_1 F_{1v}^1 = 0. \qquad (5.8)$$

We claim that the E_{2u}'s are additive in x_1. Indeed, replacing x_1 by $x_1' + x_1''$ in (5.8) it follows easily that

$$\sum_{u=0}^{a} \left(E_{2u}^2(x_1' + x_1'') - E_{2u}^2(x_1') - E_{2u}^2(x_1'') \right) x_2 t^u = 0$$

(here we have simplified the notation by neglecting $x_3, \ldots, x_m$). It follows from Lemma 5.4 (i) that $E_{2u}^2(x_1' + x_1'') = E_{2u}^2(x_1') + E_{2u}^2(x_1'')$, proving our claim. Now fix $0 \le v \le b$. Since $f\text{-}\deg(t) > b$ there exist $c_l, d_l \in \mathrm{Id}(\mathcal{A})$, $l = 1, 2, \ldots, n$, such that $\sum_{l=1}^{n} c_l t^w d_l = 0$ if $w \ne v$ and $c = \sum_{l=1}^{n} c_l t^v d_l \in \mathrm{Id}(\mathcal{A})$ is fractionable in $\mathcal{Q}$. Using (5.8) it follows that

$$0 = \sum_{l=1}^{n} c_l D(d_l x_1) = \sum_{u=0}^{a} H_{2u}^2 x_2 t^u + c x_1 F_{1v} \tag{5.9}$$

where $H_{2u}(\overline{x}_m^2) = \sum_{l=1}^{n} c_l E_{2u}(d_l x_1, x_3, \ldots, x_m)$, and so the H_{2u}'s are additive in x_1. Replacing x_1 by $y c x_1$ in (5.9) gives

$$\sum_{u=0}^{a} H_{2u}(y c x_1, x_3, \ldots, x_m) x_2 t^u + c y c x_1 F_{1v}^1 = 0.$$

Since, on the other hand, $cy(c x_1 F_{1v}^1) = -\sum_{u=0}^{a} c y H_{2u}^2 x_2 t^u$ by (5.9), it follows that

$$\sum_{u=0}^{a} \left(H_{2u}(y c x_1, x_3, \ldots, x_m) - c y H_{2u}^2 \right) x_2 t^u = 0.$$

Applying Lemma 5.4 (i) again we arrive at

$$H_{2u}(y c x_1, x_3, \ldots, x_m) = c y H_{2u}(x_1, x_3, \ldots, x_m)$$

for all $u = 0, 1, \ldots, a$ and for all $y, x_1, x_3, \ldots, x_m \in \mathcal{A}$. According to Definition 5.1 there exists a unique element $p_u(\overline{x}_m^{12}) \in \mathcal{Q}$ such that

$$H_{2u}(\overline{x}_m^2) = c x_1 p_u(\overline{x}_m^{12}).$$

Going back to (5.9) we now have

$$c x_1 \left(\sum_{u=0}^{a} p_u^{12} x_2 t^u + F_{1v}^1 \right) = 0.$$

Since c is fractionable we conclude that

$$F_{1v} = -\sum_{u=0}^{a} p_u^{12} x_2 t^u \quad \text{for all } 0 \le v \le b.$$

This means that all the F_{1v}'s are given according to (2.16) and so the same holds true for all the E_{2u}'s. The proof of the second subcase is complete.

We may now assume $|\mathcal{I}| + |\mathcal{J}| > 2$ and make the inductive step. Further we may assume $|\mathcal{J}| \geq 2$ and that $1, 2 \in \mathcal{J}$. We have

$$H(\bar{x}_m) = \sum_{i \in \mathcal{I}} \sum_{u=0}^{a} E_{iu}^i x_i t^u + \sum_{j \in \mathcal{J}} \sum_{v=0}^{b} t^v x_j F_{jv}^j = 0.$$

Applying the t-substitution operation, i.e., computing $H(tx_1) - tH(\bar{x}_m)$, it follows that

$$\sum_{i \in \mathcal{I}} \sum_{u=0}^{a} G_{iu}^i x_i t^u + \sum_{\substack{j \in \mathcal{J}, \\ j \neq 1}} x_j F_{j0}^j(tx_1)$$

$$+ \sum_{\substack{j \in \mathcal{J}, \\ j \neq 1}} \sum_{v=1}^{b} t^v x_j \{F_{jv}^j(tx_1) - F_{j,v-1}^j\} - \sum_{\substack{j \in \mathcal{J}, \\ j \neq 1}} t^{b+1} x_j F_{jb}^j = 0$$

for appropriate G_{iu}'s. Noting that $|\mathcal{J} \setminus \{1\}| + b + 1 = |\mathcal{J}| + b$ we see that the degree condition on t holds. Applying the induction assumption we get

$$F_{jb}^j = -\sum_{\substack{i \in \mathcal{I}, \\ i \neq j}} \sum_{u=0}^{a} q_{iuj\,b+1}^{ij} x_i t^u - \sum_{u=0}^{a} \mu_{ju\,b+1}^j t^u, \quad j \neq 1,$$

$$F_{jv}^j(tx_1) - F_{j,v-1}^j = -\sum_{\substack{i \in \mathcal{I}, \\ i \neq j}} \sum_{u=0}^{a} q_{iujv}^{ij} x_i t^u - \sum_{u=0}^{a} \mu_{juv}^j t^u,$$

$$j \neq 1, \quad v = 1, 2, \ldots, b,$$

and $\mu_{juv} = 0$ if $j \notin \mathcal{I}$. Beginning with F_{jb} and proceeding recursively, we see that these identities yield

$$F_{jv}^j = -\sum_{\substack{i \in \mathcal{I}, \\ i \neq j}} \sum_{u=0}^{a} p_{iujv}^{ij} x_i t^u - \sum_{u=0}^{a} \lambda_{juv}^j t^u, \quad j \neq 1$$

for appropriate p_{iujv} and λ_{juv} with $\lambda_{juv} = 0$ if $j \notin \mathcal{I}$. In a similar fashion, by computing $H(tx_2) - tH(\bar{x}_m)$, we obtain

$$F_{1v}^1 = -\sum_{\substack{i \in \mathcal{I}, \\ i \neq 1}} \sum_{u=0}^{a} p_{iu1v}^{i1} x_i t^u - \sum_{u=0}^{a} \lambda_{1uv}^1 t^u,$$

$\lambda_{1uv} = 0$ if $1 \notin \mathcal{I}$, and so all the F_{jv}'s are in the standard form. Consequently, the E_{iu}'s are of standard form as well. $\qquad\square$

5.2 A List of d-Free Subsets of Prime Rings

Up until this point in Part II we have not shown the existence of any d-free sets; the results have focused on showing that if a certain set is d-free, then a certain related set is d'-free, where d' is computed in terms of d. For instance, if we knew that a certain prime ring was d-free, then we have a host of results showing various important related sets are d'-free. In this section we will find a sufficient condition for a prime ring $\mathcal{A}$ to be a $(t; d)$-free subset of its maximal left ring of quotients

$$\mathcal{Q} = \mathcal{Q}_{ml}(\mathcal{A}),$$

and moreover a necessary and sufficient condition for $\mathcal{A}$ to be a d-free subset of $\mathcal{Q}$. These will make it possible for us to make a list of various d-free subsets of prime rings at the end of this section.

Having applications in the next section in mind we begin by considering a more general situation where $\mathcal{A}$ is a semiprime ring. We refer the reader to appendix A for various properties of the maximal left ring of quotients $\mathcal{Q}$. Since throughout the section $\mathcal{Q}$ will stand for $\mathcal{Q}_{ml}(\mathcal{A})$, we again abbreviate the notation by writing $f\text{-}\deg(\,.\,)$ instead of $f\text{-}\deg_{\mathcal{Q}}(\,.\,)$, and $\mathrm{Id}(\mathcal{A})$ instead of $\mathrm{Id}_{\mathcal{Q}}(\mathcal{A})$.

We point out that the condition that was assumed throughout Section 5.1, i.e., that the centralizer of $\mathcal{A}$ in $\mathcal{Q}$ is $\mathcal{C}$ $((\mathcal{C}(\mathcal{A}) = \mathcal{C})$, is indeed satisfied in the present context.

We begin with a lemma of crucial importance; it brings to light what is, in the present setting, really hidden behind the technical conditions of Definition 5.1.

Lemma 5.7. *If $\mathcal{A}$ is semiprime, then the conditions* (i) *and* (ii) *of Definition* 5.1 *are equivalent for every $a \in \mathrm{Id}(\mathcal{A})$. Furthermore, they are equivalent to the condition* (iii) $\ell(\mathcal{A}; \mathcal{A}a) = 0$.

Proof. Let us first show that (ii) and (iii) are equivalent. Of course it is enough to show that (iii) implies (ii). Given $\ell(\mathcal{A}; \mathcal{A}a) = 0$, suppose $q\mathcal{A}a = 0$ for some $0 \neq q \in \mathcal{Q}$. Then $0 \neq rq \in \mathcal{A}$ for some $r \in \mathcal{A}$. Thus we reach the contradiction $rq\mathcal{A}a = 0$. Therefore $\ell(\mathcal{Q}; \mathcal{A}a) = 0$. Next suppose $a\mathcal{A}q = 0$ for some $q \in \mathcal{Q}$. Lemma A.3 tells us that $q\mathcal{A}a = 0$, and hence $q = 0$ by what we have just shown. Thus, (ii) holds indeed.

Now suppose that (i) holds, and pick $b \in \ell(\mathcal{A}; \mathcal{A}a)$. Then $b \in r(\mathcal{A}; a\mathcal{A})$ by Lemma A.3, i.e., $a\mathcal{A}b = 0$. Consider the map $\varphi : \mathcal{A} \to \mathcal{Q}$ given by $\varphi(x) = xb$. We have $\varphi(xay) = xayb = 0 = axyb = ax\varphi(y)$ for all $x, y \in \mathcal{A}$, and so, by assumption, there exists $q \in \mathcal{Q}$ such that $\varphi(x) = axq$ for all $x \in \mathcal{A}$. Thus, $xb = axq$ for all $x \in \mathcal{A}$, and hence, since $ba = 0$ (by Lemma A.3), we have $(\mathcal{A}b)^2 = 0$. But then $b = 0$ by the semiprimeness of $\mathcal{A}$. This shows that (i) implies (iii).

Finally we prove the most important part of the lemma, namely that (ii) implies (i). Thus assume $\varphi : \mathcal{A} \to \mathcal{Q}$ satisfies $\varphi(xay) = ax\varphi(y)$ for all $x, y \in \mathcal{A}$. Set $\mathcal{L} = \mathcal{Q}a\mathcal{A}$. We claim that $\mathcal{L}$ is a dense left ideal of $\mathcal{Q}$. Pick $q_1, q_2 \in \mathcal{Q}$ with $q_1 \neq 0$. Let $\mathcal{J}$ be a dense left ideal of $\mathcal{A}$ such that $\mathcal{J}q_2 \subseteq \mathcal{A}$. Since $\mathcal{J}$ is dense there exists

$u \in \mathcal{J}$ such that $uq_1 \neq 0$. From $r(\mathcal{Q}; a\mathcal{A}) = 0$ it follows that $v(uq_1) \neq 0$ for some $v \in \mathcal{L}$. Consequently, $q = vu \in \mathcal{Q}$ satisfies $qq_1 \neq 0$ and $qq_2 \in \mathcal{L}$, proving that $\mathcal{L}$ is dense. Now define $f : \mathcal{L} \to \mathcal{Q}$ by

$$f\left(\sum q_i a y_i\right) = \sum q_i \varphi(y_i) \quad \text{for all} \quad q_i \in \mathcal{Q}, \; y_i \in \mathcal{A}.$$

We claim that f is well-defined. Indeed, suppose $\sum q_i a y_i = 0$ for some $q_i \in \mathcal{Q}$, $y_i \in \mathcal{A}$. Let $\mathcal{K}$ be a dense left ideal of $\mathcal{A}$ such that $\mathcal{K}q_i \subseteq \mathcal{A}$ for every i (see Corollary A.2). Then for every $y \in \mathcal{K}$ we have

$$ay \sum q_i \varphi(y_i) = \sum a y q_i \varphi(y_i) = \varphi\left(\sum y q_i a y_i\right) = \varphi\left(y \sum q_i a y_i\right) = 0.$$

Thus, $a\mathcal{K}\left(\sum q_i \varphi(y_i)\right) = 0$, and so $a\mathcal{A}\mathcal{K}\left(\sum q_i \varphi(y_i)\right) = 0$. Again using $r(\mathcal{Q}; a\mathcal{A}) = 0$ it follows that $\mathcal{K}\left(\sum q_i \varphi(y_i)\right) = 0$ which in turn implies $\sum q_i \varphi(y_i) = 0$. This proves that f is well-defined. Clearly f is a left $\mathcal{Q}$-module homomorphism. Therefore there exists $q \in \mathcal{Q}_{ml}(\mathcal{Q}) = \mathcal{Q}$ (see Corollary A.5) such that $f(x) = xq$ for all $x \in \mathcal{L}$. But then $\varphi(y) = f(ay) = ayq$ for all $y \in \mathcal{A}$. $\qquad \square$

In the prime ring case all these become extremely simple:

Lemma 5.8. *If $\mathcal{A}$ is prime, then every nonzero element in $\mathrm{Id}(\mathcal{A})$ is fractionable in $\mathcal{Q}$.*

The notion of the fractional degree can now be represented in a much simpler way:

Lemma 5.9. *If $\mathcal{A}$ is semiprime (resp. prime) and $t \in \mathrm{Id}(\mathcal{A})$, then $f\text{-}\deg(t) > n$ if and only if for every $i = 0, 1, \ldots, n$ there exists $\mathcal{E}_i \in \mathcal{M}(\mathrm{Id}(\mathcal{A}))$ such that $\mathcal{E}_i(t^j) = 0$ if $j \neq i$ and $\ell(\mathcal{A}; \mathcal{A}\mathcal{E}_i(t^i)) = 0$ (resp. $\mathcal{E}_i(t^i) \neq 0$).*

From now on we confine ourselves to the case where $\mathcal{A}$ is prime. Recall that the *extended centroid C* of $\mathcal{A}$, that is, the center of $\mathcal{Q}$, is a field (Theorem A.6). If $x \in \mathcal{Q}$ is algebraic over C, then we denote by $\deg(x)$ its *degree of algebraicity*. If x is not algebraic, then we write $\deg(x) = \infty$. So $\deg(x) \geq d$ means that either x is not algebraic over C or it is algebraic and its degree of algebraicity is $\geq d$. For a nonempty set $\mathcal{R} \subseteq \mathcal{Q}$ we define

$$\deg(\mathcal{R}) = \sup\{\deg(x) \mid x \in \mathcal{R}\}.$$

Lemma 5.10. *If $\mathcal{A}$ is prime, then $f\text{-}\deg(t) = \deg(t)$ for every $t \in \mathrm{Id}(\mathcal{A})$.*

Proof. First, $f\text{-}\deg(t) \leq \deg(t)$ follows from Lemma 5.4 (see the remarks following the proof of this lemma). Now assume that $\deg(t) > n \geq 0$. Then $1, t, \ldots, t^n$ are linearly independent over C and so by Theorem A.8, for each $i = 0, 1, \ldots, n$ there exists $\mathcal{E}_i \in \mathcal{M}(\mathcal{A})$ such that $\mathcal{E}_i(t^j) = 0$ if $j \neq i$ and $\mathcal{E}_i(t^i) \neq 0$. That is, $f\text{-}\deg(t) > n$ by Lemma 5.8. Accordingly, $f\text{-}\deg(t) \geq \deg(t)$. $\qquad \square$

Theorem 5.6 and Lemma 5.10 immediately yield the fundamental

Theorem 5.11. *Let $\mathcal{A}$ be a prime ring. If there exists $t \in \mathrm{Id}(\mathcal{A})$ such that $\deg(t) \geq d$, then $\mathcal{A}$ is a $(t;d)$-free subset of $\mathcal{Q} = \mathcal{Q}_{ml}(\mathcal{A})$; in particular, $\mathcal{A}$ is d-free.*

The last assertion can be sharpened as follows.

Corollary 5.12. *Let $\mathcal{A}$ be a prime ring, and let $d \in \mathbb{N}$. Then $\mathcal{A}$ is a d-free subset of $\mathcal{Q} = \mathcal{Q}_{ml}(\mathcal{A})$ if and only if $\deg(\mathcal{A}) \geq d$.*

Proof. The "if" part follows from Theorem 5.11. To prove the "only if" part, assume that $\deg(\mathcal{A}) < d$. Applying Theorem C.2 and Corollary 4.21 it follows that $\mathcal{A}$ is not d-free. $\qquad\square$

This clearly yields

Corollary 5.13. *A prime ring $\mathcal{A}$ is a d-free subset of $\mathcal{Q}$ for every $d \in \mathbb{N}$ if and only $\deg(\mathcal{A}) = \infty$ (i.e., $\mathcal{A}$ is not a PI-ring).*

Another slight generalization of the last assertion in Theorem 5.11 follows from Theorem 5.11, Theorem A.4 and Lemma C.5 (here one should note that a nonzero ideal of $\mathcal{A}$ is automatically a noncentral Lie ideal).

Corollary 5.14. *Let $\mathcal{I}$ be a nonzero ideal of a prime ring $\mathcal{A}$. If $\deg(\mathcal{A}) \geq d$, then $\mathcal{I}$ is a d-free subset of $\mathcal{Q}$.*

Our next goal is to consider d-freeness of Lie ideals. Here we shall apply Corollary 3.18. The assumption that $\deg(t)$ should be ≥ 3 in this result forces us to deal with a small technical problem before reaching our goal. We shall do this in the next lemma. In the proof we shall make use of Theorem 5.11, primarily to illustrate how the theory just developed works in practice. More direct approaches would also be possible.

Lemma 5.15. *Let $\mathcal{L}$ be a noncommutative Lie ideal of a prime ring $\mathcal{A}$. If $a, b \in \mathcal{Q}$ are such that $ax + xb = 0$ for all $x \in \mathcal{L}$, then $a = -b \in \mathcal{C}$.*

Proof. By assumption there are $t_1, t_2 \in \mathcal{L}$ such that $[t_1, t_2] \neq 0$. Replacing x by $[t_i, y]$ where $y \in \mathcal{A}$ we get

$$(at_i)y - ayt_i - y(t_ib) + t_iyb = 0 \quad \text{for all } y \in \mathcal{A}. \tag{5.10}$$

Since $t_i \notin \mathcal{C}$ we have of course $\deg(t_i) \geq 2$. Therefore $\mathcal{A}$ is a $(t_i;2)$-free subset of $\mathcal{Q}$ by Theorem 5.11. Note that this can be applied to (5.10); in particular it follows that there exist $\lambda_i, \mu_i \in \mathcal{C}$ such that $b = \lambda_i t_i + \mu_i$, $i = 1, 2$. Since t_1 and t_2 do not commute (and since $\mathcal{C}$ is a field!) it follows that $\lambda_1 = \lambda_2 = 0$. Consequently, $b \in \mathcal{C}$. Similarly we see that $a \in \mathcal{C}$. The initial identity now gives $(a+b)\mathcal{L} = 0$ with $a + b \in \mathcal{C}$, which clearly yields $a + b = 0$. $\qquad\square$

Corollary 5.16. *Let $\mathcal{L}$ be a noncommutative Lie ideal of a prime ring $\mathcal{A}$. If $\deg(\mathcal{A}) \geq d + 1$, then $\mathcal{L}$ is a d-free subset of $\mathcal{Q}$.*

Proof. By Lemma C.5 there exists $t \in \mathcal{L}$ such that $\deg(t) \geq d + 1$. Consequently, $\mathcal{A}$ is $(t; d+1)$-free by Theorem 5.11. Since $[t, \mathcal{A}] \subseteq \mathcal{L}$, Corollary 3.18 yields the desired conclusion, provided that $\deg(t) \geq 3$.

Thus we have to consider separately the case when $\deg(\mathcal{A}) = 2$. We first remark that then $\mathcal{Q}$ is equal to the central closure $\mathcal{AC}$ of $\mathcal{A}$ and it is a simple unital ring (in fact, a 4-dimensional central simple algebra; see Theorem C.2). The goal is to prove that $\mathcal{L}$ is a 1-free subset of $\mathcal{Q}$. In view of Lemma 5.15 it suffices to consider the FI $E_1(x_2)x_1 + x_2 F_2(x_1) = 0$ for all $x_1, x_2 \in \mathcal{L}$, where $E_1, F_2 : \mathcal{L} \to \mathcal{Q}$. Since $\mathcal{L}$ is a Lie ideal, we have $\mathcal{AL} \subseteq \mathcal{L} + \mathcal{LA}$. As $\mathcal{Q} = \mathcal{AC}$, this readily implies that $\mathcal{LQ}$ is an ideal of $\mathcal{Q}$. However, $\mathcal{Q}$ is simple and so we have $\mathcal{LQ} = \mathcal{Q}$. Therefore there exist $l_i \in \mathcal{L}$, $q_i \in \mathcal{Q}$ such that $\sum_i l_i q_i = 1$. Accordingly,

$$E_1(x_2) = \sum_i E_1(x_2) l_i q_i = -\sum_i x_2 F_2(l_i) q_i = x_2 p,$$

where $p = -\sum_i F_2(l_i) q_i \in \mathcal{Q}$. Consequently, $x_2(px_1 + F_2(x_1)) = 0$ for all $x_1, x_2 \in \mathcal{L}$, and from Lemma 5.15 we infer that $px_1 + F_2(x_1) = 0$, i.e., $F_2(x_1) = -px_1$. This proves that the FI $E_1(x_2)x_1 + x_2 F_2(x_1) = 0$ has only standard solutions on $\mathcal{L}$. □

Let us remark that noncommutative Lie ideals of a prime ring $\mathcal{A}$ coincide with noncentral ones, unless $\operatorname{char}(\mathcal{A}) = 2$ and $\deg(\mathcal{A}) = 2$. This well-known fact basically follows from Herstein's theory of Lie structures in associative rings; explicitly it is stated for example in [135, Lemma 6]. If $\mathbb{F}$ is a field with $\operatorname{char}(\mathbb{F}) = 2$, then the set $\mathcal{L}$ of all elements of the form $\begin{bmatrix} x & y \\ y & x \end{bmatrix}$, $x, y \in \mathbb{F}$, is a standard example of a commutative noncentral ideal of $\mathcal{A} = M_2(\mathbb{F})$. Of course $\mathcal{L}$ is not 1-free, although $\mathcal{A}$ is 2-free (as a subset of itself).

Let us mention another yet more simple example illustrating Corollary 5.16. Again let $\mathbb{F}$ be a field (of arbitrary characteristic), $\mathcal{A} = M_2(\mathbb{F})$, and let $\mathcal{L}$ be the set of all matrices in $\mathcal{A}$ with trace 0. Clearly $\mathcal{L}$ is a noncommutative Lie ideal of $\mathcal{A}$, $\mathcal{A}$ is 2-free, $\mathcal{L}$ is 1-free but not 2-free. The latter follows from the fact that $ax + xa$ lies in the center of $\mathcal{A}$ for all $a, x \in \mathcal{L}$.

Corollary 3.17 makes it possible for us to obtain a similar result for Jordan ideals. However, since, at least when $\operatorname{char}(\mathcal{A}) \neq 2$, a nonzero Jordan ideal always contains a nonzero ideal [114, Theorem 1.1], this result would be, in view of Corollary 5.14, hardly interesting.

We now turn to rings with involution. Recall that $\mathcal{S} = \mathcal{S}(\mathcal{A})$ (resp. $\mathcal{K} = \mathcal{K}(\mathcal{A})$) denotes the set of all symmetric (resp. skew) elements in $\mathcal{A}$. As a corollary to Theorems 5.11 and 3.28 we have

Corollary 5.17. *Let $\mathcal{A}$ be a prime ring with involution. If there exists $t \in \mathcal{S} \cup \mathcal{K}$ such that $\deg(t) \geq 2d + 1$, then $\mathcal{S}$ and $\mathcal{K}$ are $(t; d)$-free subsets of $\mathcal{Q}$.*

This corollary together with Lemma C.6 gives

Corollary 5.18. *Let $\mathcal{A}$ be a prime ring with involution. If $\operatorname{char}(\mathcal{A}) \neq 2$ and $\deg(\mathcal{A}) \geq 2d + 1$, then $\mathcal{S}$ and $\mathcal{K}$ are d-free subsets of $\mathcal{Q}$.*

Finally we consider Lie ideals of $\mathcal{K}$ (the consideration of Jordan ideals of $\mathcal{S}$ will be omitted for similar reasons as in the non-involution case). We remark that the proof of the next corollary indirectly uses a number of important results of Chapters 3 and 5.

Corollary 5.19. *Let $\mathcal{A}$ be a prime ring with involution, and let $\mathcal{L}$ be a noncentral Lie ideal of $\mathcal{K}$. If $\mathrm{char}(\mathcal{A}) \neq 2$ and $\deg(\mathcal{A}) \geq 2d+3$, then $\mathcal{L}$ is a d-free subset of $\mathcal{Q}$.*

Proof. Since $2d+3$ is certainly ≥ 5, by Lemma C.6 there exists $t \in \mathcal{L}$ such that $\deg(t) \geq 2d+3 = 2(d+1)+1$. Therefore Corollary 5.17 tells us that $\mathcal{K}$ is $(t; d+1)$-free. Since $[t, \mathcal{K}] \subseteq \mathcal{L}$ we infer from Corollary 3.18 that $\mathcal{L}$ is d-free. $\qquad\square$

We conclude this section by summarizing its most useful results into one statement. Together with Corollary 3.5 these results yield the following.

A LIST OF d-FREE SETS: *Let $\mathcal{A}$ be a prime ring (with involution) such that $\mathrm{char}(\mathcal{A}) \neq 2$ (this assumption is not needed in (a)), and let $\mathcal{R}$ be a subset of $\mathcal{Q} = \mathcal{Q}_{ml}(\mathcal{A})$. Then $\mathcal{R}$ is a d-free subset of $\mathcal{Q}$ if one of the following conditions holds:*

(a) $\deg(\mathcal{A}) \geq d$ *and $\mathcal{R}$ contains a nonzero ideal of $\mathcal{A}$;*

(b) $\deg(\mathcal{A}) \geq d+1$ *and $\mathcal{R}$ contains a noncentral Lie ideal of $\mathcal{A}$;*

(c) $\deg(\mathcal{A}) \geq 2d+1$ *and $\mathcal{R}$ contains $\mathcal{S}$;*

(d) $\deg(\mathcal{A}) \geq 2d+1$ *and $\mathcal{R}$ contains $\mathcal{K}$;*

(e) $\deg(\mathcal{A}) \geq 2d+3$ *and $\mathcal{R}$ contains a noncentral Lie ideal of $\mathcal{K}$.*

5.3 *d*-**Freeness of Semiprime Rings**

We turn now to the more involved problem of trying to show that semiprime rings are $(t; d)$-free, or, at least d-free. We keep the notation of the preceding section: $\mathcal{A}$ will be a semiprime ring, $\mathcal{Q} = \mathcal{Q}_{ml}(\mathcal{A})$ its maximal left ring of quotients, and $\mathcal{C}$ its extended centroid. Further, $f\text{-}\deg(.)$ stands for $f\text{-}\deg_{\mathcal{Q}}(.)$ and $\mathrm{Id}(.)$ for $\mathrm{Id}_{\mathcal{Q}}(.)$. Also, recall that $\ell(\mathcal{T}; \mathcal{S})$ (resp. $r(\mathcal{T}; \mathcal{S})$) denotes the left (resp. right) annihilator of $\mathcal{S}$ in $\mathcal{T}$.

We remark that the fractional degree of an element t, as introduced in Definition 5.1, does not depend only on the ring $\mathcal{Q}$ but also on the ring $\mathcal{A}$ whose idealizer contains t. In the sequel we shall consider the fractional degree of a certain element t with respect to different semiprime rings whose maximal left ring of quotients are equal to $\mathcal{Q}$ (specifically, for the orthogonal completion $\mathcal{O}$ of $\mathcal{A}$ (see Appendix B), and for an essential ideal $\mathcal{L}$ of $\mathcal{A}$). It will be clear from the context which setting we have in mind.

In contrast to the prime ring case the notion of the degree of an element over $\mathcal{C}$ is no longer easy to work with, since $\mathcal{C}$ is no longer a field. Recall, however, the result for a prime ring $\mathcal{A}$ that says that there exists an element t of degree

$> n$ (possibly infinite) if and only if $\mathcal{A}$ does not satisfy the standard polynomial identity of degree $2n$ (Theorem C.2). For semiprime rings this latter concept is easier to work with. We denote by St_{2n} the standard polynomial of degree $2n$ and, for any $\mathcal{T} \subseteq \mathcal{A}$, we denote by $(St_{2n})_{\mathcal{T}}$ the ideal of $\mathcal{A}$ generated by the set $\{St_{2n}(\overline{x}_{2n}) \,|\, \overline{x}_{2n} \in \mathcal{T}^{2n}\}$. We begin with two simple lemmas that are not of great importance for our goals, but they describe the framework in which we shall be working.

Lemma 5.20. *The ideal $(St_{2n})_{\mathcal{A}}$ is essential in $\mathcal{A}$ if and only if no nonzero ideal of $\mathcal{A}$ satisfies St_{2n}.*

Proof. Clearly $\mathcal{I} = \ell(\mathcal{A}; (St_{2n})_{\mathcal{A}})$ is an ideal of $\mathcal{A}$, and note that $((St_{2n})_{\mathcal{I}})^2 = 0$. Since $\mathcal{I}$ is also semiprime it follows that $(St_{2n})_{\mathcal{I}} = 0$, that is $\mathcal{I}$ satisfies St_{2n}. This proves the "if" part. To prove the converse, assume that $\mathcal{J}$ is a nonzero ideal of $\mathcal{A}$ satisfying St_{2n}. Then $\mathcal{J}$ is a semiprime PI-ring and so $\mathcal{J}$ has a nonzero center (Theorem C.4). Pick a nonzero element c from the center of $\mathcal{J}$. Actually c lies in the center of $\mathcal{A}$. Indeed, given $u \in \mathcal{J}$ and $x \in \mathcal{A}$ we have $[c, ux] = 0$, and hence, since c commutes with u, $u[c, x] = 0$; that is, $\mathcal{J}[c, x] = 0$ which clearly yields $[c, x] = 0$. We have $St_{2n}(cx_1, cx_2, \dots, cx_{2n}) = 0$ for all $x_i \in \mathcal{A}$. However, since c is lies in the center of $\mathcal{A}$ this can be written as $c^{2n} St_{2n}(\overline{x}_{2n}) = 0$. Nonzero central elements in semiprime rings are not nilpotent, and so this shows that $(St_{2n})_{\mathcal{A}}$ is not essential. $\qquad\square$

If $\mathcal{A}$ is prime, then Lemma 5.20 basically says that we have just two possibilities: either $\mathcal{A}$ satisfies St_{2n} or no nonzero ideal of $\mathcal{A}$ satisfies St_{2n}. This is of course not the case in the semiprime context since here we can take direct sums.

Lemma 5.21. *Let $n \in \mathbb{N}$. Then there exists an idempotent $e \in C$ such that the ring $(1 - e)\mathcal{A}$ satisfies St_{2n} and the ring $e\mathcal{A}$ does not contain nonzero ideals satisfying St_{2n}.*

Proof. Let $e = E((St_{2n})_{\mathcal{A}})$ (see Lemma B.1). Then

$$St_{2n}((1 - e)x_1, (1 - e)x_2, \dots, (1 - e)x_{2n}) = (1 - e)St_{2n}(\overline{x}_{2n}) = 0$$

for all $x_i \in \mathcal{A}$. Suppose there was a nonzero ideal of $e\mathcal{A}$ satisfying St_{2n}. By Lemma 5.20, applied to $e\mathcal{A}$, then there would exist a nonzero $a \in \ell(e\mathcal{A}; (St_{2n})_{e\mathcal{A}})$. Note that then $a \in \ell(e\mathcal{A}; (St_{2n})_{\mathcal{A}})$. However, from Lemmas A.3 and B.1 we see that $\ell(e\mathcal{A}; (St_{2n})_{\mathcal{A}})$ is contained in $(1 - e)\mathcal{Q}$. Since $a = ea \neq 0$, this is impossible. $\qquad\square$

The situation described in this lemma becomes particularly nice if $\mathcal{A}$ is centrally closed; in this case we have that $\mathcal{A}$ is the direct sum of two ideals, one of them satisfying St_{2n} and another one having no nonzero ideals satisfying St_{2n}.

We now focus our attention on the situation when $\mathcal{A}$ satisfies the conditions of Lemma 5.20 (i.e., the situation when $e = 1$ in Lemma 5.21). Our aim is to prove that $\mathcal{A}$ is $(n + 1)$-free (it does not appear easy to find an appropriate $t \in \mathrm{Id}(\mathcal{A})$ such that $\mathcal{A}$ is $(t; n+1)$-free). The method of proof we will use is based heavily on the theory of "orthogonal completions", and so at this point we strongly urge the reader to become familiar with appendix B, where we define the relevant notions and prove (in a self-contained way) those results which we require for the proofs of the results which follow. We shall therefore feel free to make use of these notions and results without detailed comment.

We start, then, with a semiprime ring $\mathcal{A}$ and a positive integer n such that $(St_{2n})_\mathcal{A}$ is an essential ideal of $\mathcal{A}$. We let $\mathcal{O}$ denote the orthogonal completion of $\mathcal{A}$ in $\mathcal{Q}$. Most of our work will take place in $\mathcal{O}$. Further, let $\mathcal{B}$ denote the set of all idempotents in $\mathcal{C}$. We note that $\mathcal{B}$ is a Boolean ring under the new addition $e \oplus f = e + f - 2ef$ and same multiplication, and that the following defines a partial order in $\mathcal{B}$: $e \leq f$ if $e = ef$. We will first concentrate our attention on a fixed maximal ideal $\mathcal{M}$ of the Boolean ring $\mathcal{B}$, i.e., $\mathcal{M} \in Spec(\mathcal{B})$, and form the ideal $\mathcal{O}\mathcal{M}$ of $\mathcal{O}$. By Theorem B.9 we know that $\mathcal{O}_\mathcal{M} = \mathcal{O}/\mathcal{O}\mathcal{M}$ is a prime ring.

At this point let us digress for a moment to give the reader a rough idea of the path ahead. To temporarily simplify matters let us assume that $\mathcal{A}$ is already orthogonally complete, i.e., $\mathcal{A} = \mathcal{O}$. Then the idea is to show that the $f\text{-}\deg(\mathcal{A}) > n$, and so by Theorem 5.6 $\mathcal{A}$ is $(t; n+1)$-free for some $t \in \mathcal{A}$. For each $\mathcal{M} \in Spec(\mathcal{B})$ it is first shown that the corresponding prime ring $\mathcal{A}/\mathcal{A}\mathcal{M}$ has fractional degree $> n$, and then using this it is shown that there is an idempotent $w = w_\mathcal{M} \in \mathcal{B} \setminus \mathcal{M}$ such that $f\text{-}\deg(w\mathcal{A}) > n$. One goes on to show that $\mathcal{A} = w_1 \mathcal{A} + w_2 \mathcal{A} + \ldots + w_k \mathcal{A}$ for a finite number of such w's, and finally one replaces these w_i's by orthogonal idempotents $e_1, \ldots, e_k$ whose sum is 1. Since the fractional degree of each ideal $e_i \mathcal{A}$ is $> n$ it follows in an obvious way that the fractional degree of $\mathcal{A}$ itself is $> n$.

In general, of course, we are not assuming that $\mathcal{A}$ is orthogonally complete, so we do the natural thing by copying the above process for the orthogonal completion $\mathcal{O}$ of $\mathcal{A}$. A minor problem arises since the various (but finite number of) elements appearing in the definition of $f\text{-}\deg(t)$ lie in $\mathcal{O}$ rather than in $\mathrm{Id}(\mathcal{A})$. Loosely speaking, these elements are "connected" to $\mathcal{A}$ by various idempotents in $\mathcal{B}$, and these idempotents lead us to an essential ideal $\mathcal{L}$ of $\mathcal{A}$ which again has the fractional degree $> n$. Thus $\mathcal{L}$ is $(t; n + 1)$-free, whence $\mathcal{L}$ is $(n + 1)$-free, and finally $\mathcal{A}$ is $(n + 1)$-free by Corollary 3.5.

We return now to our consideration of an arbitrary but fixed $\mathcal{M} \in Spec(\mathcal{B})$, where we have noted that $\mathcal{O}_\mathcal{M} = \mathcal{O}/\mathcal{O}\mathcal{M}$ is a prime ring. By $\mathcal{C}(\mathcal{O}_\mathcal{M})$ we denote its extended centroid. An element $x + \mathcal{O}\mathcal{M} \in \mathcal{O}_\mathcal{M}$ will frequently be denoted by $\bar{x}$.

In what follows we assume that $\mathcal{A}$ satisfies the conditions of Lemma 5.20.

Since the intersection of any nonzero ideal of $\mathcal{O}$ with $\mathcal{A}$ is again a nonzero ideal of $\mathcal{A}$, it is clear that $\mathcal{O}$ also satisfies the conditions of Lemma 5.20.

Lemma 5.22. $\mathcal{O}_\mathcal{M}$ *does not satisfy* St_{2n} *(and hence there exists an element* $t \in \mathcal{O}$ *such that the degree of* $\bar{t} \in \mathcal{O}_\mathcal{M}$ *over* $\mathcal{C}(\mathcal{O}_\mathcal{M})$ *is* $> n$*).*

Proof. Suppose that $\mathcal{O}_\mathcal{M}$ satisfies St_{2n}, that is the set

$$\mathcal{T} = \{St_{2n}(x_1, x_2, \ldots, x_n) \,|\, x_i \in \mathcal{O}\} \subseteq \mathcal{O}\mathcal{M}.$$

We claim that $\mathcal{T}$ is orthogonally complete. Indeed, let $\mathcal{U}$ be a dense orthogonal subset of $\mathcal{B}$ and let $\{St_{2n}(x_{1u}, x_{2u} \dots, x_{2nu}) \mid u \in \mathcal{U}\} \subseteq \mathcal{T}$. Since $\mathcal{O}$ is orthogonally complete, for each $i = 1, 2, \dots, 2n$ there exists $x_i \in \mathcal{O}$ such that $x_i u = x_{iu} u$ for every $u \in \mathcal{U}$. Then it is easily seen that $St_{2n}(x_1, \dots, x_{2n})$ is the desired element of $\mathcal{T}$. By Lemma B.7 (ii) there exists $t \in \mathcal{T}$ such that $E(\mathcal{T}) = E(t)$, whence $E(\mathcal{T}) \in \mathcal{M}$ by Lemma B.8. Therefore $e = 1 - E(\mathcal{T}) \notin \mathcal{M}$ and we have $e\mathcal{T} = 0$. Consequently, $e(St_{2n})_\mathcal{A} = 0$ which contradicts our assumption that the conditions of Lemma 5.20 are fulfilled. Thus $\mathcal{O}_\mathcal{M}$ does not satisfy St_{2n} and so Theorem C.2 tells us that some $\bar{t} \in \mathcal{O}_\mathcal{M}$ has degree over $\mathcal{C}(\mathcal{O}_\mathcal{M})$ greater than n. $\qquad \square$

For each $\mathcal{M} \in Spec(\mathcal{B})$ we fix $t = t_\mathcal{M} \in \mathcal{O}$ such that $\bar{1}, \bar{t}, \dots, \bar{t}^n$ are linearly independent over $\mathcal{C}(\mathcal{O}_\mathcal{M})$.

Lemma 5.23. *Let $\mathcal{M} \in Spec(\mathcal{B})$. Then there exists $w = w_\mathcal{M} \in \mathcal{B} \setminus \mathcal{M}$ such that for every $0 \le i \le n$ there exist $a_{ij}, b_{ij} \in \mathcal{O}$ having the following properties:*

(a) $\sum_{j=1}^{m_i} w a_{ij} t^k b_{ij} = 0$ *if $k \ne i$, $0 \le k \le n$;*

(b) $\ell(w\mathcal{O}; \mathcal{O}s_i) = 0$ *where $s_i = \sum_{j=1}^{m_i} a_{ij} t^i b_{ij}$.*

Proof. Pick $0 \le i \le n$. Since $\mathcal{O}_\mathcal{M}$ is prime and $\bar{1}, \bar{t}, \dots, \bar{t}^n$ are linearly independent, by Theorem A.8 there exist $a_{ij}, b_{ij} \in \mathcal{O}$, $1 \le j \le m_i$, such that

(a') $\sum_{j=1}^{m_i} \bar{a}_{ij} \bar{t}^k \bar{b}_{ij} = \bar{0}$ *if $k \ne i$, $0 \le k \le n$;*

(b') $\sum_{j=1}^{m_i} \bar{a}_{ij} \bar{t}^i \bar{b}_{ij} \ne \bar{0}$.

Fix $k \ne i$, set $r_k = \sum_{j=1}^{m_i} a_{ij} t^k b_{ij}$, and let $e_k = 1 - E(r_k)$. Clearly, if $v \le e_k$ we have $vr_k = 0$. From (a') we see that $r_k \in \mathcal{O}\mathcal{M}$, and so by Lemma B.8 we have $E(r_k) \in \mathcal{M}$, whence $e_k \notin \mathcal{M}$. Next, we set $s_i = \sum_{j=1}^{m_i} a_{ij} t^i b_{ij}$ and note from (b') that $s_i \notin \mathcal{O}\mathcal{M}$. Therefore $e_i = E(s_i) \notin \mathcal{M}$ by Lemma B.8. Suppose that $x \in \mathcal{O}$ is such that $e_i x \mathcal{O} s_i = 0$. In view of Lemmas A.3 and B.1 this yields $e_i x \in (1 - e_i)Q$, and so $e_i x = 0$. Thus we proved that $\ell(e_i \mathcal{O}; \mathcal{O}s_i) = 0$. Now set $w_i = e_0 e_1 \dots e_n \notin \mathcal{M}$ and note that $w_i r_k = 0$, and, since $w_i \le e_i$, $\ell(w_i \mathcal{O}; \mathcal{O}s_i) = 0$. Consequently $w = w_0 w_1 \dots w_n \notin \mathcal{M}$ satisfies the conditions (a) and (b). $\qquad \square$

Let us add to the above statement that without loss of generality we may assume that for each $\mathcal{M}$ the upper limits m_i are all equal $m_\mathcal{M}$ (just add zeros in appropriate places if necessary).

Lemma 5.24. $f\text{-}\deg(\mathcal{O}) > n$.

Proof. For each $\mathcal{M} \in Spec(\mathcal{B})$ we have the idempotent $w_\mathcal{M} \in \mathcal{B} \setminus \mathcal{M}$, provided by Lemma 5.23. By Lemma B.10 there exist $\mathcal{M}_1, \dots, \mathcal{M}_q \in Spec(\mathcal{B})$ and pairwise orthogonal idempotents $e_1, \dots, e_q \in \mathcal{B}$ whose sum is 1 such that $e_p \le w_p = w_{\mathcal{M}_p}$ for $p = 1, 2, \dots, q$. Similarly as above we may assume without loss of generality that $m_{\mathcal{M}_1} = \dots = m_{\mathcal{M}_q} = m$. Set $t_p = t_{\mathcal{M}_p}$. Now, for any fixed $1 \le p \le q$ and fixed $0 \le i \le n$ there are $a_{ijp}, b_{ijp} \in \mathcal{O}$, $j = 1, \dots, m$, which in view of $e_p \le w_p$ satisfy

(a) $\sum_{j=1}^{m} e_p a_{ijp} t_p^k b_{ijp} = 0$ if $k \neq i$, $0 \leq k \leq n$;

(b) $\ell(e_p \mathcal{O}; \mathcal{O} s_{ip}) = 0$ where $s_{ip} = \sum_{j=1}^{m} a_{ijp} t_p^i b_{ijp}$.

We define

$$t = \sum_{p=1}^{q} e_p t_p, \quad a_{ij} = \sum_{p=1}^{q} e_p a_{ijp}, \quad b_{ij} = \sum_{p=1}^{q} e_p b_{ijp}.$$

Note that $t, a_{ij}, b_{ij} \in \mathcal{O}$ by Lemma B.7 (i). By the orthogonality of the e_p's it is clear from (a) that $\sum_{j=1}^{m} a_{ij} t^k b_{ij} = 0$ if $k \neq i$. Suppose that $x \in \mathcal{O}$ is such that $x\mathcal{O}\left(\sum_{j=1}^{m} a_{ij} t^i b_{ij}\right) = 0$. Multiplying this identity by e_p we get $e_p x \mathcal{O} s_{ip} = 0$ and hence $e_p x = 0$ by (b). Since the sum of e_p's is 1 it follows that $x = 0$. In view of Lemma 5.7 (and Theorem A.4) we may conclude that $\sum_{j=1}^{m} a_{ij} t^i b_{ij}$ is fractionable in $\mathcal{Q}$. Accordingly, $f\text{-}\deg(t) > n$. $\square$

Lemma 5.25. *There exists an essential ideal $\mathcal{L}$ of $\mathcal{A}$ such that $f\text{-}\deg(\mathcal{L}) > n$.*

Proof. So far we know (from the proof of Lemma 5.24) that there are $t, a_{ij}, b_{ij} \in \mathcal{O}$, $0 \leq i \leq n$, $1 \leq j \leq m$, such that for every i we have

(a) $\sum_{j=1}^{m} a_{ij} t^k b_{ij} = 0$ if $k \neq i$;

(b) $\ell(\mathcal{O}; \mathcal{O} s_i) = 0$ where $s_i = \sum_{j=1}^{m} a_{ij} t^i b_{ij}$.

By Lemma B.6 there exist dense orthogonal subsets $\mathcal{U}, \mathcal{V}_{ij}, \mathcal{W}_{ij}$ of $\mathcal{B}$ and elements $t_u, a_{ijv}, b_{ijw} \in \mathcal{A}$ such that $t = \sum_{u \in \mathcal{U}}^{\perp} t_u u$, $a_{ij} = \sum_{v \in \mathcal{V}_{ij}}^{\perp} a_{ijv} v$, and $b_{ij} = \sum_{w \in \mathcal{W}_{ij}}^{\perp} b_{ijw} w$. Then

$$\mathcal{Z} = \mathcal{U} \prod_{\substack{0 \leq i \leq n \\ 1 \leq j \leq m}} \mathcal{V}_{ij} \mathcal{W}_{ij}$$

is again a dense orthogonal subset of $\mathcal{B}$ (Lemma B.3). Given $z \in \mathcal{Z}$ we set $\mathcal{L}_z = \{x \in \mathcal{A} \mid zx \in \mathcal{A}\}$. Clearly $z\mathcal{L}_z$ is an ideal of $\mathcal{A}$ and moreover, $\mathcal{L} = \sum_{z \in \mathcal{Z}} z\mathcal{L}_z$ is an essential ideal of $\mathcal{A}$ by Lemma B.4. We claim that $t, a_{ij}, b_{ij} \in \mathrm{Id}(\mathcal{L})$. Let us show, for example, that $t\mathcal{L} \subseteq \mathcal{L}$ (the other conditions can be checked similarly). It suffices to show that for any given $z \in \mathcal{Z}$ and $x \in \mathcal{L}_z$ we have $tzx \in \mathcal{L}$. Clearly there exists $u_0 \in \mathcal{U}$ such that $u_0 z = z$. Therefore

$$tzx = (\sum_{u \in \mathcal{U}}^{\perp} t_u u) u_0 z x = t_{u_0} u_0 z x = t_{u_0} z x \in z\mathcal{L}_z \subseteq \mathcal{L},$$

proving our claim. Finally, making use of Lemma A.3 we see that (b) yields $\ell(\mathcal{L}; \mathcal{L} s_i) = 0$. Since the maximal left ring of quotients of $\mathcal{L}$ is $\mathcal{Q}$ (Theorem A.4) we now see that the fractional degree of t (here considered as an element of $\mathrm{Id}(\mathcal{L})$) is indeed $> n$. $\square$

Theorem 5.26. *Let $\mathcal{A}$ be a semiprime ring, and let n be a positive integer such that $\mathcal{A}$ does not contain nonzero ideals satisfying St_{2n}. Then $\mathcal{A}$ is an $(n+1)$-free subset of $\mathcal{Q} = \mathcal{Q}_{ml}(\mathcal{A})$.*

Proof. Lemma 5.25 and Theorem 5.6 tell us that there is an essential ideal $\mathcal{L}$ of $\mathcal{A}$ which is $(t; n+1)$-free for some $t \in \mathrm{Id}(\mathcal{L})$. In particular, $\mathcal{L}$ is $(n+1)$-free (Lemma 3.13), and so $\mathcal{A}$ is $(n+1)$-free by Corollary 3.5. $\square$

We remark that applying Theorem 3.14 instead of Corollary 3.5 we get that $\mathcal{A}$ is actually $(t; n+1)$-free, however with t lying in $\mathrm{Id}(\mathcal{L})$ rather than in $\mathrm{Id}(\mathcal{A})$.

If $\mathcal{A}$ is prime, then the condition that $\mathcal{A}$ does not contain nonzero ideals satisfying St_{2n} is equivalent to the condition that $\mathcal{A}$ itself does not satisfy St_{2n} (see e.g., Lemma 5.20), and is further equivalent to the condition that $\deg(\mathcal{A}) \geq n+1$ (see Theorem C.2). Thus in this case Theorem 5.26 gives the same assertion as Theorem 5.11 (just take $d = n+1$).

One might consider Theorem 5.26 as a basis for analyzing d-freeness of various subsets of semiprime rings. However, we shall not investigate these somewhat technical questions here, since the results for subsets of prime rings from the preceding section are sufficient for our further purposes in this book.

5.4   Commuting Maps on (Semi)prime Rings

In this section we will consider the problem of characterizing the trace of an n-additive map $T : \mathcal{A} \to \mathcal{Q}$ that is commuting, i.e.,

$$[T(x), x] = 0 \quad \text{for all } x \in \mathcal{A}.$$

We are facing a particular example of an FI, but one that is of special interest, as has been indicated several times so far. In particular the case where $n = 2$ deserves a special attention.

As in the preceding sections we shall be interested in the case where $\mathcal{A}$ is a (semi)prime ring and $\mathcal{Q} = \mathcal{Q}_{ml}(\mathcal{A})$ is its maximal left quotient ring. In practice we are mostly interested in the situation where the range of T lies in $\mathcal{A}$ itself, but the greater level of generality when T maps in $\mathcal{Q}$ will not cause any problems for us (and in fact in view of the arguments it is more natural to work in that setting). By $\mathcal{C}$ we, of course, again denote the extended centroid of $\mathcal{A}$.

Let us first of all record a result that follows immediately from the general theory, more precisely from Corollary 4.16 and Theorem 5.26.

Theorem 5.27. *Let $\mathcal{A}$ be a semiprime ring and let $T : \mathcal{A} \to \mathcal{Q}$ be the trace of an n-additive map, $n \geq 1$. Suppose that T is commuting. If $n!$ is invertible in $\mathcal{C}$ and $\mathcal{A}$ does not contain nonzero ideals satisfying St_{2n}, then there exist traces of i-additive maps $\lambda_i : \mathcal{A} \to \mathcal{C}$, $i = 0, 1, \ldots, n$ such that $T(x) = \sum_{i=0}^{n} \lambda_i(x) x^{n-i}$ for all $x \in \mathcal{A}$.*

For $n = 1$ the assumptions concerning $\mathcal{A}$ can be simply read as that $\mathcal{A}$ does not contain nonzero commutative ideals. But even this assumption is superfluous.

Corollary 5.28. *Let $\mathcal{A}$ be a semiprime ring and let $f : \mathcal{A} \rightarrow \mathcal{Q}$ be an additive commuting map. Then there exist $\lambda \in \mathcal{C}$ and an additive map $\mu : \mathcal{A} \rightarrow \mathcal{C}$ such that $f(x) = \lambda x + \mu(x)$ for all $x \in \mathcal{A}$.*

Proof. By Lemma 5.21 there is an idempotent $e \in \mathcal{C}$ such that the ring $e\mathcal{A}$ does not contain commutative ideals and the ring $(1-e)\mathcal{A}$ is commutative. We remark that the latter implies that $(1 - e)\mathcal{A} \subseteq \mathcal{C}$ since the elements in $(1 - e)\mathcal{A}$ commute with all elements from $\mathcal{A}$, and for a similar reason this further yields that $(1-e)\mathcal{Q} \subseteq \mathcal{C}$.

Let $\mathcal{L} = \mathcal{L}_e$ be the ideal of $\mathcal{A}$ from Lemma B.2, i.e., $\mathcal{L}$ consists of all elements $x \in \mathcal{A}$ such that $ex \in \mathcal{A}$. Just using the facts that $\mathcal{L}$ is an ideal and $e\mathcal{A}$ does not contain nonzero commutative ideals, it is easy to see that $e\mathcal{L}$, which is clearly a semiprime ring, also does not contain nonzero commutative ideals. Now define $f_e : e\mathcal{L} \rightarrow e\mathcal{Q} = \mathcal{Q}_{ml}(e\mathcal{L})$ (see Lemma B.2) by $f_e(ex) = ef(ex)$. Note that f_e is an additive commuting map, and moreover, all the conditions of Theorem 5.27 are fulfilled. The extended centroid of $e\mathcal{L}$ is equal to $e\mathcal{C}$ (this also follows from Lemma B.2) and so there exists $\lambda \in e\mathcal{C} \subseteq \mathcal{C}$ such that $f_e(ex) - \lambda ex \in e\mathcal{C} \subseteq \mathcal{C}$ for all $x \in \mathcal{L}$; that is, $e(f(ex) - \lambda x) \in \mathcal{C}$.

Linearizing $[f(x), x] = 0$ we get $[f(x), y] = [x, f(y)]$ for all $x, y \in \mathcal{A}$. This clearly implies that f maps the center of $\mathcal{A}$ into $\mathcal{C}$; in particular, $f((1 - e)\mathcal{L}) \subseteq \mathcal{C}$. Accordingly,

$$e(f(x) - \lambda x) = e(f(ex) - \lambda x) + ef((1 - e)x) \subseteq \mathcal{C}$$

for every $x \in \mathcal{L}$. Since $(1 - e)\mathcal{Q} \subseteq \mathcal{C}$ this clearly implies that in fact $f(x) - \lambda x \in \mathcal{C}$ for every $x \in \mathcal{L}$. Again using $[f(x), y] = [x, f(y)]$ it follows that $[x, f(y) - \lambda y] = 0$ for all $x \in \mathcal{L}$ and all $y \in \mathcal{A}$. Since $\mathcal{L}$ is an essential ideal of $\mathcal{A}$ (see Lemma B.2) it follows that $\mu(y) = f(y) - \lambda y \in \mathcal{C}$ for all $y \in \mathcal{A}$. $\square$

The proof of Corollary 5.28 demonstrates how the machinery that we have developed works. It should be mentioned that there are more direct and simpler approaches available to prove this result. Perhaps the most suitable one is via the notion of a biderivation: one has to argue similarly as when treating Example 1.5, and then apply a certain version of Theorem A.7 for semiprime rings.

Let us mention that the appearance of the extended centroid in Corollary 5.28 is necessary even in the case when one assumes that f maps into $\mathcal{A}$, that is, in this case we cannot, in general, replace the role of the extended centroid by the center (or at least by the ordinary centroid in case $\mathcal{A}$ is not unital). Of course in many important cases the center coincides with the extended centroid, for example this is true for simple unital rings. In the following example the ring in question is in some sense close to a simple one, but still the involvement of the extended centroid is necessary.

Example 5.29. Let $\mathcal{V}$ be an infinite dimensional vector space over a field $\mathcal{C}$ and let $\mathcal{A}_0$ be the algebra of all finite rank $\mathcal{C}$-linear operators on $\mathcal{V}$. Incidentally we remark that $\mathcal{A}_0$ is a simple ring with trivial center. Now let $\mathcal{Z}$ be any proper unital subring of $\mathcal{C}$, and consider the ring consisting of all linear operators of the form $a_0 + z$ where $a_0 \in \mathcal{A}_0$ and $z \in \mathcal{Z}$; here, z is identified by the corresponding scalar operator $v \mapsto zv$. Clearly $\mathcal{A}$ is a unital ring with center $\mathcal{Z}$, and one can check that the extended centroid of $\mathcal{A}$ is $\mathcal{C}$. Let $\lambda \in \mathcal{C} \setminus \mathcal{Z}$ and define $f : \mathcal{A} \to \mathcal{A}$ by $f(a_0 + z) = \lambda a_0$. Then f is an additive commuting map, and obviously there is no $\alpha \in \mathcal{Z}$ such that $f(a) - \alpha a \in \mathcal{Z}$ for all $a \in \mathcal{A}$. On the other hand, clearly $f(a) - \lambda a \in \mathcal{C}$ for all $a \in \mathcal{A}$. More precisely, we can write f as $f(a) = \lambda a + \mu(a)$ where $\mu : \mathcal{A} \to \mathcal{C}$ is defined by $\mu(a_0 + z) = -\lambda z$.

From now on we confine ourselves to the case where $\mathcal{A}$ is prime. Let us first restate Theorem 5.27 for this case. The condition that $\mathcal{A}$ does not contain nonzero ideals satisfying St_{2n} now transforms into the condition that $\deg(\mathcal{A}) \geq n + 1$, and the condition that $n!$ is invertible in $\mathcal{C}$ transforms into the condition that $\mathrm{char}(\mathcal{C}) = 0$ or $\mathrm{char}(\mathcal{C}) \geq n + 1$. Since $\mathcal{C}$ and $\mathcal{A}$ have the same characteristic, we may state

Corollary 5.30. *Let $\mathcal{A}$ be a prime ring and let $T : \mathcal{A} \to \mathcal{Q}$ be the trace of an n-additive map, $n \geq 1$. Suppose that T is commuting. If $\deg(\mathcal{A}) \geq n + 1$ and either $\mathrm{char}(\mathcal{A}) = 0$ or $\mathrm{char}(\mathcal{A}) \geq n + 1$, then there exist traces of i-additive maps $\lambda_i : \mathcal{A} \to \mathcal{C}$, $i = 0, 1, \ldots, n$ such that $T(x) = \sum_{i=0}^{n} \lambda_i(x) x^{n-i}$ for all $x \in \mathcal{A}$.*

Let us mention that using various results in Section 5.2 one can get similar characterizations of commuting traces of multiadditive maps on various subsets of prime rings. We shall not state them explicitly since it is obvious what they give in connection with Corollary 4.16. Just as an example, which however is motivated by the classical Lie isomorphism problem (see Section 1.4), we mention that commuting traces of 3-additive maps on $\mathcal{K}$, the set of skew elements of a prime ring $\mathcal{A}$ with involution, are of standard form $x \mapsto \sum_{i=0}^{3} \lambda_i(x) x^{n-i}$ provided that $\deg(\mathcal{A}) \geq 9$ and $\mathrm{char}(\mathcal{A}) \neq 2, 3$. This follows immediately from Corollaries 4.16 and 5.18.

Are the assumptions concerning $\mathcal{A}$ in Corollary 5.30 really necessary? We shall not bother with assumptions concerning the characteristic, but what about the assumption on $\deg(\mathcal{A})$? One can ask a similar question with respect to Theorem 5.27. As we saw, at least for $n = 1$ the assumption that $\mathcal{A}$ does not contain nonzero commutative ideals (i.e., ideals satisfying St_2) can be removed. What about for an arbitrary n? It is rather typical of FI theory that questions like this arise. Theorem 5.27 and Corollary 5.30 were derived as by-products of the general theory. On the one hand we have thereby nicely illustrated the power of the general theory; on the other hand, however, we have also faced its limitations. The theory just stops working in rings of small degrees. We know of course that there are good reasons for this (see the "only if" part of Corollary 5.12), but when considering some special FI's these restrictions may no longer be justifiable. In particular, there does not seem to exist a good reason for assuming that $\deg(\mathcal{A})$ should be $\geq n + 1$ in Corollary 5.30. The problem whether this assumption can be removed is still open. Our main goal in the remainder of this section is to solve this problem for the important special case where $n = 2$.

Our next result basically shows that in any case, also when $\deg(\mathcal{A}) \leq n$, we can represent T as $T(x) = \sum_{i=0}^{n} \lambda_i(x) x^{n-i}$. However, unfortunately we are unable to show that each λ_i is the trace of an i-additive map.

Corollary 5.31. *Let $\mathcal{A}$ be a prime ring and let $T : \mathcal{A} \to \mathcal{Q}$ be the trace of an n-additive map, $n \geq 1$. If T is commuting and either $\mathrm{char}(\mathcal{A}) = 0$ or $\mathrm{char}(\mathcal{A}) \geq n+1$, then $T(x) \in \sum_{j=0}^{n} \mathcal{C}x^j$ for all $x \in \mathcal{A}$.*

Proof. In view of Corollary 5.30 we may assume that $\deg(\mathcal{A}) \leq n$.

By assumption there is an n-additive map $B : \mathcal{A}^n \to \mathcal{Q}$ such that $T(x) = B(x, x, \ldots, x)$ for every $x \in \mathcal{A}$. Without loss of generality we may assume that B is symmetric, i.e., $B(x_{\pi(1)}, x_{\pi(2)}, \ldots, x_{\pi(n)}) = B(x_1, x_2, \ldots, x_n)$ for every permutation $\pi \in S_n$. Namely, we may replace B by the map

$$(x_1, x_2, \ldots, x_n) \mapsto \frac{1}{n!} \sum_{\pi \in S_n} B(x_{\pi(1)}, x_{\pi(2)}, \ldots, x_{\pi(n)}),$$

which is symmetric and has T as its trace. A linearization of $[B(x, x, \ldots, x), x] = 0$ now gives

$$\sum_{i=1}^{n+1} [B(x_1, \ldots, x_{i-1}, x_{i+1}, \ldots, x_n), x_i] = 0$$

for all $x_i \in \mathcal{A}$. Setting $x_1 = \ldots = x_n = x$ and $x_{n+1} = y$ it follows that

$$n[B(y, x, \ldots, x), x] + [B(x, x, \ldots, x), y] = 0$$

for all $x, y \in \mathcal{A}$. We now fix a nonzero $x \in \mathcal{A}$, and set $q = B(x, x, \ldots, x)$. The last identity can now be written as

$$n[f(y), x] + [q, y] = 0 \quad \text{for all } y \in \mathcal{A},$$

where $f(y) = [B(y, x, \ldots, x), x]$.

By our assumption we have $\deg(x) = s \leq n$. Thus $\sum_{i=0}^{s} \alpha_i x^i = 0$ for some $\alpha_i \in \mathcal{C}$ with $\alpha_s = 1$. Hence we have

$$\sum_{i=1}^{s} \alpha_i [f(y), x^i] = [f(y), \sum_{i=1}^{s} \alpha_i x^i] = -[f(y), \alpha_0] = 0. \tag{5.11}$$

Since

$$[f(y), x^i] = [f(y), x] x^{i-1} + x[f(y), x] x^{i-2} + \ldots + x^{i-1}[f(y), x],$$

and since $[f(y), x] = -\frac{1}{n}[q, y]$ it follows from (5.11), by appropriately collecting the terms and using $[q, x] = 0$, that

$$\sum_{j=0}^{s-1} q_j y x^j = \sum_{j=0}^{s-1} x^j y q_j \quad \text{for all } y \in \mathcal{A}$$

and some $q_j \in C$; these q_j's can be easily explicitly expressed through α_i's and powers of x, but the only thing that matters for us is that $q_{s-1} = q$. Since $1, x, \ldots, x^{s-1}$ are linearly independent over C it follows from Theorem A.7 that each q_j, and so in particular q, is a C-linear combination of $1, x, \ldots, x^{s-1}$. □

Making use of both preceding corollaries we will now be able to obtain the definitive result for $n = 2$. Just an elementary linear algebraic consideration is still needed.

Theorem 5.32. *Let A be a prime ring with* $\mathrm{char}(A) \neq 2$ *and let $T : A \to Q$ be the trace of a biadditive map. If T is commuting, then there exist $\lambda \in C$, an additive map $\mu : A \to C$ and the trace of a biadditive map $\nu : A \to C$ such that $T(x) = \lambda x^2 + \mu(x)x + \nu(x)$ for all $x \in A$. In case $\deg(A) = 2$ we may assume that $\lambda = 0$.*

Proof. In view of Corollary 5.30 we only need to prove the theorem for the case where $\deg(A) \leq 2$. The case where $\deg(A) = 1$, i.e., A is commutative, is trivial (just take $\lambda = 0$, $\mu = 0$ and $\nu = T$). So we may assume that $\deg(A) = 2$. We shall prove the desired conclusion with $\lambda = 0$.

Corollary 5.31 tells us that, for every $x \in A$, $T(x)$ lies in $C + Cx + Cx^2$; however, since x^2 itself is a C-linear combination of 1 and x it follows that $T(x)$ actually lies in $C + Cx$. It suffices to show that there exists an additive map $\mu : A \to C$ such that $T(x) - \mu(x)x \in C$. Indeed, then we define ν by $\nu(x) = T(x) - \mu(x)x$, and clearly ν is the trace of a biadditive map having its range in C.

The proof that follows is similar to the proof of Lemma 3.3 (vii), but necessarily it is somewhat more complicated. For every $x \in A \setminus C$ there is a uniquely determined $\mu(x) \in C$ such that $T(x) - \mu(x)x \in C$. We note that T, as the trace of a biadditive map, satisfies

$$T(x + y) + T(x - y) = 2T(x) + 2T(y) \quad \text{for all } x, y \in A.$$

If x and y are such that none of $x, y, x + y, x - y$ lies in C, then this shows that

$$\mu(x + y)(x + y) + \mu(x - y)(x - y) - 2\mu(x)x - 2\mu(y)y \in C,$$

which in turn implies

$$\mu(x + y) = \mu(x) + \mu(y) \quad \text{if } x, y, 1 \text{ are } C\text{-independent.} \tag{5.12}$$

We next claim that

$$\mu(c + y) - \mu(y) = \mu(c + z) - \mu(z) \quad \text{if } c \in A \cap C, \ y, z \in A \setminus C. \tag{5.13}$$

Indeed, if $z \notin C + Cy$, then we get (5.13) by applying (5.12) for $\mu((c + y) + z) = \mu((c + z) + y)$. So let $z \in C + Cy$. Since A is noncommutative there exists $w \in A$ such that $w \notin C + Cy$, and hence also $w \notin C + Cz$. Therefore, on the one hand we have $\mu(c + w) - \mu(w) = \mu(c + y) - \mu(y)$, and on the other hand we have $\mu(c + w) - \mu(w) = \mu(c + z) - \mu(z)$. Comparing both relations results in (5.13).

From (5.13) we see that μ may be extended to a well-defined map on $\mathcal{A}$ by setting, for every $c \in \mathcal{A} \cap \mathcal{C}$, $\mu(c) = \mu(c+y) - \mu(y)$ for any $y \in \mathcal{A} \backslash \mathcal{C}$. We claim that μ is additive. If $c \in \mathcal{A} \cap \mathcal{C}$ and $y \in \mathcal{A} \backslash \mathcal{C}$, then by definition $\mu(c+y) = \mu(c) + \mu(y)$. If $c, d \in \mathcal{A} \cap \mathcal{C}$, then by definition

$$\begin{aligned} \mu(c+d) &= \mu(c+d+y) - \mu(y) = \mu(c) + \mu(d+y) - \mu(y) \\ &= \mu(c) + \mu(d) + \mu(y) - \mu(y) = \mu(c) + \mu(d). \end{aligned}$$

Therefore, in view of (5.12) it remains to consider only one case: $x, y \in \mathcal{A} \backslash \mathcal{C}$ and $x \in \mathcal{C} + \mathcal{C}y$. Pick $w \notin \mathcal{C} + \mathcal{C}y$. Then $\mu(y+w) = \mu(y) + \mu(w)$ and $\mu(x+y+w) = \mu(x) + \mu(y+w)$ by (5.12). Further, we claim that $\mu(x+y+w) = \mu(x+y) + \mu(w)$. Indeed, if $x+y \in \mathcal{C}$, then this is clear, and if not, then this follows from $w \notin \mathcal{C} + \mathcal{C}(x+y)$. Comparing the last three identities we arrive at $\mu(x+y) = \mu(x) + \mu(y)$. Thus μ is additive. $\square$

Remark 5.33. If $\mathcal{A}$ is a $\mathcal{C}$-algebra and T is the trace of a bilinear map, then a standard argument shows that μ is linear and ν is the trace of a bilinear map.

5.5 Generalized Functional Identities

Throughout this section $\mathcal{A}$ will be a prime ring with maximal left quotient ring $\mathcal{Q} = \mathcal{Q}_{ml}(\mathcal{A})$ and extended centroid $\mathcal{C}$. We consider $\mathcal{Q}$ as a vector space over $\mathcal{C}$.

We have seen that if we have a basic functional identity on $\mathcal{A}$:

$$\sum_{i \in \mathcal{I}} E_i(\bar{x}_m^i) x_i + \sum_{j \in \mathcal{J}} x_j F_j(\bar{x}_m^j) = 0, \tag{5.14}$$

then either $\mathcal{A}$ is PI-ring or (5.14) has only the so-called standard solutions. In the former case there is then Posner's theorem which says that the central closure $\mathcal{AC}$ of $\mathcal{A}$ is a finite dimensional central simple algebra (Theorem C.1). It is therefore natural to conjecture that we can obtain an analogous conclusion in case $\mathcal{A}$ satisfies a so-called basic "generalized" functional identity (a notion presently to be made precise). The purpose of this section is to answer this conjecture in the affirmative: either $\mathcal{A}$ is a GPI-ring or the generalized functional identity has only the "obvious" solutions. In the former case there is then Martindale's theorem that says that $\mathcal{AC}$ is a primitive ring with a minimal idempotent e such that $e\mathcal{AC}e$ is a finite dimensional division algebra over $\mathcal{C}$ (Theorem D.1).

We now proceed to define what we mean by a basic *generalized functional identity* (GFI).

Let $m \geq 1$, let $\mathcal{I}$, $\mathcal{J}$ be subsets of $\{1, 2, \ldots, m\}$, let s_i, $i \in \mathcal{I}$ and t_j, $j \in \mathcal{J}$ be given positive integers, and let $\mathcal{V}$ be a finite dimensional subspace of $\mathcal{Q}$. Let $E_{iu}, F_{jv} : \mathcal{A}^{m-1} \to \mathcal{Q}$ be maps such that

$$\sum_{i \in \mathcal{I}} \sum_{u=1}^{s_i} E_{iu}(\bar{x}_m^i) x_i a_{iu} + \sum_{j \in \mathcal{J}} \sum_{v=1}^{t_j} b_{jv} x_j F_{jv}(\bar{x}_m^j) \in \mathcal{V} \tag{5.15}$$

for all $x_m \in \mathcal{A}$. But this is a contradiction to $\mathcal{A}$ being non-GPI. Namely, (5.19) implies that there is a nonzero element of the multiplication ring of $\mathcal{A}$ (see Theorem A.7) which maps $\mathcal{AC}$ into the finite dimensional space $b\mathcal{V}$; by Theorem A.9 this is impossible.

Now suppose $|\mathcal{I}| > 1$. Again without loss of generality it suffices to show that $E_{m1} = 0$, and so we may assume that each $a_{mu} \in \mathcal{A}$.

First suppose $s_m = 1$ and set $a = a_{m1} \neq 0$. By Lemma D.2 (ii) there exists $y \in \mathcal{A}$ such that the set

$$\{a_{i1}, a_{i2}, \ldots, a_{is_i}, a_{i1}ya, a_{i2}ya, \ldots, a_{is_i}ya\}$$

is $\mathcal{C}$-independent for all $i \neq m$. Set

$$K(\bar{x}_m) = H(\bar{x}_{m-1}, x_m ay) - H(\bar{x}_m)ya.$$

It is clear that, setting $\mathcal{I}' = \mathcal{I} \setminus \{m\}$, we have

$$K(\bar{x}_m) = \sum_{i \in \mathcal{I}'} \sum_{u=1}^{s_i} G_{iu}(\bar{x}_m^i)x_i a_{iu} - \sum_{i \in \mathcal{I}'} \sum_{u=1}^{s_i} E_{iu}(\bar{x}_m^i)x_i(a_{iu}ya) \in \mathcal{V} + \mathcal{V}ya$$

for all $\bar{x}_m \in \mathcal{A}^m$, where $G_{iu}(\bar{x}_m^i) = E_{iu}(\bar{x}_{m-1}^i, x_m ay)$. Since $\mathcal{V} + \mathcal{V}ya$ is also finite dimensional, by the induction assumption we have in particular that each $E_{iu} = 0$ for $i \neq m$. Therefore (5.17) becomes

$$E_{m1}(\bar{x}_{m-1})x_m a \in \mathcal{V}$$

for all $\bar{x}_m \in \mathcal{A}^m$. That is, again we have arrived at the situation where the induction assumption can be used; thus $E_{m1} = 0$.

Now suppose $s_m > 1$. By Theorem A.8 there exists $\mathcal{E} = \sum_{j=1}^{n} b_j M_{c_j} \in \mathcal{M}(\mathcal{A})$ such that $\mathcal{E}(a_{m1}) = a \neq 0$ and $\mathcal{E}(a_{mu}) = 0$ for all $u = 2, 3, \ldots, s_m$. Set

$$L(\bar{x}_m) = \sum_{j=1}^{n} H(\bar{x}_{m-1}, x_m b_j)c_j.$$

Clearly $L(\bar{x}_m) \in \sum_{j=1}^{n} \mathcal{V}c_j$. Furthermore, after expanding $L(\bar{x}_m)$ in full, we can write $L(\bar{x}_m)$ in the form

$$L(\bar{x}_m) = \sum_{i \in \mathcal{I}'} \sum_{u=1}^{r_i} G_{iu}(\bar{x}_m^i)x_i d_{iu} + E_{m1}(\bar{x}_{m-1})x_m a, \tag{5.20}$$

where $\mathcal{I}' \subseteq \mathcal{I} \setminus \{m\}$, for some maps G_{iu}, some $r_i \geq 1$, and some $d_{iu} \in \mathcal{Q}$. Since we are only interested in the term $E_{m1}(\bar{x}_{m-1})x_m a$ there is no loss of generality in assuming that each of the sets $\{d_{i1}, d_{i2}, \ldots, d_{ir_i}\}$ is $\mathcal{C}$-independent. Namely, otherwise one can pick its maximal linearly independent subset, express all other elements as linear combinations of elements from this set, and this clearly yields the desired form. We are thus back to the situation where $s_m = 1$ and so we conclude that $E_{m1} = 0$. This concludes the proof of the first assertion.

The proof of the second assertion is similar. The important difference is only in the first step, which is in this case in fact simpler. Using Lemma D.2 (i) we can find an element $a \in \mathcal{A}$ such that each of the sets $\{ab_{j1}, ab_{j2}, \ldots, ab_{jt_j}\}$, $j \in \mathcal{J}$, is $\mathcal{C}$-independent and each $ab_{jv} \in \mathcal{A}$. Multiplying (5.18) on the left by a we are thus in the situation where we can assume the b_{jv}'s lie in $\mathcal{A}$ to begin with. The rest of the proof is just a modification of the proof of the first assertion and so we omit it. □

Lemma 5.34 makes it possible for us to state a GFI version of results we have often met when dealing with FI's.

Corollary 5.35. *Let $\mathcal{A}$ be a non-GPI prime ring. Then:*

(i) *Any standard solution of (5.15) is unique.*

(ii) *If all E_{iu}'s and F_{jv}'s are $(m-1)$-additive, then all p_{iujv}'s and λ_{iuv}'s (from (5.16)) are $(m-2)$-additive and $(m-1)$-additive, respectively.*

Proof. (i) Suppose $p_{iujv}, q_{iujv} : \mathcal{A}^{m-2} \to \mathcal{Q}$ and $\lambda_{kuv}, \mu_{kuv} : \mathcal{A}^{m-1} \to \mathcal{C}$ are such that for each $i \in \mathcal{I}$, $1 \le u \le s_i$,

$$E_{iu}(\overline{x}^i_m) = \sum_{\substack{j \in \mathcal{J} \\ j \neq i}} \sum_{v=1}^{t_j} b_{jv} x_j p_{iujv}(\overline{x}^{ij}_m) + \sum_{v=1}^{t_i} \lambda_{iuv}(\overline{x}^i_m) b_{iv}$$

$$= \sum_{\substack{j \in \mathcal{J} \\ j \neq i}} \sum_{v=1}^{t_j} b_{jv} x_j q_{iujv}(\overline{x}^{ij}_m) + \sum_{v=1}^{t_i} \mu_{iuv}(\overline{x}^i_m) b_{iv}.$$

Then

$$\sum_{\substack{j \in \mathcal{J} \\ j \neq i}} \sum_{v=1}^{t_j} b_{jv} x_j \left(p_{iujv}(\overline{x}^{ij}_m) - q_{iujv}(\overline{x}^{ij}_m) \right) \in \sum_{v=1}^{t_i} \mathcal{C} b_{iv}.$$

By Lemma 5.34 we have $p_{iujv} = q_{iujv}$. Thus

$$\sum_{v=1}^{t_i} \left(\lambda_{iuv}(\overline{x}^i_m) - \mu_{iuv}(\overline{x}^i_m) \right) b_{iv} = 0,$$

whence $\lambda_{iuv} = \mu_{iuv}$ by the $\mathcal{C}$-independence of the b_{iv}'s.

(ii) The proof is just as simple as the proof of (i); let us just outline it. We substitute $x'_m + x''_m$ for x_m in (5.16). Since E_{iu}, $i \neq m$, is additive in x_m this results, after rearrangement of terms, in a GFI of the type (5.18). Applying Lemma 5.34 then one easily arrives at the desired conclusion. □

Theorem 5.36. *Let $\mathcal{A}$ be a prime ring satisfying (5.15). Then either $\mathcal{A}$ is a GPI-ring or (5.15) has only a unique standard solution (5.16).*

Proof. We assume throughout the proof that $\mathcal{A}$ is a non-GPI ring.

Let (5.15) hold. By Lemma 5.34 we may assume that both $\mathcal{I}$ and $\mathcal{J}$ are nonempty. Our goal is to show that E_{iu}, F_{jv} are of the form (5.16).

We claim that it suffices to show only that all F_{jv}'s are of this standard form (with $\lambda_{juv} = 0$ if $j \notin \mathcal{I}$). Indeed, knowing that the F_{jv}'s are of this form, it follows by substituting the expressions for F_{jv}'s in (5.15) that

$$\sum_{i\in\mathcal{I}}\sum_{u=1}^{s_i}\left(E_{iu}(\overline{x}_m^i) - \sum_{\substack{j\in\mathcal{J}\\j\neq i}}\sum_{v=1}^{t_j}b_{jv}x_jp_{iujv}(\overline{x}_m^{ij}) - \sum_{v=1}^{t_i}\lambda_{iuv}(\overline{x}_m^i)b_{iv}\right)x_ia_{iu} \in \mathcal{V}.$$

Lemma 5.34 now says that the E_{iu}'s are of the form (5.16). The uniqueness was already established in Corollary 5.35 (i).

So, our task is to prove that the F_{jv}'s are standard, i.e., they are of standard form as given in (5.16). The proof is by induction on $|\mathcal{J}|$. We shall defer until a little later on the proof for the case $|\mathcal{J}| = 1$. Rather, we will proceed by a series of reductions of an inductive nature.

(A) We may assume that $b_{jv} \in \mathcal{A}$. Indeed by Lemma D.2 (i) there exists $a \in \mathcal{A}$ such that each of the sets $\{ab_{j1}, ab_{j2}, \ldots, ab_{jt_j}\}$, $j \in \mathcal{J}$, is $\mathcal{C}$-independent and each $ab_{jv} \in \mathcal{A}$. Multiplying (5.15) from left by a gives us

$$\sum_{i\in\mathcal{I}}\sum_{u=1}^{s_i}(aE_{iu})(\overline{x}_m^i)x_ia_{iu} + \sum_{j\in\mathcal{J}}\sum_{v=1}^{t_j}ab_{jv}x_jF_{jv}(\overline{x}_m^j) \in \mathcal{V}.$$

Clearly, as long as only the functions F_{jv}'s are concerned, this GFI and (5.15) have the same standard solutions.

(B) Given any fixed $k \in \mathcal{J}$ it is enough to show that each F_{kv}, $1 \leq v \leq t_k$, is standard. Indeed, for simplicity we may assume $k = 1$. For $v = 1, 2, \ldots, t_1$ we have

$$F_{1v}(\overline{x}_m^1) = -\sum_{\substack{i\in\mathcal{I}\\i\neq1}}\sum_{u=1}^{s_i}p_{iu1v}(\overline{x}_m^{i1})x_ia_{iu} - \sum_{u=1}^{s_1}\lambda_{1uv}(\overline{x}_m^1)a_{1u}. \qquad (5.21)$$

We set

$$\overline{E}_{iu}(\overline{x}_m^i) = E_{iu}(\overline{x}_m^i) + \sum_{v=1}^{t_1}b_{1v}x_1p_{iu1v}(\overline{x}_m^{i1}), \quad i \neq 1,$$

$$\overline{E}_{1u}(\overline{x}_m^1) = E_{1u}(\overline{x}_m^1) + \sum_{v=1}^{t_1}\lambda_{1uv}(\overline{x}_m^1)b_{1v}$$

(of course if $1 \notin \mathcal{I}$, then the last summation in (5.21) does not appear and in this case there is no need to introduce $\overline{E}_{1u}$). Substitution of (5.21) in (5.15) results in

$$\sum_{i\in\mathcal{I}}\sum_{u=1}^{s_i}\overline{E}_{iu}(\overline{x}_m^i)x_ia_{iu} + \sum_{j\in\mathcal{J}'}\sum_{v=1}^{t_j}b_{jv}x_jF_{jv}(\overline{x}_m^j) \in \mathcal{V} \qquad (5.22)$$

for suitable $\overline{E}_{iu}$, where $\mathcal{J}' = \mathcal{J} \setminus \{1\}$. Since $|\mathcal{J}'| < |\mathcal{J}|$ by induction applied to (5.22) we may conclude that the remaining F_{jv}'s, $j \neq 1$ are also standard.

(C) Without loss of generality we may assume that $m \in \mathcal{J}$ and $t_m = 1$, and it is enough to prove that F_{m1} is standard. The latter of course follows from (B); the new information here is that we may assume that $t_m = 1$.

So assume that the theorem is true if $m \in \mathcal{J}$ and $t_m = 1$, and let us prove that then it is true in any case. In view of (B) our task is to show that each F_{mv}, $1 \leq v \leq t_m$, is standard. We proceed by induction on t_m. There is nothing to prove if $t_m = 1$, so let $t_m > 1$. We write (5.15) as

$$H(\overline{x}_m) = \sum_{i \in \mathcal{I}} \sum_{u=1}^{s_i} E_{iu}(\overline{x}_m^i) x_i a_{iu} + \sum_{v=1}^{t_m} b_{mv} x_m F_{mv}(\overline{x}_{m-1})$$

$$+ \sum_{\substack{j \in \mathcal{J} \\ j \neq m}} \sum_{v=1}^{t_j} b_{jv} x_j F_{jv}(\overline{x}_m^j) \in \mathcal{V}. \qquad (5.23)$$

By Theorem A.8 there exists $\mathcal{E} = \sum_{j=1}^{n} c_j M_{d_j} \in \mathcal{M}(\mathcal{A})$ such that $\mathcal{E}(b_{mt_m}) = b \neq 0$ and $\mathcal{E}(b_{mv}) = 0$ for all $v = 1, 2, \ldots, t_m - 1$. Then

$$\overline{H}(\overline{x}_m) = \sum_{j=1}^{n} c_j H(\overline{x}_{m-1}, d_j x_m) \in \sum_{j=1}^{n} c_j \mathcal{V}$$

and we may write

$$\overline{H}(\overline{x}_m) = \sum_{i \in \mathcal{I}} \sum_{u=1}^{s_i} \overline{E}_{iu}(\overline{x}_m^i) x_i a_{iu} + b x_m F_{mt_m}(\overline{x}_{m-1}) + \sum_{\substack{j \in \mathcal{J} \\ j \neq m}} \sum_{l=1}^{h_j} d_{jl} x_j \overline{F}_{jl}(\overline{x}_m^j) \in \mathcal{V}$$

where $0 \neq b \in \mathcal{A}$, $\overline{E}_{iu}$ and $\overline{F}_{jl}$, $j \neq m$, are suitably chosen maps, and without loss of generality (cf. the argument following (5.20) in the proof of Lemma 5.34) the sets $\{d_{j1}, d_{j2}, \ldots, d_{jh_j}\}$, $j \neq m$ are $\mathcal{C}$-independent. Therefore, we are in a position to use the assumption we are making in (C), and hence it follows, in particular, that F_{mt_m} is standard (since the s_i's and a_{iu}'s have not changed). Thus

$$F_{mt_m}(\overline{x}_{m-1}) = - \sum_{\substack{i \in \mathcal{I} \\ i \neq m}} \sum_{u=1}^{s_i} p_{iumt_m}(\overline{x}_m^{im}) x_i a_{iu} - \sum_{u=1}^{s_m} \lambda_{mut_m}(\overline{x}_{m-1}) a_{mu}.$$

Inserting this expression in (5.23) and rearranging terms in an obvious way we end up with

$$\sum_{i \in \mathcal{I}} \sum_{u=1}^{s_i} \widetilde{E}_{iu}(\overline{x}_m^i) x_i a_{iu} + \sum_{v=1}^{t_m-1} b_{mv} x_m F_{mv}(\overline{x}_{m-1}) + \sum_{\substack{j \in \mathcal{J} \\ j \neq m}} \sum_{v=1}^{t_j} b_{jv} x_j F_{jv}(\overline{x}_m^j) \in \mathcal{V}$$

for some $\widetilde{E}_{iu}$'s. By the induction assumption on t_m we may conclude that all F_{mv} are standard.

(D) At this point we choose to prove the theorem for the initial case where $|\mathcal{J}| = 1$. So, in view of (C) we may assume that $\mathcal{J} = \{m\}$ and $t_m = 1$. We begin by writing

$$H(\bar{x}_m) = \sum_{i\in\mathcal{I}}\sum_{u=1}^{s_i} E_{iu}(\bar{x}_m^i)x_i a_{iu} + bx_m F_{m1}(\bar{x}_{m-1}) \in \mathcal{V} \tag{5.24}$$

where, in view of (A), we have $b \in \mathcal{A}$. We compute

$$H(\bar{x}_{m-1}, x_m' + x_m'') - H(\bar{x}_{m-1}, x_m') - H(\bar{x}_{m-1}, x_m'')$$

obtaining

$$\sum_{\substack{i\in\mathcal{I}\\i\neq m}}\sum_{u=1}^{s_i} \left(E_{iu}(\bar{x}_{m-1}^i, x_m' + x_m'') - E_{iu}(\bar{x}_{m-1}^i, x_m') - E_{iu}(\bar{x}_{m-1}^i, x_m'')\right) x_i a_{iu} \in \mathcal{V}.$$

Lemma 5.34 now implies that each E_{iu}, $i \neq m$, is additive in x_m. Let $y \in \mathcal{A}$ and note that

$$H(\bar{x}_{m-1}, ybx_m) - byH(\bar{x}_m)$$

$$= \sum_{\substack{i\in\mathcal{I}\\i\neq m}}\sum_{u=1}^{s_i} \left(E_{iu}(\bar{x}_{m-1}^i, ybx_m) - byE_{iu}(\bar{x}_m^i)\right) x_i a_{iu}$$

$$+ \sum_{u=1}^{s_m} \left(E_{mu}(\bar{x}_{m-1})yb - byE_{mu}(\bar{x}_{m-1})\right) x_m a_{mu} \in \mathcal{V} + by\mathcal{V}.$$

Using Lemma 5.34 in an obvious way, we draw two conclusions from this identity. First we see that

$$E_{mu}(\bar{x}_{m-1})yb - byE_{mu}(\bar{x}_{m-1}) = 0$$

for all $y \in \mathcal{A}$. By Theorem A.7 we then see that

$$E_{mu}(\bar{x}_{m-1}) = \lambda_{mu1}(\bar{x}_{m-1})b \tag{5.25}$$

where $\lambda_{mu1} : \mathcal{A}^{m-1} \to \mathcal{C}$. Secondly we conclude that for $i \neq m$,

$$E_{iu}(\bar{x}_{m-1}^i, ybx_m) = byE_{iu}(\bar{x}_m^i)$$

for all $y \in \mathcal{A}$. Since we have previously shown that each E_{iu} is additive in x_m for $i \neq m$ we see that we can now make use of the fact that b, as a nonzero element in $\mathcal{Q}$, is fractionable in $\mathcal{Q}$ (Lemma 5.8). Accordingly, there exists a (unique) element $p_{ium1}(\bar{x}_m^{im}) \in \mathcal{Q}$ such that

$$E_{iu}(\bar{x}_m^i) = bx_m p_{ium1}(\bar{x}_m^{im}). \tag{5.26}$$

Substituting (5.25) and (5.26) in (5.24) we obtain

$$\sum_{u=1}^{s_m} \lambda_{mu1}(\overline{x}_{m-1})bx_m a_{mu} + \sum_{\substack{i \in \mathcal{I} \\ i \neq m}} \sum_{u=1}^{s_i} bx_m p_{ium1}(\overline{x}_m^{im})x_i a_{iu} + bx_m F_{m1}(\overline{x}_{m-1}) \in \mathcal{V},$$

that is

$$bx_m \left(F_{m1}(\overline{x}_{m-1}) + \sum_{\substack{i \in \mathcal{I} \\ i \neq m}} \sum_{u=1}^{s_i} p_{ium1}(\overline{x}_m^{im})x_i a_{iu} + \sum_{u=1}^{s_m} \lambda_{mu1}(\overline{x}_{m-1})a_{mu} \right) \in \mathcal{V}.$$

Applying Lemma 5.34 we see that

$$F_{m1}(\overline{x}_{m-1}) = -\sum_{\substack{i \in \mathcal{I} \\ i \neq m}} \sum_{u=1}^{s_i} p_{ium1}(\overline{x}_m^{im})x_i a_{iu} - \sum_{u=1}^{s_m} \lambda_{mu1}(\overline{x}_{m-1})a_{mu},$$

that is, F_{m1} is standard.

(E) It remains to consider the case where $|\mathcal{J}| > 1$, and make the inductive step. Of course we are still assuming that $m \in \mathcal{J}$ with $t_m = 1$, and our goal is to show that F_{m1} is standard.

As usual we begin by writing down

$$H(\overline{x}_m) = \sum_{i \in \mathcal{I}} \sum_{u=1}^{s_i} E_{iu}(\overline{x}_m^i)x_i a_{iu} + bx_m F_{m1}(\overline{x}_{m-1}) + \sum_{\substack{j \in \mathcal{J} \\ j \neq m}} \sum_{v=1}^{t_j} b_{jv} x_j F_{jv}(\overline{x}_m^j) \in \mathcal{V}.$$

By Lemma D.2 (iii) there is $z \in \mathcal{A}$ such that for each $j \neq m$,

$$\{b_{j1}, b_{j2}, \dots, b_{jt_j}, bz b_{j1}, bz b_{j2}, \dots, bz b_{jt_j}\}$$

is a $\mathcal{C}$-independent set. We set $\overline{H}(\overline{x}_m) = H(\overline{x}_{m-1}, zbx_m) - bzH(\overline{x}_m)$ and expand $\overline{H}$, obtaining

$$\sum_{i \in \mathcal{I}} \sum_{u=1}^{s_i} \overline{E}_{iu}(\overline{x}_m^i)x_i a_{iu} + \sum_{\substack{j \in \mathcal{J} \\ j \neq m}} \sum_{v=1}^{t_j} b_{jv} x_j F_{jv}(\overline{x}_{m-1}^j, zbx_m)$$

$$- \sum_{\substack{j \in \mathcal{J} \\ j \neq m}} \sum_{v=1}^{t_j} bz b_{jv} x_j F_{jv}(\overline{x}_m^j) \in \mathcal{V} + bz\mathcal{V}$$

where the $\overline{E}_{iu}$'s are suitable maps. Note that this identity is of the proper form; in particular the new index set $\mathcal{J}' = \mathcal{J} \setminus \{m\}$ allows us to apply induction to conclude that each F_{jv} is standard, for $j \neq m$. Substituting these standard expressions for F_{jv} into (5.15) and rearranging terms in the obvious way we are then left with

$$H(\bar{x}_m) = \sum_{i \in \mathcal{I}} \sum_{u=1}^{s_i} \widetilde{E}_{iu}(\bar{x}_m^i)x_i a_{iu} + bx_m F_{m1}(\bar{x}_{m-1}) \in \mathcal{V}$$

for suitable maps $\widetilde{E}_{iu}$. But this case has already been taken care of in (D). Thus F_{m1} is standard and the proof is now complete.   □

As a final remark we note that Theorem 5.36 affords another proof that a non-GPI prime ring is a $(t;d)$-free subset of $\mathcal{Q}$ for all d. To see this one simply takes the $\mathcal{C}$-independent powers of a suitable element t for the coefficients.

Let us conclude this chapter by comparing the result of this section by the results of Section 5.2. A rough summary of this section can be stated as follows: If $\mathcal{A}$ is a prime ring, then the GFI's (5.15) have only standard solutions if and only if $\mathcal{A}$ is not a GPI-ring. This is analogous to Corollary 5.13 which basically says that the basic FI's through which d-freeness is defined have only standard solutions on a prime ring $\mathcal{A}$ if and only if $\mathcal{A}$ is not a PI-ring.

Literature and Comments. The history of the subject of Chapter 5 is much richer than that of the preceding chapters, so these concluding notes must necessary be somewhat longer.

The order of topics presented in this chapter is almost opposite to the order of their historic developments. Let us therefore first discuss the topic of Section 5.4, i.e., commuting maps, since these are the roots of FI theory. For a complete account of commuting maps we refer the reader to Brešar's survey paper [66]. Here we shall give just a very brief and rough summary.

As already mentioned above, the study of commuting maps originated in Posner's theorem [182] from 1957 on centralizing derivations on prime rings (we recall that the definition of a centralizing map is just slightly more general than that of a commuting map). There is a vast literature on extensions of Posner's result treating more general conditions with derivations and some related maps, such as ring homomorphisms (see [66] for references). As a curiosity we mention that some connections of these ring-theoretic results to certain problems studied in the theory of Banach algebras have been discovered (see for example surveys [66, 158]). The first results on commuting maps in which the role of a derivation or some other special map was replaced by an arbitrary additive map were obtained by Brešar [54, 56] at the beginning of the 90's. In particular in [56] he obtained Corollary 5.28 for the case where the ring in question is prime, what can be considered as the first result on FI's (in the sense of this book). The extension to semiprime rings was obtained somewhat later by Ara and Mathieu [6], which was followed by a paper by Brešar [59] giving a shorter proof based on biderivations. Theorem 5.32 was also proved by Brešar [58] in 1993, however under the assumption that $\deg(\mathcal{A}) \neq 2$. The fact that this assumption is redundant was observed only ten years later in [84]. In [58] it was also shown for the first time that commuting maps are applicable to many areas, which made the subject interesting and gave a good motivation for their further investigation. Corollaries 5.30 and 5.31 were obtained by Lee, Lin, Wang and Wong [140] in 1997.

All these results were originally proved by a self-contained and direct approach (usually making use of some versions of Theorem A.7). So far we have mentioned just a minor part of numerous results on commuting maps on (semi)prime rings (and their various subsets) that were obtained before the general theory of FI's was created, i.e., before Beidar's seminal paper [16]. Here is a list of some of the papers studying this subject: [12, 35, 39, 55, 60, 62, 77, 81, 137, 139, 141, 142]. Some of the results from these papers were rather important at that time, particularly the one from [39] leading to the solution of Herstein's problem on Lie isomorphisms of skew elements. However, from the present perspective they are largely overshadowed by the general theory. Further applications of the general theory to commuting (and related) maps that were not yet discussed can be found in [16, 19, 38].

There are many results on commuting (and related) maps that are not considered in this book and are not superseded by the general theory. We list some of them:

(a) Commuting maps on operator algebras [6, 54, 79]; a complete account of this subject is given in the book of Ara and Mathieu [7].

(b) Commuting maps on triangular rings [18, 48, 99]; these are typical examples of rings that are not d-free (so the general theory fails), but commuting maps can nevertheless be described.

(c) Associating maps in Jordan algebras [69, 73]; in special Jordan algebras the notion of an associating map coincides with the notion of a centralizing map, and these results give appropriate Jordan algebra generalizations of Corollary 5.28 and Theorem 5.32.

(d) Range-inclusive maps [67, 128]; these are additive maps $f : \mathcal{A} \to \mathcal{A}$ satisfying $[f(x), \mathcal{A}] \subseteq [x, \mathcal{A}]$, so this notion generalizes the notion of a commuting additive map.

(e) Appropriate extensions of commuting additive maps on ordinary rings to associative superalgebras [108].

The next topic we are going to discuss are GFI's. Paradoxically, the fundamental results on GFI's were discovered earlier than similar results on FI's. For some time this area had appeared as the most promising one; only later, after [16], did the FI's prevail.

The first result on GFI's was obtained in 1995 by Brešar [61]. Neglecting some technical details he basically proved Theorem 5.36 for $m = 2$. Incidentally, this is the paper where the phrase "functional identity" was introduced (although later an adjective "generalized" was added to identities such as those treated in [61]). Some of the arguments from [61] may nowadays seem somewhat clumsy, but on the other hand this paper brought some fundamental ideas important for further development. In particular, a version of the key Lemma 5.8 appeared there. In 1998 Chebotar [88] generalized the result of [61] to an arbitrary m, i.e., he proved a version of Theorem 5.36, and simultaneously he simplified the proof from [61]. Certainly this paper of Chebotar was one of the milestones in the development of the theory treated in this book. It initiated the direction in which the general theory was later created. In particular, in his fundametal paper [16] Beidar used several ideas from [88]. Another paper that needs to be pointed out is [17]. Its main result is considerably more general than Theorem 5.36. It treats GFI's which also involve automorphisms, antiautomorphisms and derivations of $\mathcal{A}$, and so it connects and in some sense unifies two theories: the theory of GFI's and the theory of rings with generalized identities with automorphisms, antiautomorphisms and derivations (see [40] for a full account). Next we list a few papers [89, 102, 103, 195] that treat some special questions concerned with GFI's. Finally we say a few words about yet another type of "generalized" FI's. It is a challenging problem to study identities involving summands such as $E(x_1, \ldots, x_{i-1})x_i F(x_{i+1}, \ldots, x_m)$ where of course E and F are the unknown functions. The papers [71, 72] show that even some simple and rather special identities of this type are difficult to handle, but on the other hand they also show that nevertheless there is some hope here for getting interesting results.

Now about FI's on prime rings. As already mentioned at the end of Chapter 1, in the early 1990s numerous authors studied some special FI's, mostly on prime rings and their subsets. Among them we point out the paper by Brešar [60] which is the closest one to the present conception of FI's. Namely, one of its results basically discovers 2-freeness of noncommutative prime rings. Motivated by this result and the aforementioned paper by Chebotar, in his 1998 paper [16] Beidar established the foundations of the advanced FI theory. The notions of d-freeness and $(t; d)$-freeness are not yet introduced in [16], but they are hidden in the results. An almost equivalent version of the fundamental Theorem 5.11 is proved, as well as a version of Corollary 5.16. The next important step was the study of FI's in rings with involution, started in [38] and continued in [22], which in particular led to Corollaries 5.17 and 5.18.

There are some other relevant results on FI's in prime rings that, however, are not treated in this book. For instance, the results concerning FI's on one-sided ideals of prime rings [27]. In [20] an alternative approach to the study of FI's in prime rings is proposed; the coefficients are no longer necessarily powers of a fixed elements t (as in the definition of $(t; d)$-freeness), but elements satisfying certain conditions. These conditions are admittedly somewhat technical, but the results from [20] have really turned out to be useful. They were used in solving certain problems [20, 198] for which $(t; d)$-freeness is not sufficient.

After the theory of FI's in prime rings was more or less completed, Beidar and Chebotar introduced the concept of a d-free set in an arbitrary ring [29, 30]. Regarding FI's from this more abstract aspect made it possible to obtain a new insight even in the classical prime and semiprime ring context. In [19] the fractional degree was introduced which led to establishing d-freeness of semiprime rings (Theorem 5.26).

Part III

Applications

Chapter 6

Lie Maps and Related Topics

Every associative ring $\mathcal{A}$ can be turned into a Lie ring by introducing a new product $[x, y] = xy - yx$. So we may regard $\mathcal{A}$ simultaneously as an associative ring and as a Lie ring. What is the connection between the associative and the Lie structure of $\mathcal{A}$? This question has been studied for more than fifty years by numerous authors, most notably by Herstein and many of his students (see, for example, [113, 114, 115]). One of the first questions that one might ask in this context is: If rings $\mathcal{B}$ and $\mathcal{A}$ are isomorphic as Lie rings, are they then also isomorphic (or at least antiisomorphic) as associative rings? In more technical terms one can rephrase this question as whether a Lie isomorphism $\alpha : \mathcal{B} \to \mathcal{A}$ always "arises" from an (anti)isomorphism. This is just the simplest question that one can ask in this setting. More general (and from the point of view of the theory of Lie algebras also more natural) questions concern the structure of Lie homomorphisms between various Lie subrings of associative rings. Analogous problems can be formulated for Lie derivations.

As usual we leave historic details for the end of the chapter. Let us just mention here that unlike for most of the other basic questions concerning the Lie structure of associative rings, Herstein and his school did not obtain definitive answers for questions about Lie homomorphisms and Lie derivations. In his 1961 "AMS Hour Talk" (which was published in [113]) Herstein formulated several conjectures about Lie homomorphisms and Lie derivations of "simple (or, perhaps, even of prime) rings". Until rather recently these conjectures had only been settled under the assumption that the rings in question contain nontrivial idempotents. Making use of advanced FI theory, all conjectures have now been completely settled. Most of FI theory was actually developed when searching for suitable tools for settling these conjectures.

The main purpose of this chapter is to present solutions of some of Herstein's conjectures (Sections 6.1-6.3). Only *some* of them will be given; a systematic treatment of *all* conjectures in full detail would require tackling a number of tedious technical problems, which might overshadow what is our intention in Part III, which is to indicate the applicability of the general FI theory.

We will also obtain analogous results for Jordan homomorphisms and Jordan derivations (Section 6.4). This topic is equally interesting; however, there are other powerful methods, based on Zelmanov's approach [199], that can be used for describing Jordan maps. Therefore the results that we obtain using FI's cannot really be considered as such a breakthrough as in the Lie map case.

At the end of the chapter, in Section 6.5, we will explore considerably more general types of maps that act as homomorphisms (or derivations) with respect to an arbitrary fixed polynomial in $\mathbb{Z}\langle X \rangle$, rather than just to the Lie or Jordan product.

Our main results will be stated in terms of d-free sets, and then we will derive as applications results for the classical prime ring case. One of the advantages of the approach based on d-freeness is that the results on derivations are obtained as byproducts of results on Lie homomorphisms (rather than requiring separate independent proofs).

6.1   Lie Maps on Rings

Let $\mathcal{B}$ and $\mathcal{Q}$ be rings, and as usual we assume that $\mathcal{Q}$ is unital and we denote by $\mathcal{C}$ its center. A Lie homomorphism is an additive map $\alpha : \mathcal{B} \to \mathcal{Q}$ satisfying $[x, y]^\alpha = [x^\alpha, y^\alpha]$ for all $x, y \in \mathcal{B}$. One obvious possibility when this is fulfilled is that α satisfies $(xy)^\alpha = x^\alpha y^\alpha$, and another one is that α satisfies $(xy)^\alpha = -y^\alpha x^\alpha$. In the first case α is of course a (ring) homomorphism, and in the second case it is the negative of an antihomomorphism. A more general example of a Lie homomorphism is a *direct sum* of a homomorphism and the negative of an antihomomorphism. By this we mean a map $\sigma : \mathcal{B} \to \mathcal{Q}$ such that for some idempotent $\varepsilon \in \mathcal{C}$, $x \mapsto \varepsilon x^\sigma$ is a homomorphism and $x \mapsto (1 - \varepsilon)x^\sigma$ is the negative of an antihomomorphism. Another relevant example is of an entirely different nature: any additive mapping $\tau : \mathcal{B} \to \mathcal{C}$ sending commutators to 0 is a Lie homomorphism. Furthermore, the sum, $\sigma + \tau$, of these two types of examples yields another example of a Lie homomorphism. When can a Lie homomorphism $\alpha : \mathcal{B} \to \mathcal{Q}$ be represented as such a sum? This question is the main issue of this section.

The topic of Lie homomorphisms was roughly introduced in Section 1.4, where it was shown how this notion leads to a certain FI which can be interpreted in terms of commuting maps. Let us recall the main idea of this approach. One just has to replace x by y^2 in $[x, y]^\alpha = [x^\alpha, y^\alpha]$, which gives a relatively simple FI

$$[(y^2)^\alpha, y^\alpha] = 0 \tag{6.1}$$

(and if α is bijective, then we can regard this as that $x \mapsto \left(\left(x^{\alpha^{-1}} \right)^2 \right)^\alpha$ is a commuting trace of a biadditive map). The approach based on (6.1) is very simple and easy to memorize, which is the main reason for pointing it out in the introductory chapter. We shall return to it at the end of the section. At the beginning, however, we will use a slightly different way, based on the observation that

$$[xy, z] + [zx, y] + [yz, x] = 0 \tag{6.2}$$

holds for any three elements x, y, z in an (associative) ring. This identity readily implies that every Lie homomorphism $\alpha : \mathcal{B} \to \mathcal{Q}$ satisfies the FI

$$[(xy)^\alpha, z^\alpha] + [(zx)^\alpha, y^\alpha] + [(yz)^\alpha, x^\alpha] = 0. \tag{6.3}$$

Unlike (6.1), (6.3) is applicable equally well in rings of characteristic 2. This is basically the only advantage of (6.3); otherwise the approaches based on (6.1) and (6.3) are equivalent.

The FI (6.3) is a simple example of an identity for which the results of Chapter 4 are applicable. To make this more clear we rewrite it as

$$B(x, y)z^\alpha + B(z, x)y^\alpha + B(y, z)x^\alpha$$
$$- z^\alpha B(x, y) - y^\alpha B(z, x) - x^\alpha B(y, z) = 0, \tag{6.4}$$

where $B : \mathcal{B}^2 \to \mathcal{Q}$ is defined by $B(x, y) = (xy)^\alpha$. Assuming that $\mathcal{B}^\alpha$ is a 3-free subset of $\mathcal{Q}$ we are then in a position to apply Theorem 4.13 (with $m = 3$, $n = 2$, $P = 0$ and $c = \pm 1$). Hence it follows that $B(x, y) = (xy)^\alpha$ is a quasi-polynomial, meaning that

$$B(x, y) = \varepsilon x^\alpha y^\alpha + \varepsilon' y^\alpha x^\alpha + \mu_1(x)y^\alpha + \mu_2(y)x^\alpha + \nu(x, y) \tag{6.5}$$

for some $\varepsilon, \varepsilon' \in \mathcal{C}$, $\mu_1, \mu_2 : \mathcal{B} \to \mathcal{C}$, and $\nu : \mathcal{B}^2 \to \mathcal{C}$. We have thereby found out how α acts on the associative product xy, and so it is no longer surprising that we are able to express α through homomorphisms and antihomomorphisms. We will do this in the proof of the next theorem. Before proceeding with this, let us use the opportunity for warning the reader about some possible mistakes in using the results on quasi-polynomials. We referred to Theorem 4.13 in order to conclude that B is a quasi-polynomial. But since (6.4) involves only one function, B, one might be inclined to use Corollary 4.14 instead. However, formally this would be a mistake. Namely, one has to take care about the order of variables on which B acts. If B was symmetric, then we could replace $B(z, x)$ by $B(x, z)$ in (6.4) and Corollary 4.14 would be applicable. But in our case B is not symmetric. Still, even in our case we can use this corollary, but only after noticing that $B(z, x) = B(x, z) + [z^\alpha, x^\alpha]$, so that (6.4) can be rewritten as

$$B(x, y)z^\alpha + B(x, z)y^\alpha + B(y, z)x^\alpha$$
$$- z^\alpha B(x, y) - y^\alpha B(x, z) - x^\alpha B(y, z) = [y^\alpha, [z^\alpha, x^\alpha]].$$

This form is suitable for applying Corollary 4.14 (now with P being equal to $[y^\alpha, [z^\alpha, x^\alpha]]$). Anyway, in one way or another, (6.5) is established.

Theorem 6.1. *Let $\mathcal{B}$ be any ring and let $\mathcal{Q}$ be a unital ring with center $\mathcal{C}$. If $\alpha : \mathcal{B} \to \mathcal{Q}$ is a Lie homomorphism such that $\mathcal{B}^\alpha$ is a 3-free subset of $\mathcal{Q}$, then $\alpha = \sigma + \tau$, where $\sigma : \mathcal{B} \to \mathcal{Q}$ is a direct sum of a homomorphism and the negative of an antihomomorphism, and $\tau : \mathcal{B} \to \mathcal{C}$ is an additive map which vanishes on commutators.*

Proof. We have already seen that $B(x, y) = (xy)^\alpha$ is a quasi-polynomial according to (6.5). Since B is a biadditive function, it follows from Lemma 4.6 that its coefficients are multiadditive functions. This means that μ_1 and μ_2 are additive and ν is biadditive. Substituting (6.5) in (6.4) we obtain

$$(\mu_1 - \mu_2)(x)[y^\alpha, z^\alpha] + (\mu_1 - \mu_2)(y)[z^\alpha, x^\alpha] + (\mu_1 - \mu_2)(z)[x^\alpha, y^\alpha] = 0.$$

Applying Lemma 4.4 it follows that $\mu_1 = \mu_2$. So we have

$$(xy)^\alpha = \varepsilon x^\alpha y^\alpha + \varepsilon' y^\alpha x^\alpha + \mu(x)y^\alpha + \mu(y)x^\alpha + \nu(x, y), \qquad (6.6)$$

where $\mu = \mu_1 = \mu_2$.

We shall now compute $(xyz)^\alpha$ in two different ways. On the one hand we have

$$
\begin{aligned}
((xy)z)^\alpha ={}& \varepsilon(xy)^\alpha z^\alpha + \varepsilon' z^\alpha (xy)^\alpha + \mu(xy)z^\alpha + \mu(z)(xy)^\alpha + \nu(xy, z) \\
={}& \varepsilon^2 x^\alpha y^\alpha z^\alpha + \varepsilon\varepsilon' y^\alpha x^\alpha z^\alpha + \varepsilon\mu(x)y^\alpha z^\alpha + \varepsilon\mu(y)x^\alpha z^\alpha + \varepsilon\nu(x, y)z^\alpha \\
& + \varepsilon\varepsilon' z^\alpha x^\alpha y^\alpha + \varepsilon'^2 z^\alpha y^\alpha x^\alpha + \varepsilon'\mu(x)z^\alpha y^\alpha + \varepsilon'\mu(y)z^\alpha x^\alpha + \varepsilon'\nu(x, y)z^\alpha \\
& + \mu(xy)z^\alpha + \varepsilon\mu(z)x^\alpha y^\alpha + \varepsilon'\mu(z)y^\alpha x^\alpha + \mu(x)\mu(z)y^\alpha + \mu(y)\mu(z)x^\alpha \\
& + \mu(z)\nu(x, y) + \nu(xy, z).
\end{aligned}
$$

On the other hand,

$$
\begin{aligned}
(x(yz))^\alpha ={}& \varepsilon x^\alpha (yz)^\alpha + \varepsilon'(yz)^\alpha x^\alpha + \mu(x)(yz)^\alpha + \mu(yz)x^\alpha + \nu(x, yz) \\
={}& \varepsilon^2 x^\alpha y^\alpha z^\alpha + \varepsilon\varepsilon' x^\alpha z^\alpha y^\alpha + \varepsilon\mu(y)x^\alpha z^\alpha + \varepsilon\mu(z)x^\alpha y^\alpha + \varepsilon\nu(y, z)x^\alpha \\
& + \varepsilon\varepsilon' y^\alpha z^\alpha x^\alpha + \varepsilon'^2 z^\alpha y^\alpha x^\alpha + \varepsilon'\mu(y)z^\alpha x^\alpha + \varepsilon'\mu(z)y^\alpha x^\alpha + \varepsilon'\nu(y, z)x^\alpha \\
& + \varepsilon\mu(x)y^\alpha z^\alpha + \varepsilon'\mu(x)z^\alpha y^\alpha + \mu(x)\mu(y)z^\alpha + \mu(x)\mu(z)y^\alpha + \mu(x)\nu(y, z) \\
& + \mu(yz)x^\alpha + \nu(x, yz).
\end{aligned}
$$

Comparing both expressions we obtain

$$\varepsilon\varepsilon'[y^\alpha, [x^\alpha, z^\alpha]] + \omega(x, y)z^\alpha - \omega(y, z)x^\alpha \in \mathcal{C} \qquad (6.7)$$

for suitable $\omega : \mathcal{B}^2 \to \mathcal{C}$. Our goal is to show that $\varepsilon\varepsilon' = 0$. If $\mathcal{B}^\alpha$ was 4-free this would follow immediately by applying Lemma 4.4 to (6.7). However, only 3-freeness is assumed, so a somewhat more careful analysis of (6.7) is necessary. Suppose $\varepsilon\varepsilon' \neq 0$. Then there is $b \in \mathcal{B}$ such that $a = \varepsilon\varepsilon' b^\alpha \notin \mathcal{C}$ (2-freeness of $\mathcal{B}^\alpha$ is enough for establishing this). Set $y = b$ in (6.7), and note that the identity which we obtain can be written as

$$E_1(z)x^\alpha + E_2(x)z^\alpha + x^\alpha F_1(z) + z^\alpha F_2(x) \in \mathcal{C},$$

where $E_1(z) = -az^\alpha$, $E_2(x) = ax^\alpha$, $F_1(z) = -z^\alpha a - \omega(b, z)$, $F_2(x) = x^\alpha a + \omega(x, b)$. Since $\mathcal{B}^\alpha$ is 3-free it follows from Theorem 4.3 that, in particular, there are $p \in \mathcal{Q}$ and $\lambda : \mathcal{B} \to \mathcal{C}$ such that $E_1(z) = z^\alpha p + \lambda(z)$; accordingly, $az^\alpha + z^\alpha p \in \mathcal{C}$. But this contradicts $a \notin \mathcal{C}$ (cf. observation 5 following Definition 3.1). Therefore $\varepsilon\varepsilon'$ must be 0.

Another relation involving ε and ε' can be derived easily from (6.6) and the fact that α is a Lie homomorphism. Indeed, we have

$$[x^\alpha, y^\alpha] = (xy)^\alpha - (yx)^\alpha = \varepsilon x^\alpha y^\alpha + \varepsilon' y^\alpha x^\alpha - \varepsilon y^\alpha x^\alpha - \varepsilon' x^\alpha y^\alpha + \nu(x, y) - \nu(y, x),$$

and hence

$$(1 - \varepsilon + \varepsilon')[x^\alpha, y^\alpha] \in \mathcal{C}.$$

Again applying Lemma 4.4 we get $1 - \varepsilon + \varepsilon' = 0$, i.e., $\varepsilon' = -(1 - \varepsilon)$. From $\varepsilon\varepsilon' = 0$ we now see that ε is an idempotent.

Now define $\sigma : \mathcal{B} \to \mathcal{Q}$ by

$$x^\sigma = x^\alpha - (1 - 2\varepsilon)\mu(x).$$

We claim that $x \mapsto \varepsilon x^\sigma$ is a homomorphism and $x \mapsto (1-\varepsilon)x^\sigma$ is the negative of an antihomomorphism. Using (6.6) one can easily check that $\rho(x, y) = \varepsilon(xy)^\sigma - \varepsilon x^\sigma y^\sigma$ always lies in $\varepsilon\mathcal{C} \subseteq \mathcal{C}$. Computing $\varepsilon(xyz)^\sigma$ in two different ways, as before by using $(xy)z = x(yz)$, we are left with $\varepsilon\rho(x, y)z^\sigma - \varepsilon\rho(y, z)x^\sigma \in \mathcal{C}$, which in turn implies

$$\varepsilon\rho(x, y)z^\alpha - \varepsilon\rho(y, z)x^\alpha \in \mathcal{C}. \tag{6.8}$$

We are now in a position to apply Lemma 4.4 and conclude that $\rho(x, y) = \varepsilon\rho(x, y) = 0$ for all $x, y \in \mathcal{B}$. Apparently 4-freeness of $\mathcal{B}^\alpha$ would be necessary for using this lemma at this point, but regarding (6.8) for an arbitrary but fixed y we see that 3-freeness is again sufficient. Thus $\rho(x, y) = 0$, meaning that $x \mapsto \varepsilon x^\sigma$ is a homomorphism. Similarly we see that $(1 - \varepsilon)(xy)^\sigma = -(1 - \varepsilon)y^\sigma x^\sigma$, i.e., $x \mapsto (1 - \varepsilon)x^\sigma$ is the negative of an antihomomorphism. This proves that σ is the direct sum of a homomorphism and the negative of an antihomomorphism.

Finally we set $x^\tau = (1 - 2\varepsilon)\mu(x)$. Since $\alpha = \sigma + \tau$ and τ maps into $\mathcal{C}$ it is clear that $[x^\alpha, y^\alpha] = [x^\sigma, y^\sigma]$. Both α and σ are Lie homomorphisms so this can be written as $[x, y]^\alpha = [x, y]^\sigma$, so that $[x, y]^\tau = 0$. The proof is thus complete. $\square$

At this point we shall take the opportunity to generalize the notion of a Lie homomorphism as follows. Let $\mathcal{L}$ be a Lie subring of a ring $\mathcal{B}$ and (as usual) let $\mathcal{Q}$ be a unital ring with center $\mathcal{C}$. An additive map $\alpha : \mathcal{L} \to \mathcal{Q}$ is said to be a *weak Lie homomorphism* if $[x, y]^\alpha - [x^\alpha, y^\alpha] \in \mathcal{C}$ for all $x, y \in \mathcal{L}$. This is not intended as a fundamental concept which is an end in itself, but rather as a useful technical notion which will be seen to arise naturally in the remaining sections of this chapter.

In particular we will have need for an analogue of Theorem 6.1 which holds for weak Lie homomorphisms of a ring $\mathcal{B}$ into $\mathcal{Q}$. If one inspects the proof of Theorem 6.1 one will notice that only minor changes are necessary if α is assumed to be a weak Lie homomorphism. First of all, the right-hand side of (6.4) may be a central element instead of just 0. Thus 4-freeness is needed to obtain (6.6). Also it has been pointed out that in two places ((6.7) and (6.8)) of the proof, the assumption of 4-freeness allows one to considerably shorten the proof. Of course it is clear that the map $\tau : \mathcal{B} \to \mathcal{C}$ may no longer vanish on commutators. With these observations in mind we leave it for the reader to verify that the proof of Theorem 6.1 essentially carries over to establishing

Remark 6.2. Let $\mathcal{B}$ be a ring, let $\mathcal{Q}$ be a unital ring with center $\mathcal{C}$, and let $\alpha : \mathcal{B} \to \mathcal{Q}$ be a weak Lie homomorphism such that $\mathcal{B}^\alpha$ is 4-free in $\mathcal{Q}$. Then $\alpha = \sigma + \tau$, where $\sigma : \mathcal{B} \to \mathcal{Q}$ is the direct sum of a homomorphism and the negative of an antihomomorphism and $\tau : \mathcal{B} \to \mathcal{C}$ is an additive map.

Combining Theorem 6.1 with concrete examples of d-free sets that were presented in Sections 2.4, 5.2 and 5.3, one gets the description of Lie homomorphisms in various instances. For example, from Theorem 5.26 we immediately get the following result.

Corollary 6.3. *Let $\mathcal{B}$ be any ring, let $\mathcal{A}$ be a semiprime ring with extended centroid $\mathcal{C}$, and let α be a Lie homomorphism from $\mathcal{B}$ onto $\mathcal{A}$. If $\mathcal{A}$ does not contain nonzero ideals satisfying St_4, then $\alpha = \sigma + \tau$, where $\sigma : \mathcal{B} \to \mathcal{AC} + \mathcal{C}$ is the direct sum of a homomorphism and the negative of an antihomomorphism and $\tau : \mathcal{B} \to \mathcal{C}$ is an additive map which vanishes on commutators.*

Actually, Theorem 6.1 tells us that σ has its image in $\mathcal{Q} = \mathcal{Q}_{ml}(\mathcal{A})$, but since α by assumption maps into $\mathcal{A}$ and τ maps into $\mathcal{C}$, it is clear that σ must map into $\mathcal{A} + \mathcal{C}$, which is an additive subgroup of the subring $\mathcal{AC} + \mathcal{C}$ of $\mathcal{Q}$. In general the ranges of σ and τ do not lie in $\mathcal{A}$ (see Example 6.10 below).

In case $\mathcal{A}$ is a prime ring, Corollary 6.3 gets a somewhat simpler form. The extended centroid $\mathcal{C}$ is then a field and hence it does not contain nontrivial idempotents. Accordingly, σ is either a homomorphism or the negative of an antihomomorphism (namely, ε is 1 or 0). Recall that the condition that $\mathcal{A}$ does not contain nonzero ideals satisfying St_4 is in the prime case equivalent to the condition that $\deg(\mathcal{A})$, the degree of algebraicity of $\mathcal{A}$ over the extended centroid $\mathcal{C}$ of $\mathcal{A}$, is ≥ 3. So we get the following result.

Corollary 6.4. *Let $\mathcal{B}$ be any ring, let $\mathcal{A}$ be a prime ring with extended centroid $\mathcal{C}$, and let α be a Lie homomorphism from $\mathcal{B}$ onto $\mathcal{A}$. If $\deg(\mathcal{A}) \geq 3$, then $\alpha = \sigma + \tau$, where $\sigma : \mathcal{B} \to \mathcal{AC} + \mathcal{C}$ is a either a homomorphism or the negative of an antihomomorphism and $\tau : \mathcal{B} \to \mathcal{C}$ is an additive map which vanishes on commutators.*

Is the condition $\deg(\mathcal{A}) \geq 3$ really vital? Rings $\mathcal{A}$ such that $\deg(\mathcal{A}) < 3$ are indeed very special and their structure is simple (cf. Theorem C.2), but nevertheless the answer to this question does not come so easily. It is often so that the lower the degree of a prime ring is, the more difficult it is to apply FI's. So, paradoxically, our problems may become more involved when the rings become simpler. The Lie homomorphism problem serves as a good illustration for this. We already saw in the proof of Theorem 6.1 that the case when $\deg(\mathcal{A}) = 3$ (which corresponds to the case where $\mathcal{B}^{\alpha} = \mathcal{A}$ is 3-free) requires some extra effort in comparison with the case when $\deg(\mathcal{A}) \geq 4$ (i.e., $\mathcal{B}^{\alpha} = \mathcal{A}$ is 4-free). Still, we were able to handle this case using the general FI machinery. The case when $\deg(\mathcal{A}) = 2$ is more complicated. First of all, not every Lie automorphism of such a ring is of the form $\sigma + \tau$; see Example 6.11 below. However, this example concerns rings with $\text{char}(\mathcal{A}) = 2$. If $\text{char}(\mathcal{A}) \neq 2$, then we can still get the desired description of Lie homomorphisms (Corollary 6.5 below), but under the additional assumption of bijectivity and by using a result which is not a part of the general FI theory based on the notion of d-freeness. Going the final step downwards, i.e., considering the $\deg(\mathcal{A}) = 1$ case, there is just nothing of interest that can be said. Of course, setting $\sigma = 0$ the standard conclusion formally still holds, but there is no reason to believe that there exist nontrivial homomorphisms from $\mathcal{B}$ into $\mathcal{A}$.

Let us mention, just as a curiosity, that every prime ring $\mathcal{A}$ with $\deg(\mathcal{A}) = 2$ satisfies the identity of the type (6.7) with $\varepsilon\varepsilon' \neq 0$, so that there is no way to handle such rings by the method of the proof of Theorem 6.1. Indeed, $\deg(\mathcal{A}) = 2$ implies that there exists an additive map $\tau : \mathcal{A} \to \mathcal{C}$ such that $u^2 - \tau(u)u \in \mathcal{C}$ for every $u \in \mathcal{A}$ (see Theorem C.2). Linearizing we then get $u \circ v = \tau(u)v + \tau(v)u + \zeta(u, v)$ where $\zeta(u, v) \in \mathcal{C}$. Using the identity $[u, [v, w]] = (u \circ v) \circ w - (u \circ w) \circ v$ it then easily follows that

$$[u, [v, w]] + \omega(v, u)w - \omega(u, w)v \in \mathcal{C}$$

for all $u, v, w \in \mathcal{A}$, where $\omega(v, u) = \omega(u, v) = -\tau(u)\tau(v) - 2\zeta(u, v)$. Thus, another method is needed for handling the $\deg(\mathcal{A}) = 2$ case.

Corollary 6.5. *Let $\mathcal{B}$ be any ring, let $\mathcal{A}$ be a noncommutative prime ring with extended centroid $\mathcal{C}$, and let α be a Lie isomorphism from $\mathcal{B}$ onto $\mathcal{A}$. If $\text{char}(\mathcal{A}) \neq 2$, then $\alpha = \sigma + \tau$, where $\sigma : \mathcal{B} \to \mathcal{A} + \mathcal{C}$ is either a homomorphism or the negative of an antihomomorphism and $\tau : \mathcal{B} \to \mathcal{C}$ is an additive map which vanishes on commutators. Moreover, if $\mathcal{B}$ is prime, then σ is injective.*

Proof. By Corollary 6.4 we may assume that $\deg(\mathcal{A}) = 2$. We now use the approach based on (6.1), noticing that the map $x \mapsto \left(\left(x^{\alpha^{-1}} \right)^2 \right)^{\alpha}$ is a commuting trace of a biadditive map of the prime ring $\mathcal{A}$ into itself. By Theorem 5.32 we have

$$\left(\left(x^{\alpha^{-1}} \right)^2 \right)^{\alpha} = \mu(x)x + \nu(x),$$

where μ and ν map $\mathcal{A}$ into $\mathcal{C}$ and μ is additive. Setting $y = x^{\alpha^{-1}}$, we thus have

$$(y^2)^{\alpha} - \mu'(y)y^{\alpha} \in \mathcal{C} \tag{6.9}$$

Proof. By Theorem 6.6 we know that δ is of the form $x^\delta = \lambda x + x^d + x^\tau$. Using this in $[x, y]^\delta = [x^\delta, y] + [x, y^\delta]$ we obtain $\lambda[x, y] - [x, y]^\tau = 0$. Applying Lemma 4.4 it follows that $\lambda = 0$ and $[x, y]^\tau = 0$, which proves the corollary. $\qquad\square$

Corollary 6.7 could also be easily derived from Theorem 3.7.

Remark 6.8. As for weak Lie homomorphisms, an analogous result holds for *weak Lie derivations*, i.e., additive maps $\delta : A \to Q$ such that $[x, y]^\delta - [x^\delta, y] - [x, y^\delta] \in C$ for all $x, y \in A$. That is, Corollary 6.7 holds also in case δ is a weak Lie derivation; just two modifications are necessary: we must assume that A is a 4-free, and we cannot claim that τ vanishes on commutators.

It is clear that Corollary 6.7 implies derivation analogues of Corollaries 6.3 and 6.4. We shall omit stating them, but rather establish only an analogue of Corollary 6.5.

Corollary 6.9. *Let A be a prime ring with extended centroid C. Then every Lie derivation $\delta : A \to A$ is of the form $\delta = d + \tau$ where $d : A \to AC + C$ is a derivation and $\tau : A \to C$ is an additive map vanishing on commutators, unless $\deg(A) = 2$ and $\mathrm{char}(A) = 2$.*

Proof. In view of Corollaries 6.7 and 5.12 we may assume that $\deg(A) \le 2$. If $\deg(A) = 1$, then we just set $d = 0$ and $\tau = \delta$. So we may assume that $\deg(A) = 2$ and $\mathrm{char}(A) \ne 2$. As already mentioned in Section 1.4, every Lie derivation δ satisfies

$$[(x^2)^\delta - x^\delta x - x x^\delta, x] = 0, \qquad (6.15)$$

which enables one to apply results on commuting traces of biadditive maps. Proving (6.15) is easy. Just note that

$$[(r^2)^\delta, r] - [x^2, x]^\delta - [x^2, x^\delta] - [x^\delta, x^2] - [x^\delta x \mid x x^\delta, x].$$

An application of Theorem 5.32 now shows that there exists an additive map $\mu : A \to C$ such that

$$(x^2)^\delta - x^\delta x - x x^\delta - \mu(x) x \in C.$$

Linearizing this identity, and comparing the resulting relation with $[x, y]^\delta = [x^\delta, y] + [x, y^\delta]$, we get

$$2\left((xy)^\delta - x^\delta y - x y^\delta\right) - \mu(x) y - \mu(y) x \in C.$$

Setting $x^\tau = -\frac{\mu(x)}{2}$ and $d = \delta - \tau$ we see that

$$\epsilon(x, y) = (xy)^d - x^d y - x y^d \in C.$$

Repeatedly applying d to $(xy)z = x(yz)$ now yields

$$\epsilon(x, y)z - \epsilon(y, z)x \in C.$$

This is reminiscent of the relation (6.12); arguing as in the proof of Corollary 6.5 we can show that $\epsilon(x, y) = 0$ for all $x, y \in A$. Thus d is a derivation. Since $\tau = \delta - d$

is a Lie derivation with values in $\mathcal{C}$ it is clear that τ must vanish on commutators. This completes the proof. $\qquad\square$

We conclude this section with three examples. In the first one we show that the images of σ (in Corollaries 6.4 and 6.5) and d (in Corollary 6.9) may not be contained in $\mathcal{A}$.

Example 6.10. Let $\mathcal{Z}$ be a subfield of a field $\mathcal{C}$. Let $\mathcal{A}$ be the set of all (countably infinite) matrices of the form $a + \lambda I$, where a is an $n \times n$ upper left corner matrix over $\mathcal{C}$ with n varying, and λI is a scalar matrix with $\lambda \in \mathcal{Z}$. It is easy to check that $\mathcal{A}$ is a prime ring (in fact a primitive ring) with center $\mathcal{Z}$ and extended center $\mathcal{C}$.

We first let $\mathcal{C}$ be the field of rational functions in two variables x and y over a field $\mathbb{F}$, and let $\mathcal{Z}$ be its subfield of rational functions in x. If $a = (a_{ij})$, with $a_{ij} \in \mathcal{C}$, then we set $\widehat{a} = (a_{ij}^\theta)$ where θ is the automorphism of $\mathcal{C}$ determined by interchanging x and y. Now let $\alpha : \mathcal{A} \to \mathcal{A}$ be given according to $(a+\lambda I)^\alpha = \widehat{a}+\lambda I$. One checks that α is a Lie automorphism of $\mathcal{A}$. Suppose $\alpha = \sigma + \tau$ (according to Corollaries 6.4 and 6.5), where $\mathcal{A}^\sigma \subseteq \mathcal{A}$ (and hence $\mathcal{A}^\tau \subseteq \mathcal{Z}$). We take the case that σ is a homomorphism; a similar argument will work in case σ is the negative of an antihomomorphism. Consider the element $b = xe_{12}$, where e_{12} is of course a matrix unit. Since b is a commutator, we have $b^\sigma = b^\alpha$. On the one hand $b^\alpha = ye_{12}$. On the other hand $b^\sigma = (xI)^\sigma e_{12}^\sigma = (f(x)I)e_{12} = f(x)e_{12}$ for some $f(x) \in \mathcal{Z}$ (since σ, being a homomorphism, must map the center $\mathcal{Z}$ into itself). Thus we have reached a contradiction. Of course, there is no conflict with Corollaries 6.4 and 6.5: the correct choices for σ and τ are $(a+f(x)I)^\sigma = \widehat{a}+f(y)I$ and $(a + f(x)I)^\tau = (f(x) - f(y))I$.

We next let $\mathcal{C} = \mathbb{F}(x)$ and $\mathcal{Z} = \mathbb{F}(x^2)$, and let Δ denote the ordinary derivative in $\mathbb{F}(x)$. If $a = (a_{ij})$ with $a_{ij} \in \mathcal{C}$, then we set $\widetilde{a} = (a_{ij}^\Delta)$. Now let $\delta : \mathcal{A} \to \mathcal{A}$ be given by $(a + \lambda I)^\delta = \widetilde{a} + \lambda I$. One verifies that δ is a Lie derivation. Suppose $\delta = d + \tau$ (according to Corollary 6.9), with $\mathcal{A}^d \subseteq \mathcal{A}$ (and hence $\mathcal{A}^\tau \subseteq \mathcal{A}$). Consider the commutator $c = x^2 e_{12}$; thus $c^d = c^\delta$. Now $c^\delta = 2xe_{12}$. On the other hand $c^d = (x^2 I)^d e_{12} + x^2 e_{12}^d = f(x^2)e_{12}$ for some $f(x^2) \in \mathcal{Z}$ (since a derivation must map $\mathcal{Z}$ into itself), and a contradiction is reached. The correct choices for d and τ to satisfy Corollary 6.9 are of course $(a + \lambda I)^d = \widetilde{a} + \lambda^\Delta I$ and $(a + \lambda I)^\tau = (\lambda - \lambda^\Delta)I$.

The next example justifies the exclusion of rings with characteristic 2 in Corollaries 6.5 and 6.9.

Example 6.11. Let $\mathbb{F}$ be a field with $\mathrm{char}(\mathbb{F}) = 2$, and let $\mathcal{A} = M_2(\mathbb{F})$ (so $\deg(\mathcal{A}) = 2$ and $\mathrm{char}(\mathcal{A}) = 2$). For $a = (a_{ij}) \in \mathcal{A}$ we define $\delta, \alpha : \mathcal{A} \to \mathcal{A}$ by $a^\delta = a_{21}e_{12}$ and $a^\alpha = a + a^\delta$. One can check that δ is a Lie derivation and α is a Lie automorphism. If α was of a standard form $\alpha = \sigma + \tau$, then it would follow that $e_{21}^\sigma = e_{21}^\alpha$ (since e_{21} is a commutator) and so $e_{11} + e_{22} = (e_{12} + e_{21})^2 = (e_{21}^\sigma)^2 = (e_{21}^\sigma)^2 = (e_{21}^2)^\sigma = 0$, a contradiction. Similarly we see that δ is not of a standard form $d + \tau$.

The last example shows that in the setting of Corollary 6.5, in general σ need not be injective nor must $\mathcal{B}$ be prime.

Example 6.12. Let $\mathcal{A} = \mathbb{F}\langle X \rangle$ be the free noncommutative algebra over a field $\mathbb{F}$, and let $\mathcal{T}$ be the subalgebra of $\mathcal{A}$ consisting of all elements with constant term 0. We let $\mathcal{B}$ be the ring theoretic direct sum $\mathbb{F} \oplus \mathcal{T}$ and define $\alpha : \mathcal{B} \to \mathcal{A}$ according to $(\lambda \oplus t)^\alpha = \lambda + t$. One checks that α is a Lie isomorphism, but clearly $\mathcal{B}$ is not prime and σ maps $\mathbb{F}$ to 0.

6.2   Lie Maps on Skew Elements

In Section 6.1 we saw how d-freeness can be used to study a Lie homomorphism α in the simplest situation, i.e., $\alpha : \mathcal{B} \to \mathcal{Q}$ where $\mathcal{B}$ and $\mathcal{Q}$ are associative rings. In contrast, our goal in Sections 6.2 and 6.3 is to study Lie maps in a far more general context, which we indicate as follows. Let $\mathcal{B}$ be an associative ring, $\mathcal{L}$ a Lie subring of $\mathcal{B}$, and $\mathcal{S}$ a Lie ideal of $\mathcal{L}$. Let $\mathcal{Q}$ be a unital ring with center $\mathcal{C}$ and let $\overline{\mathcal{Q}}$ denote the factor Lie ring $\mathcal{Q}/\mathcal{C}$. For every $y \in \mathcal{Q}$ we shall write $\overline{y} = y + \mathcal{C} \in \overline{\mathcal{Q}}$, and for every set $\mathcal{R} \subseteq \mathcal{Q}$ we shall write $\overline{\mathcal{R}} = \{\overline{r} \mid r \in \mathcal{R}\}$. We will be studying a Lie homomorphism $\alpha : \mathcal{S} \to \overline{\mathcal{Q}}$, where $\mathcal{S}^\alpha = \overline{\mathcal{R}}$, with $\mathcal{R}$ a d-free subset of $\mathcal{Q}$ for appropriate choice of d. There are various reasons motivating this degree of generality. The situation where $\mathcal{S} = \mathcal{L}$ includes the important case where $\mathcal{S}$ is the Lie ring of skew elements of a ring $\mathcal{B}$ with involution. The reason for considering Lie ideals stems from Herstein's Lie structure theory of simple associative rings with and without involution (more details about this are given at the beginning of Section 6.3).

The fact that factor Lie rings show up in a natural way when concerned with Lie simplicity is why we want to have the range of α lying in $\overline{\mathcal{Q}}$ instead of just in $\mathcal{Q}$. At any rate any Lie homomorphism $\alpha' : \mathcal{S} \to \mathcal{Q}$ induces in an obvious way a Lie homomorphism $\alpha : \mathcal{S} \to \overline{\mathcal{Q}}$ according to $x^\alpha = \overline{x^{\alpha'}}$. The results that we will obtain therefore immediately imply similar results on Lie homomorphisms having their ranges in $\mathcal{Q}$.

In this section we study Lie homomorphisms $\alpha : \mathcal{L} \to \overline{\mathcal{Q}}$ where $\mathcal{L}$ is the Lie ring of skew elements of $\mathcal{B}$ (Theorem 6.15). This in turn is used in Section 6.3 to study Lie homomorphisms $\alpha : \mathcal{S} \to \overline{\mathcal{Q}}$ where $\mathcal{S}$ is a Lie ideal of the Lie ring $\mathcal{L}$ of skew elements of $\mathcal{B}$ (Theorem 6.18). Also in Section 6.3 we use the results of Section 6.1 to study Lie homomorphisms $\alpha : \mathcal{S} \to \overline{\mathcal{Q}}$ where $\mathcal{S}$ is a Lie ideal of $\mathcal{B}$ (Theorem 6.19).

The proofs of these results are rather demanding and we feel that we can alleviate, if only very slightly, the burdens of these proofs by establishing right now the first step they all have in common. Namely, since it is awkward to work in the factor Lie ring $\mathcal{Q}$, the common first step is to "replace" the given Lie homomorphism by a weak Lie homomorphism (as defined in Section 6.1 preceding Remark 6.2). Therefore let $\alpha : \mathcal{S} \to \overline{\mathcal{Q}}$ be a Lie homomorphism such that $\mathcal{S}^\alpha = \overline{\mathcal{R}}$ where $\mathcal{R}$ is d-free in $\mathcal{Q}$. It is at this point that we must make the additional assumption that $\mathcal{C}$ is an additive direct summand of $\mathcal{Q}$; thus there is an additive subgroup $\mathcal{W}$ of $\mathcal{Q}$ such that $\mathcal{Q} = \mathcal{C} \oplus \mathcal{W}$. Let $\theta : \overline{\mathcal{Q}} \to \mathcal{W}$ be the additive isomorphism induced by the projection of $\mathcal{Q}$ onto $\mathcal{W}$ and set $x^\beta = x^{\alpha\theta}$. It is easy to see that β is a weak Lie homomorphism from $\mathcal{S}$ into $\mathcal{W}$. Since there is no reason to believe that $\mathcal{S}^\beta$ is d-free, we solve this problem by forming the ring-theoretic direct sum $\mathcal{B}_\mathcal{C} = \mathcal{C} \oplus \mathcal{B}$, noting that $\mathcal{L}_\mathcal{C} = \mathcal{C} \oplus \mathcal{L}$ is a Lie subring of $\mathcal{B}_\mathcal{C}$ and $\mathcal{S}_\mathcal{C} = \mathcal{C} \oplus \mathcal{S}$ is a Lie ideal of $\mathcal{L}_\mathcal{C}$. We then define a map $\gamma : \mathcal{S}_\mathcal{C} \to \mathcal{Q}$ according to $(\lambda + s)^\gamma = \lambda + s^\beta$, and note that γ is also a weak Lie homomorphism. Furthermore, for $r \in \mathcal{R}$ we have $\bar{r} = x^\alpha$ for some $x \in \mathcal{S}$. Writing $r = \lambda + w$ we see by definition that $w = x^\beta = x^\gamma$. Thus $r = (\lambda + x)^\gamma$, whence $\mathcal{R}$ is contained in $\mathcal{S}_\mathcal{C}^\gamma$ and so by Corollary 3.5 $\mathcal{S}_\mathcal{C}^\gamma$ is again d-free. The map γ given above will be called the *weak Lie homomorphism associated with* α.

The remainder of this section will now be devoted to the study of Lie homomorphisms $\alpha : \mathcal{L} \to \overline{\mathcal{Q}}$, where $\mathcal{L}$ is the Lie ring of skew elements of a ring with involution. From the technical point of view this is considerably more difficult than studying a Lie homomorphism defined on a ring. The main ideas upon which this study is based, however, are similar to those from Section 6.1.

We begin by introducing some necessary notions. We will say that an additive group $\mathcal{L}$ *admits the operator* $\frac{1}{2}$ if for every $a \in \mathcal{L}$, the equation $2x = a$ has a unique solution $\frac{1}{2}a$ in $\mathcal{L}$. For example, every vector space over a field with characteristic different from 2 has this property. The condition that a unital ring admits the operator $\frac{1}{2}$ is clearly equivalent to the condition that the element $1+1$ is invertible. Anyhow, using $\frac{1}{2}$ is just something that is difficult to avoid in our context.

We now introduce a more important (in fact the key) notion of this section: we will refer to elements in a ring that are of the form $xyz + zyx$ as *triads*. We shall say that a subset $\mathcal{L}$ of a ring is closed under triads if $xyz + zyx \in \mathcal{L}$ for all $x, y, z \in \mathcal{L}$. Finally we recall that by $\langle \mathcal{L} \rangle$ we are denoting the subring generated by $\mathcal{L}$.

Lemma 6.13. *Let $\mathcal{L}$ be a Lie subring of a ring $\mathcal{B}$. Suppose that $\mathcal{L}$ is closed under triads and $\mathcal{L}$ admits the operator $\frac{1}{2}$. Then $\langle \mathcal{L} \rangle = \mathcal{L} + \mathcal{L} \circ \mathcal{L}$.*

Proof. We have to show that $u_1 u_2 \ldots u_n \in \mathcal{L} + \mathcal{L} \circ \mathcal{L}$ for all $n \geq 2$ and all $u_i \in \mathcal{L}$. For $n = 2$ this is clear, just write $u_1 u_2$ as $\frac{1}{2}[u_1, u_2] + \frac{1}{2}u_1 \circ u_2$. Let $n = 3$. Noting that

$$u_1 u_2 u_3 - u_3 u_2 u_1 = \frac{1}{2}\Big([u_1, u_2] \circ u_3 + [u_1, u_3] \circ u_2 + [u_2, u_3] \circ u_1\Big) \in \mathcal{L} \circ \mathcal{L}$$

it follows, in view of our assumption that $\mathcal{L}$ is closed under triads, that

$$u_1 u_2 u_3 = \frac{1}{2}\Big(u_1 u_2 u_3 + u_3 u_2 u_1\Big) + \frac{1}{2}\Big(u_1 u_2 u_3 - u_3 u_2 u_1\Big) \in \mathcal{L} + \mathcal{L} \circ \mathcal{L}.$$

For $n > 3$, we just consider $u_1 u_2 \ldots u_n$ as $(u_1 u_2 u_3)u_4 \ldots u_n$ and use the induction argument. $\square$

The reason for our interest in Lie rings closed under triads is obvious: the skew elements of a ring with involution do form such a Lie ring. But so do all elements in an associative ring. There is another property, given in the next lemma, which makes it possible for us to clearly distinguish between these two types of Lie rings.

Lemma 6.14. *Let $\mathcal{L}$ be a Lie subring of a ring $\mathcal{B}$ such that $\mathcal{L}$ admits the operator $\frac{1}{2}$. Then the following two conditions are equivalent:*

(i) *$\mathcal{L}$ is closed under triads and $\mathcal{L} \cap (\mathcal{L} \circ \mathcal{L}) = 0$;*

(ii) *$\langle \mathcal{L} \rangle$ has an involution such that $\mathcal{L}$ is the set of all skew elements with respect to this involution.*

Proof. Trivially (ii) implies (i). Suppose (i) holds. By Lemma 6.13 we know that $\langle \mathcal{L} \rangle = \mathcal{L} \oplus \mathcal{L} \circ \mathcal{L}$. Define $*$ by $(u + \sum_i v_i \circ w_i)^* = -u + \sum_i v_i \circ w_i$. Adapting the arguments from the proof of Lemma 6.13 one can easily check that $(u_1 u_2 \ldots u_n)^* = (-1)^n u_n \ldots u_2 u_1$ for all $u_i \in \mathcal{L}$, which in turn implies that $*$ is an involution. Clearly $\mathcal{L}$ is the set of skew elements under $*$. $\square$

We shall refer to a Lie ring $\mathcal{L}$ satisfying Lemma 6.14 as a *Lie ring of skew elements*, so we will avoid mentioning the associative ring with involution such that $\mathcal{L}$ is its set of skew elements. This is simply because this associative ring shall play no role in our arguing. Later, in applications to the classical prime ring situation, we will consider the skew elements of prime rings with involution. Then these rings will be of course pointed out.

Our intention is to show that under appropriate assumptions a Lie homomorphism $\alpha : \mathcal{L} \to \overline{\mathcal{Q}}$, where $\mathcal{L}$ is a Lie ring of skew elements, arises from an (associative) homomorphism. More precisely, we wish to show that there exists a homomorphism $\sigma : \langle \mathcal{L} \rangle \to \mathcal{Q}$ such that $x^\alpha = \overline{x^\sigma}$ for all $x \in \mathcal{L}$.

The reader might wonder why now, unlike in the previous section, antihomomorphisms do not appear in the expected conclusion. After all, the negative of an antihomomorphism $\psi : \langle \mathcal{L} \rangle \to \mathcal{Q}$ certainly induces a Lie homomorphism from $\mathcal{L}$ into $\overline{\mathcal{Q}}$. The answer is hidden in the existence of an involution on $\langle \mathcal{L} \rangle$, which is an antihomomorphism acting as the negative of the identity on $\mathcal{L}$. Namely, we can replace ψ by a homomorphism σ defined by $x^\sigma = -(x^*)^\psi$; indeed, σ agrees with ψ on $\mathcal{L}$, so that σ and ψ induce the same Lie homomorphism on $\mathcal{L}$.

The sole goal of this section is to establish the theorem that follows. We could easily derive various applications of this theorem to Lie derivations and Lie maps in prime rings with involution. However, in the next section we shall prove a substantially more general result (with a slightly strengthened d-freeness assumption), and so we postpone applications until then.

Theorem 6.15. *Let $\mathcal{L}$ be a Lie ring of skew elements, let $\mathcal{Q}$ be a unital ring with center $\mathcal{C}$, and let $\alpha : \mathcal{L} \to \overline{\mathcal{Q}}$ be a Lie homomorphism. Suppose that both $\mathcal{L}$ and $\mathcal{Q}$ admit the operator $\frac{1}{2}$, suppose that $\mathcal{C}$ is a direct summand of the additive group $\mathcal{Q}$, and suppose there exists an 8-free subset $\mathcal{R}$ of $\mathcal{Q}$ such that $\overline{\mathcal{R}} = \mathcal{L}^\alpha$. Then there exists a homomorphism $\sigma : \langle \mathcal{L} \rangle \to \mathcal{Q}$ such that $x^\alpha = \overline{x^\sigma}$ for all $x \in \mathcal{L}$.*

Proof. With reference to the discussion at the beginning of this section, let $\gamma :$ $\mathcal{L}_C \to \mathcal{Q}$ be the weak Lie homomorphism associated with α, with the reminder that $\mathcal{L}_C^\gamma$ is again an 8-free subset of $\mathcal{Q}$.

We remark that $\mathcal{L}_C$ is also closed under triads. Our first goal is to find out how γ acts on triads. Therefore we let $B : \mathcal{L}_C^3 \to \mathcal{Q}$ be the triadditive map given by

$$B(x, y, z) = (xyz + zyx)^\gamma.$$

If $p, q \in \mathcal{Q}$ are such that $\bar{p} = \bar{q}$, i.e., if $p - q \in \mathcal{C}$, then we shall write $p \equiv q$. Applying γ to the identity

$$[xyz + zyx, t] + [txy + yxt, z] + [ztx + xtz, y] + [yzt + tzy, x] = 0$$

we have

$$[B(x, y, z), t^\gamma] + [B(t, x, y), z^\gamma] + [B(z, t, x), y^\gamma] + [B(y, z, t), x^\gamma] \equiv 0.$$

Since $\mathcal{L}_C^\gamma$ is 8-free in $\mathcal{Q}$ (and hence 5-free) we see by Theorem 4.13 that B is a quasi-polynomial of degree ≤ 3, which we write as follows:

$$\begin{aligned}
B(x, y, z) = {}& \lambda_1 x^\gamma y^\gamma z^\gamma + \lambda_2 x^\gamma z^\gamma y^\gamma + \lambda_3 y^\gamma x^\gamma z^\gamma \\
&+ \lambda_4 y^\gamma z^\gamma x^\gamma + \lambda_5 z^\gamma x^\gamma y^\gamma + \lambda_6 z^\gamma y^\gamma x^\gamma \\
&+ \nu_1(x) y^\gamma z^\gamma + \nu_2(x) z^\gamma y^\gamma + \nu_3(y) x^\gamma z^\gamma \\
&+ \nu_4(y) z^\gamma x^\gamma + \nu_5(z) x^\gamma y^\gamma + \nu_6(z) y^\gamma x^\gamma \\
&+ \mu_1(x, y) z^\gamma + \mu_2(x, z) y^\gamma + \mu_3(y, z) x^\gamma \\
&+ \omega(x, y, z),
\end{aligned}$$

where $\lambda_i \in \mathcal{C}$, $\nu_i : \mathcal{L}_C \to \mathcal{C}$, $\mu_i : \mathcal{L}_C^2 \to \mathcal{C}$, and $\omega : \mathcal{L}_C^3 \to \mathcal{C}$. Moreover, by Lemma 4.6 ν_i is additive, μ_i is biadditive, and ω is triadditive.

Many of the calculations we are about to make arise from noting that if two quasi-polynomials are equal to each other, then we may conclude from Lemma 4.4 that the corresponding coefficients are equal. Frequently we will indicate what is to be done and leave it for the reader to provide the details.

Since $B(x, y, z) = B(z, y, x)$ we write out $B(z, y, x)$ in the same form as above and conclude that $\lambda_1 = \lambda_6$, $\lambda_2 = \lambda_5$, $\lambda_3 = \lambda_4$, $\nu_1 = \nu_6$, $\nu_2 = \nu_5$, $\nu_3 = \nu_4$, $\mu_1(x, y) = \mu_3(y, x)$, and $\mu_2(x, z) = \mu_2(z, x)$. So the above formula now reads

$$\begin{aligned}
B(x, y, z) = {}& \lambda_1(x^\gamma y^\gamma z^\gamma + z^\gamma y^\gamma x^\gamma) \\
&+ \lambda_2(x^\gamma z^\gamma y^\gamma + z^\gamma x^\gamma y^\gamma) + \lambda_3(y^\gamma x^\gamma z^\gamma + y^\gamma z^\gamma x^\gamma) \\
&+ \nu_1(x) y^\gamma z^\gamma + \nu_2(x) z^\gamma y^\gamma + \nu_3(y) x^\gamma z^\gamma \\
&+ \nu_3(y) z^\gamma x^\gamma + \nu_2(z) x^\gamma y^\gamma + \nu_1(z) y^\gamma x^\gamma \\
&+ \mu_1(x, y) z^\gamma + \mu_2(x, z) y^\gamma + \mu_1(z, y) x^\gamma \\
&+ \omega(x, y, z).
\end{aligned}$$

Using the fact that γ is a weak Lie homomorphism we see that

$$B(x, y, z) - B(y, x, z) = [[x, y], z]^\gamma \equiv [[x, y]^\gamma, z^\gamma] = [[x^\gamma, y^\gamma], z^\gamma].$$

On the other hand we have

$$
\begin{aligned}
B(x, y, z) &- B(y, x, z) \\
&\equiv (\lambda_1 - \lambda_3)[x^\gamma, y^\gamma]z^\gamma - (\lambda_1 - \lambda_2)z^\gamma[x^\gamma, y^\gamma] + (\lambda_2 - \lambda_3)(x^\gamma z^\gamma y^\gamma - y^\gamma z^\gamma x^\gamma) \\
&\quad + (\nu_1 - \nu_3)(x)y^\gamma z^\gamma + (\nu_2 - \nu_3)(x)z^\gamma y^\gamma + (\nu_3 - \nu_1)(y)x^\gamma z^\gamma \\
&\quad + (\nu_3 - \nu_2)(y)z^\gamma x^\gamma + (\nu_2 - \nu_1)(z)x^\gamma y^\gamma + (\nu_1 - \nu_2)(y)y^\gamma x^\gamma \\
&\quad + (\mu_1(x, y) - \mu_1(y, x))z^\gamma + (\mu_2(x, z) - \mu_1(z, x))y^\gamma + (\mu_1(z, y) - \mu_2(y, z))x^\gamma.
\end{aligned}
$$

Comparing both expressions we may then conclude that $\lambda_2 = \lambda_3$, $\lambda_1 - \lambda_2 = 1$, $\nu_1 = \nu_2 = \nu_3$ and $\mu_1 = \mu_2$ is a symmetric map. Setting $\lambda = \lambda_2$, $\nu = \nu_i$, and $\mu = \mu_i$, we thus have

$$
\begin{aligned}
B(x, y, z) &= (\lambda + 1)\big(x^\gamma y^\gamma z^\gamma + z^\gamma y^\gamma x^\gamma\big) \\
&\quad + \lambda\big(x^\gamma z^\gamma y^\gamma + z^\gamma x^\gamma y^\gamma + y^\gamma x^\gamma z^\gamma + y^\gamma z^\gamma x^\gamma\big) \\
&\quad + \nu(x)y^\gamma \circ z^\gamma + \nu(y)x^\gamma \circ z^\gamma + \nu(z)x^\gamma \circ y^\gamma \\
&\quad + \mu(x, y)z^\gamma + \mu(x, z)y^\gamma + \mu(z, y)x^\gamma \\
&\quad + \omega(x, y, z).
\end{aligned}
$$

The next computations are based on the identity

$$B(x, y, xyx) = 2(xyxyx)^\gamma = B(x, yxy, x).$$

In order to express $B(x, y, xyx)$ one first has to replace z by xyx in the above formula. The new formula that we get involves the expression $(xyx)^\gamma$, which is equal to $\frac{1}{2}B(x, y, x)$ and so it can be further expanded. At the end we thus obtain $B(x, y, xyx)$ expressed as a sum of products involving only x^γ, y^γ and central coefficients arising from λ, ν, μ and ω. The identity that we obtain is not multilinear, but fortunately we have Lemma 4.20 (the nonlinear counterpart of Lemma 4.4) to fall back upon in order to draw conclusions about the coefficients. In order to use this lemma there is no need to keep track of all coefficients. For our immediate purposes it is enough to observe that the procedure we just described leads to

$$B(x, y, xyx) = 2\lambda^2 y^\gamma x^\gamma x^\gamma x^\gamma y^\gamma + R$$

where R denotes the sum of all remaining terms, that is, R consists of summands none of which is equal to a central multiple by $y^\gamma x^\gamma x^\gamma x^\gamma y^\gamma$. Repeating the same procedure with $B(x, yxy, x)$ one easily observes that the expression that we get does not involve the term $y^\gamma x^\gamma x^\gamma x^\gamma y^\gamma$ at all. Since $B(x, y, xyx) = B(x, yxy, x)$ we can compare both expressions and arrive at a situation where Lemma 4.20 is applicable. Therefore $2\lambda^2 = 0$, and hence $\lambda^2 = 0$. But actually we want to show that $\lambda = 0$. Let us still consider the expanded forms of $B(x, y, xyx) = B(x, yxy, x)$, but this time we compare coefficients at $x^\gamma y^\gamma x^\gamma x^\gamma y^\gamma$. First note that in the expression of $B(x, yxy, x)$ this term does not appear at all; in other words, its coefficient is 0. On the other hand, in the expression of $B(x, y, xyx)$ it does appear and one can check that its coefficient is, in view of $\lambda^2 = 0$, equal to 2λ. Therefore $2\lambda = 0$ and so also $\lambda = 0$.

So now we have simplified the formula for B to

$$B(x, y, z) = x^\gamma y^\gamma z^\gamma + z^\gamma y^\gamma x^\gamma$$
$$+ \nu(x)y^\gamma \circ z^\gamma + \nu(y)x^\gamma \circ z^\gamma + \nu(z)x^\gamma \circ y^\gamma$$
$$+ \mu(x, y)z^\gamma + \mu(x, z)y^\gamma + \mu(z, y)x^\gamma$$
$$+ \omega(x, y, z).$$

Let us pause for a moment to see where we are heading. We would like to slightly alter γ in the following sense: is there an additive map $\tau : \mathcal{L}_\mathcal{C} \to \mathcal{C}$ such that $\phi = \gamma + \tau$ preserves triads? The formula above is a step in the right direction and in fact tells us that the candidate for τ must be ν. Indeed, expansion of

$$(x^\gamma + \nu(x))(y^\gamma + \nu(y))(z^\gamma + \nu(z)) + (z^\gamma + \nu(z))(y^\gamma + \nu(y))(x^\gamma + \nu(x))$$

yields

$$x^\gamma y^\gamma z^\gamma + z^\gamma y^\gamma x^\gamma + \nu(x)(y^\gamma \circ z^\gamma) + \nu(y)(x^\gamma \circ z^\gamma) + \nu(z)(x^\gamma \circ y^\gamma)$$
$$+ 2\nu(x)\nu(y)z^\gamma + 2\nu(x)\nu(z)y^\gamma + 2\nu(z)\nu(x)x^\gamma + 2\nu(x)\nu(y)\nu(z).$$

This strongly suggests that we first try to prove that $\mu(x, y) = 2\nu(x)\nu(y)$. To this end we shall again rely on the identity $B(x, y, xyx) = B(x, yxy, x)$, but this time we shall compare coefficients at $x^\gamma y^\gamma x^\gamma$. Making use of

$$(xyx)^\gamma = \frac{1}{2} B(x, y, x)$$

$$= x^\gamma y^\gamma x^\gamma + \nu(x)y^\gamma \circ x^\gamma + \nu(y)x^\gamma x^\gamma + \mu(x, y)x^\gamma + \frac{\mu(x, x)}{2} y^\gamma + \frac{\omega(x, y, x)}{2}$$

one can verify that the coefficient in the expansion of $B(x, y, xyx)$ at $x^\gamma y^\gamma x^\gamma$ is $3\mu(x, y) + 2\nu(x)\nu(y)$, while the coefficient in the expansion of $B(x, yxy, x)$ is $2\mu(x, y) + 4\nu(x)\nu(y)$. Comparing we thus get $\mu(x, y) = 2\nu(x)\nu(y)$, as desired.

Let us now define $\phi : \mathcal{L}_\mathcal{C} \to \mathcal{Q}$ as $\phi = \gamma + \nu$, and observe that ϕ is a weak Lie homomorphism, i.e., $[x, y]^\phi \equiv [x^\phi, y^\phi]$, and it also satisfies $(xyz + zyx)^\phi \equiv x^\phi y^\phi z^\phi + z^\phi y^\phi x^\phi$. Therefore, the maps

$$\epsilon(x, y) = [x, y]^\phi - [x^\phi, y^\phi],$$

$$\zeta(x, y, z) = \frac{1}{2} \left((xyz + zyx)^\phi - x^\phi y^\phi z^\phi - z^\phi y^\phi x^\phi \right)$$

have their images in $\mathcal{C}$. We note that there is no reason to believe that $\mathcal{L}_{\mathcal{C}}^{\phi}$ is 8-free in $\mathcal{Q}$, so some care must be taken in this matter.

Our next objective is to show that both ε and ζ must in fact be zero. We begin by noting that

$$[xyz + zyx, w]^{\phi} \equiv [(xyz + zyx)^{\phi}, w^{\phi}] = [x^{\phi}y^{\phi}z^{\phi} + z^{\phi}y^{\phi}x^{\phi}, w^{\phi}].$$

On the other hand, from the identity

$$[xyz + zyx, w] = \big([x, w]yz + zy[x, w]\big) + \big(x[y, w]z + z[y, w]x\big) + \big([z, w]yx + xy[z, w]\big)$$

it follows that

$$
\begin{aligned}
[xyz + zyx, w]^{\phi} &\equiv [x, w]^{\phi}y^{\phi}z^{\phi} + z^{\phi}y^{\phi}[x, w]^{\phi} + x^{\phi}[y, w]^{\phi}z^{\phi} + z^{\phi}[y, w]^{\phi}x^{\phi} \\
&\quad + [z, w]^{\phi}y^{\phi}x^{\phi} + x^{\phi}y^{\phi}[z, w]^{\phi} \\
&= ([x^{\phi}, w^{\phi}] + \epsilon(x, w))y^{\phi}z^{\phi} + z^{\phi}y^{\phi}([x^{\phi}, w^{\phi}] + \epsilon(x, w)) \\
&\quad + x^{\phi}([y^{\phi}, w^{\phi}] + \epsilon(y, w))z^{\phi} + z^{\phi}([y^{\phi}, w^{\phi}] + \epsilon(y, w))x^{\phi} \\
&\quad + ([z^{\phi}, w^{\phi}] + \epsilon(z, w))y^{\phi}x^{\phi} + x^{\phi}y^{\phi}([z^{\phi}, w^{\phi}] + \epsilon(z, w)) \\
&= [x^{\phi}y^{\phi}z^{\phi} + z^{\phi}y^{\phi}x^{\phi}, w^{\phi}] \\
&\quad + \epsilon(x, w)y^{\phi} \circ z^{\phi} + \epsilon(y, w)x^{\phi} \circ z^{\phi} + \epsilon(z, w)y^{\phi} \circ x^{\phi}.
\end{aligned}
$$

Comparing we thus get

$$\epsilon(x, w)y^{\phi} \circ z^{\phi} + \epsilon(y, w)x^{\phi} \circ z^{\phi} + \epsilon(z, w)y^{\phi} \circ x^{\phi} \equiv 0.$$

One now rewrites (in one's mind) this identity by substituting $\gamma + \nu$ for ϕ (in order to take advantage of $\mathcal{L}_{\mathcal{C}}^{\gamma}$ being 8-free (and hence 5-free) in $\mathcal{Q}$). Lemma 4.4 then implies that $\epsilon = 0$, i.e., ϕ is a Lie homomorphism.

Let us show that ζ is also 0. From the definition of ζ we see that $(xyx)^{\phi} = x^{\phi}y^{\phi}x^{\phi} + \zeta(x, y, x)$, and note that it is enough to show that $\zeta(x, y, x) = 0$. To this end we will write $(x^3yx^3)^{\phi}$ in two ways. On the one hand we have

$$(x(xyx)x)^{\phi} = x^{\phi}(x^{\phi}y^{\phi}x^{\phi} + \zeta(x, y, x))x^{\phi} + \zeta(x, xyx, x)$$

and consequently

$$
\begin{aligned}
(x(x^2yx^2)x)^{\phi} &= (x^{\phi})^3 y^{\phi}(x^{\phi})^3 + \zeta(x, y, x)(x^{\phi})^4 \\
&\quad + \zeta(x, xyx, x)(x^{\phi})^2 + \zeta(x, x^2yx^2, x).
\end{aligned}
$$

On the other hand,

$$
\begin{aligned}
(x^3yx^3)^{\phi} &= ((x^{\phi})^3 + \zeta(x, x, x))y^{\phi}((x^{\phi})^3 + \zeta(x, x, x)) + \zeta(x^3, y, x^3) \\
&= (x^{\phi})^3 y^{\phi}(x^{\phi})^3 + \zeta(x, x, x)(x^{\phi})^3 \circ y^{\phi} + \zeta(x, x, x)^2 y^{\phi} + \zeta(x^3, y, x^3).
\end{aligned}
$$

Consequently,

$$\zeta(x,y,x)(x^\phi)^4 - \zeta(x,x,x)(x^\phi)^3 \circ y^\phi + \zeta(x,xyx,x)(x^\phi)^2 - \zeta(x,x,x)^2 y^\phi \equiv 0.$$

Replace ϕ by $\gamma+\nu$ in order to make applying Lemma 4.20 possible. Focusing on the summand $\zeta(x,y,x)(x^\gamma)^4$ we are in a situation regarding Lemma 4.20 where $l_1 = 2$, $l_2 = 1$. Hence it follows that $2\zeta(x,y,x) = 0$, and so $\zeta(x,y,x) = 0$. Incidentally, this is the only point in the proof where 8-freeness is used in full generality; at all other places d-freeness with $d \le 6$ is sufficient.

So we know now that ϕ is a Lie homomorphism that also preserves triads. We are now ready to complete the proof. By Lemma 6.13 we have $\langle \mathcal{L} \rangle = \mathcal{L} + \mathcal{L} \circ \mathcal{L}$. We then define a map $\sigma : \langle \mathcal{L} \rangle \to \mathcal{Q}$ according to the rule

$$\left(x + \sum_i y_i \circ z_i\right)^\sigma = x^\phi + \sum_i y_i^\phi \circ z_i^\phi$$

for $x, y_i, z_i \in \mathcal{L}$. We first show that σ is well-defined. We are given that $\mathcal{L} \cap (\mathcal{L} \circ \mathcal{L}) = 0$, so it is enough to show that $\sum_i y_i \circ z_i = 0$ implies $\sum_i y_i^\phi \circ z_i^\phi = 0$. Let $t \in \mathcal{L}_C$. Since ϕ preserves triads we have

$$\left(\sum_i y_i^\phi \circ z_i^\phi\right) \circ t^\phi = \left(\left(\sum_i y_i \circ z_i\right) \circ t\right)^\phi = 0.$$

Setting $a = \sum_i y_i^\phi \circ z_i^\phi$, we see in particular that $a \circ [t,u]^\phi = 0$ for all $t, u \in \mathcal{L}_C$. As ϕ is a Lie homomorphism we have

$$0 = a \circ [t^\phi, u^\phi] = a \circ [t^\gamma, u^\gamma].$$

Since $\mathcal{L}_C^\gamma$ is 8-free (and hence 2-free) we conclude first that $a \in \mathcal{C}$ and then that $a = 0$. Thus σ is well-defined. We now verify that σ is a homomorphism. For $x, y \in \mathcal{L}$ we may write $xy = \frac{1}{2}(x \circ y + [x,y])$. Applying σ, we have

$$(xy)^\sigma = \frac{1}{2}\left(x^\phi \circ y^\phi + [x^\phi, y^\phi]\right) = x^\phi y^\phi = x^\sigma y^\sigma.$$

For $x, y, z \in \mathcal{L}$ we have the identity

$$(x \circ z)y = \frac{1}{2}(x \circ [z,y] + z \circ [x,y] + (x \circ z) \circ y).$$

Since ϕ is a Lie homomorphism preserving triads it follows that

$$\begin{aligned}
((x \circ z)y)^\sigma &= \frac{1}{2}\left(x^\phi \circ [z,y]^\phi + z^\phi \circ [x,y]^\phi + ((x \circ z) \circ y)^\phi\right) \\
&= \frac{1}{2}\left(x^\phi \circ [z^\phi, y^\phi] + z^\phi \circ [x^\phi, y^\phi] + ((x^\phi \circ z^\phi) \circ y^\phi)\right) \\
&= (x^\phi \circ z^\phi)y^\phi \\
&= (x \circ z)^\sigma y^\sigma.
\end{aligned}$$

Since $\mathcal{L}+\mathcal{L}\circ\mathcal{L} = \langle\mathcal{L}\rangle$ we may now conclude that $(uy)^\sigma = u^\sigma y^\sigma$ for all $u \in \langle\mathcal{L}\rangle$ and $y \in \mathcal{L}$. However, this clearly implies that σ is a homomorphism on $\langle\mathcal{L}\rangle$.

Finally, let us return to the original Lie homomorphism $\alpha : \mathcal{L} \to \overline{\mathcal{Q}}$. It is connected to σ via γ and ϕ. Indeed, we have $x^\alpha = \overline{x^\gamma} = \overline{x^\phi} = \overline{x^\sigma}$. This completes the proof. $\square$

6.3   Lie Maps on Lie Ideals

If a ring $\mathcal{A}$ has a nice structure from the ring-theoretic point of view, there is no reason to believe that it has also a nice structure when regarded as a Lie ring. For example, if $\mathcal{A}$ is a simple ring, then it is quite likely that its center $\mathcal{Z}$ and its derived Lie ideal $[\mathcal{A}, \mathcal{A}]$ are proper Lie ideals of $\mathcal{A}$ (just think of $\mathcal{A} = M_n(\mathbb{F})$). On the other hand, Herstein showed that the simplicity of the ring $\mathcal{A}$ implies the simplicity of the Lie ring $[\mathcal{A}, \mathcal{A}]/\mathcal{Z} \cap [\mathcal{A}, \mathcal{A}]$, unless $\mathrm{char}(\mathcal{A}) = 2$ and $\deg(\mathcal{A}) = 2$ (see e.g., [113] or [114]). Similarly, if $\mathcal{A}$ is a simple ring with involution, the set of its skew elements $\mathcal{K} = \mathcal{K}(\mathcal{A})$ is not necessarily a simple Lie ring, not even when $\mathcal{Z} \cap \mathcal{K} = 0$. Such examples are not obvious, but as shown by Lee [138] they exist even when $\mathcal{A}$ is infinite dimensional over $\mathcal{Z}$. However, $[\mathcal{K}, \mathcal{K}]/\mathcal{Z} \cap [\mathcal{K}, \mathcal{K}]$ is a simple Lie ring provided that $\mathrm{char}(\mathcal{A}) \neq 2$ and $\deg(\mathcal{A}) > 4$ (see the paper by Baxter [13] or Herstein's surveys [113, 114]). All these suggest that, from the point of view of the theory of Lie rings, studying $\mathcal{A}$ and $\mathcal{K}$, which was done in the preceding two sections, is not entirely sufficient. We will now consider Lie maps on Lie ideals of $\mathcal{A}$ and $\mathcal{K}$ ($[\mathcal{A}, \mathcal{A}]$ and $[\mathcal{K}, \mathcal{K}]$ serve as prototypes). It is quite clear that the methods used for $\mathcal{A}$ and $\mathcal{K}$ fail in this context. The idea now is to reduce these more involved situations to those that were successfully resolved in Sections 6.1 and 6.2.

The setting in which we shall work is more general than the one just outlined. We begin by fixing a multilinear polynomial $f = f(x_1, \ldots, x_m) \in \mathbb{Z}\langle X \rangle$ of degree m. The important polynomials for us are $f_1 = x_1 x_2$ and $f_2 = x_1 x_2 x_3 + x_3 x_2 x_1$, but the subsequent arguments are no harder if f is just any polynomial. In fact, notation-wise it is even easier to work in this more general situation. If f is a Lie polynomial, then the results that follow are formally correct, but meaningless. So it is better to think of f as being a "non-Lie" polynomial.

We shall say that a Lie subring $\mathcal{L}$ of an associative ring $\mathcal{B}$ is f-closed if $f(u_1, \ldots, u_m) \in \mathcal{L}$ whenever all $u_1, \ldots, u_m \in \mathcal{L}$. Since $\mathcal{B}$ plays only a background role in what follows (i.e., the presence of $\mathcal{B}$ is needed since the associative product is involved in $[x, y]$ and other polynomials), we shall refer to $\mathcal{L}$ simply as to an f-closed Lie ring. In the preceding two sections we were able to describe Lie maps on certain f_1-closed and f_2-closed Lie rings. Now we wish to consider Lie maps on Lie ideals of f-closed Lie subrings. Of course, these Lie ideals may not be f-closed anymore. This is in fact the core of the problem which we are facing.

Recalling the explanation of the common "first part" given at the beginning of Section 6.2, we now let $\mathcal{S}$ be a Lie ideal of an f-closed Lie ring $\mathcal{L}$ and let $\mathcal{Q}$ be a unital ring with center $\mathcal{C}$ such that $\mathcal{Q} = \mathcal{C} \oplus \mathcal{W}$ for some additive subgroup $\mathcal{W}$

of $\mathcal{Q}$. We consider a Lie homomorphism $\alpha : \mathcal{S} \to \overline{\mathcal{Q}}$ such that $\mathcal{S}^\alpha = \overline{\mathcal{R}}$ with $\mathcal{R}$ a d-free subset of $\mathcal{Q}$. Clearly $\mathcal{L}_\mathcal{C}$ is f-closed and, letting $\gamma : \mathcal{L}_\mathcal{C} \to \mathcal{Q}$ be the weak Lie homomorphism associated with α, we know that $\mathcal{L}_\mathcal{C}^\gamma$ is d-free in $\mathcal{Q}$.

Consider the following identity, an extension of (6.3):

$$[x_1 x_2 \ldots x_m, x_{m+1}] + [x_{m+1} x_1 \ldots x_{m-1}, x_m] + \ldots + [x_2 x_3 \ldots x_{m+1}, x_1] = 0.$$

Since f is a multilinear polynomial of degree m this identity has an immediate repercussion:

$$[f(x_1, x_2, \ldots, x_m), x_{m+1}] + [f(x_{m+1}, x_1, \ldots, x_{m-1}), x_m]$$
$$+ \ldots + [f(x_2, x_3, \ldots, x_{m+1}), x_1] = 0 \quad (6.16)$$

(incidentally, for $f = f_2$ this identity was already stated at the beginning of the proof of Theorem 6.15). Since $f(\overline{x}_m) \in \mathcal{L}_\mathcal{C}$ for all $\overline{x}_m \in \mathcal{S}_\mathcal{C}^m$ and $\mathcal{S}_\mathcal{C}$ is a Lie ideal of $\mathcal{L}_\mathcal{C}$ it follows that $[f(\overline{x}_m), x_{m+1}] \in \mathcal{S}_\mathcal{C}$ for all $x_1, \ldots, x_{m+1} \in \mathcal{S}_\mathcal{C}$. Therefore, since we may apply γ to this quantity, we then define an $(m+1)$-additive map $B : \mathcal{S}_\mathcal{C}^{m+1} \to \mathcal{Q}$ according to

$$B(x_1, \ldots, x_{m+1}) = [f(\overline{x}_m), x_{m+1}]^\gamma \quad (6.17)$$

for all $x_1, \ldots, x_{m+1} \in \mathcal{S}_\mathcal{C}$. Applying γ to (6.16) we obtain

$$B(x_1, x_2, \ldots, x_{m+1}) + B(x_{m+1}, x_1, \ldots, x_m)$$
$$+ \ldots + B(x_2, \ldots, x_{m+1}, x_1) = 0. \quad (6.18)$$

Another identity that we need is just a version of the Jacobi identity:

$$[f(\overline{x}_m), [u, v]] = [[f(\overline{x}_m), u], v] + [u, [f(\overline{x}_m), v]].$$

Since γ is a weak Lie homomorphism it follows, by applying γ to this identity, that

$$B(\overline{x}_m, [u, v]) \equiv [B(\overline{x}_m, u), v^\gamma] + [u^\gamma, B(\overline{x}_m, v)] \quad (6.19)$$

for all $x_1, \ldots, x_m, u, v \in \mathcal{S}_\mathcal{C}$. Recall that $u \equiv v$ means that $u - v \in \mathcal{C}$.

Now we abstract this situation as follows.

Lemma 6.16. *Let $\mathcal{U}$ be a Lie ring, let $\mathcal{Q}$ be a unital ring, let $\gamma : \mathcal{U} \to \mathcal{Q}$ be a weak Lie homomorphism with $\mathcal{U}^\gamma$ a $(2m+3)$-free subset of $\mathcal{Q}$, and let $B : \mathcal{U}^{m+1} \to \mathcal{Q}$ be an $(m+1)$-additive map satisfying (6.18) and (6.19). Then there exists a quasi-polynomial P (with respect to γ) of degree $\leq m$ such that $B(x_1, \ldots, x_{m+1}) \equiv [P(\overline{x}_m), x_{m+1}^\gamma]$ for all $x_1, \ldots, x_{m+1} \in \mathcal{U}$.*

Proof. We first claim that B is a quasi-polynomial (with respect to γ). We shall begin by settling the claim in the case when each of the x_i's, for $i = 1, 2, \ldots, m$, is a commutator. This will be the first step in an inductive process in which one by one the commutators are replaced by arbitrary elements. To this end let $\mathcal{I}$ be a subset of $\{1, 2, \ldots, m\}$ and let $r = |\mathcal{I}|$. We define $y_1, \ldots, y_m$ as follows:

$$y_i = \begin{cases} [u_i, v_i] & \text{if } i \in \mathcal{I} \\ u_i & \text{otherwise} \end{cases}$$

where $u_i, v_i \in \mathcal{U}$, $i = 1, 2, \ldots, m$. By (6.18) we may write (with y_{m+1} still to be determined):

$$B(y_1, y_2, \ldots, y_{m+1}) + \ldots + B(y_{i+1}, \ldots, y_{m+1}, y_1, \ldots, y_i)$$
$$+ \ldots + B(y_2, \ldots, y_{m+1}, y_1) = 0. \quad (6.20)$$

We claim that $B(y_1, \ldots, y_m, u_{m+1})$ is a quasi-polynomial of degree $\leq m + r + 1$, and we proceed to prove this claim by induction on $m + 1 - r$.

We first let $r = m$ and set each $y_i = [u_i, v_i]$, $1 \leq i \leq m + 1$. For each i we see by (6.19) that

$$B(y_{i+1}, \ldots, y_{m+1}, y_1, \ldots, y_i)$$
$$\equiv [B(y_{i+1}, \ldots, y_{i-1}, u_i), v_i^\gamma] + [u_i^\gamma, B(y_i, \ldots, y_{i-1}, v_i)]. \quad (6.21)$$

Note that (6.20) and (6.21) together form a core functional identity in $2m + 2$ variables u_i^γ, v_i^γ, $1 \leq i \leq m + 1$. A glance at (6.21) shows that, neglecting the sign, every "middle" function also appears as a "leftmost" function. The conditions of Theorem 4.13 are thus satisfied, whence we conclude in particular that $B(y_1, \ldots, y_m, u_{m+1})$ is a quasi-polynomial of degree $\leq 2m + 1$ in $u_1^\gamma, \ldots, u_{m+1}^\gamma$, $v_1^\gamma \ldots, v_m^\gamma$. We remark that this is the one place in which the full force of $(2m+3)$-freeness is needed.

Now we suppose the claim is true for $r + 1$ and try to show it is true for r, where $0 \leq r \leq m - 1$. We focus on (6.20), where we are assuming that r of the $y_1, \ldots, y_m$ are commutators, and we set $y_{m+1} = [u_{m+1}, v_{m+1}]$. Without loss of generality we may assume that $\mathcal{I} = \{1, 2, \ldots, r\}$. Then $m - r$ summands of (6.20), namely, those that have one of $y_{r+1}, \ldots, y_m$ as the last variable, are quasi-polynomials of degree $\leq m + r + 1$ in $u_1^\gamma, \ldots, u_{m+1}^\gamma, v_1^\gamma, \ldots, v_r^\gamma$. Consequently we may rewrite (6.20) in the form:

$$B(y_1, \ldots, y_{m+1}) + B(y_{r+1}, \ldots, y_r) + \ldots + B(y_2, \ldots, y_1) = Q \quad (6.22)$$

where Q is a quasi-polynomial. Together (6.22) and (6.21), for $i = 1, \ldots, r, m + 1$, form a core functional identity (similar to (6.20) and (6.21) above). Therefore again we conclude from Theorem 4.13 that, in particular, $B(y_1, \ldots, y_m, u_{m+1})$ is a quasi-polynomial, thus completing the inductive step and establishing the claim. Setting $r = 0$, we then have that $B(u_1, \ldots, u_{m+1})$ is a quasi-polynomial of degree $\leq m + 1$. Therefore $B = B(x_1, \ldots, x_{m+1})$ may be represented as

$$\sum_{M,N} \lambda_{M x_{m+1}^\gamma N} M x_{m+1}^\gamma N + \sum_K \lambda_K K + \lambda_1 \quad (6.23)$$

where x_{m+1}^γ does not appear in K and λ_1 is a central coefficient. We replace x_{m+1} by $[u, v]$ in (6.23) and use the fact that γ is a weak Lie homomorphism. Note that this substitution does not affect the coefficient $\lambda_{Mx_{m+1}^\gamma N}$, whereas the value of the coefficient λ_K is affected (which will be indicated, with some abuse of notation, by writing $\lambda_K([u, v])$ in its place). Using the property (6.19) we then have

$$\sum_{M,N} \lambda_{Mx_{m+1}^\gamma N} M[u^\gamma, v^\gamma]N + \sum_K \lambda_K([u, v])K$$

$$\equiv [\sum_{M,N} \lambda_{Mx_{m+1}^\gamma N} Mu^\gamma N + \sum_K \lambda_K(u)K, v^\gamma] \qquad (6.24)$$

$$+ [u^\gamma, \sum_{M,N} \lambda_{Mx_{m+1}^\gamma N} Mv^\gamma N + \sum_K \lambda_K(v)K].$$

By Lemma 4.4 we may equate coefficients in (6.24). In particular, since the term Kv^γ only appears once, we conclude that $\lambda_K(u) = 0$ for all $u \in \mathcal{U}$, i.e., each $\lambda_K = 0$. Furthermore, if $M \neq 1$ and $N \neq 1$, the term $Mu^\gamma Nv^\gamma$ only appears once, and so in this case $\lambda_{Mx_{m+1}^\gamma N} = 0$. Therefore B may be rewritten in a simplified form as

$$\sum_N \lambda_{x_{m+1}^\gamma N} x_{m+1}^\gamma N + \sum_M \lambda_{Mx_{m+1}^\gamma} Mx_{m+1}^\gamma + \lambda_1,$$

and we may assume without loss of generality that all $\lambda_{x_{m+1}^\gamma N}$ and $\lambda_{Mx_{m+1}^\gamma}$ are nonzero. Accordingly (6.24) now reads

$$\sum_N \lambda_{x_{m+1}^\gamma N}[u^\gamma, v^\gamma]N + \sum_M \lambda_{Mx_{m+1}^\gamma} M[u^\gamma, v^\gamma]$$

$$\equiv [\sum_N \lambda_{x_{m+1}^\gamma N} u^\gamma N + \sum_M \lambda_{Mx_{m+1}^\gamma}^\gamma Mu^\gamma, v^\gamma]$$

$$+ [u^\gamma, \sum_N \lambda_{x_{m+1}^\gamma N} v^\gamma N + \sum_M \lambda_{Mx_{m+1}^\gamma} Mv^\gamma].$$

Suppose a monomial M appears in this relation which is not equal to any N which appears. Then the term $u^\gamma Mv^\gamma$ appears alone, whence we have the contradiction that $\lambda_{Mx_{m+1}^\gamma} = 0$. But for $M = N$ the term $u^\gamma Mv^\gamma$ appears twice, whence we conclude that $\lambda_{Mx_{m+1}^\gamma} = -\lambda_{x_{m+1}^\gamma M}$. Setting $\lambda_M = \lambda_{x_{m+1}^\gamma M}$, we now see that B is equal to $[\sum \lambda_M M, x_{m+1}^\gamma] + \lambda_1$, which proves the lemma. $\square$

We are now in a position to solve the problem introduced at the beginning of the section.

Lemma 6.17. *Let $f \in \mathbb{Z}\langle X \rangle$ be a multilinear polynomial of degree m, let $\mathcal{L}$ be an f-closed Lie ring, and let $\mathcal{S}$ be a Lie ideal of $\mathcal{L}$. Let $\mathcal{Q}$ be a unital ring such that its center $\mathcal{C}$ is a direct summand of the $\mathcal{C}$-module $\mathcal{Q}$, and let $\alpha : \mathcal{S} \to \overline{\mathcal{Q}}$ be a Lie homomorphism. Suppose there exists an $(2m + 3)$-free subset $\mathcal{R}$ of $\mathcal{Q}$ such that*

$\overline{\mathcal{R}} = \mathcal{S}^{\alpha}$. Then $\mathcal{S}$ is contained in an f-closed Lie ideal $\mathcal{M}$ of $\mathcal{L}$ such that α can be extended to a Lie homomorphism from $\mathcal{M}$ into $\overline{Q}$.

Proof. Consider the set of all extensions of α to Lie homomorphisms on Lie ideals of $\mathcal{L}$ that contain $\mathcal{S}$. By Zorn's Lemma there is an extension $\alpha_1 : \mathcal{M} \to \overline{Q}$ of $\alpha : \mathcal{S} \to \overline{Q}$ which is maximal in the sense that if $\alpha_2 : \mathcal{T} \to \overline{Q}$ extends $\alpha_1 : \mathcal{M} \to \overline{Q}$, then $\mathcal{T} = \mathcal{M}$. To avoid proliferation of notation we may without loss of generality suppose that $\mathcal{M} = \mathcal{S}$ and $\alpha_1 = \alpha$.

Our goal is to show that, under the maximality condition that we just imposed, $\mathcal{S}$ is f-closed. To this end we define $\mathcal{U}$ to be the additive subgroup of $\mathcal{L}$ generated by $\mathcal{S}$ and all elements of the form $f(x_1, \ldots, x_m)$ where $x_i \in \mathcal{S}$. We want to show of course that $\mathcal{U} = \mathcal{S}$. It is easy to see that $\mathcal{U}$ is a Lie ideal of $\mathcal{L}$. Indeed, $\mathcal{U}$ is first of all a subset of $\mathcal{L}$ since $\mathcal{L}$ is f-closed. Next, from the identity

$$[x_{i_1} x_{i_2} \ldots x_{i_m}, t] = [x_{i_1}, t] x_{i_2} \ldots x_{i_m}$$
$$+ x_{i_1}[x_{i_2}, t] x_{i_3} \ldots x_{i_m} + \ldots + x_{i_1} \ldots x_{i_{m-1}}[x_{i_m}, t]$$

and the multilinearity of f we infer that

$$[f(x_1, x_2, \ldots, x_m), t] = f([x_1, t], x_2, \ldots, x_m)$$
$$+ f(x_1, [x_2, t], x_3, \ldots, x_m) + \ldots + f(x_1, \ldots, x_{m-1}, [x_m, t]).$$

Using this for the case when $x_i \in \mathcal{S}$ and $t \in \mathcal{L}$, we see that $[\mathcal{S}, \mathcal{L}] \subseteq \mathcal{S}$ forces $[\mathcal{U}, \mathcal{L}] \subseteq \mathcal{U}$. The idea now is to extend α to a Lie homomorphism on $\mathcal{U}$. The maximality of $\mathcal{S}$ then will imply $\mathcal{U} = \mathcal{S}$.

We let $\gamma : \mathcal{S}_{\mathcal{C}} \to Q$ be the weak Lie homomorphism associated with α. Defining $B : \mathcal{S}_{\mathcal{C}}^{m+1} \to Q$ according to (6.17) we now conclude from Lemma 6.16 that there exists a quasi-polynomial $P : \mathcal{S}_{\mathcal{C}}^m \to Q$ (with respect to γ) such that

$$[f(x_1, \ldots, x_m), t]^{\gamma} \equiv [P(x_1, \ldots, x_m), t^{\gamma}] \qquad (6.25)$$

for all $x_1, \ldots, x_m, t \in \mathcal{S}_{\mathcal{C}}$.

We now set $\mathcal{U}_{\mathcal{C}} = C \oplus \mathcal{U}$. Clearly $\mathcal{U}_{\mathcal{C}}$ is a Lie ideal of $\mathcal{L}_{\mathcal{C}}$ and $\mathcal{U}_{\mathcal{C}} \supseteq \mathcal{S}_{\mathcal{C}}$. Note that a typical element in $\mathcal{U}_{\mathcal{C}}$ is of the form $x + \sum f(x_{i_1}, x_{i_2}, \ldots, x_{i_m})$ where x and all x_{i_j} are from $\mathcal{S}_{\mathcal{C}}$. Let π be the projection of Q onto W, and define a map $\phi : \mathcal{U}_{\mathcal{C}} \to Q$ according to the rule

$$\left(x + \sum f(x_{i_1}, x_{i_2}, \ldots, x_{i_m})\right)^{\phi} = \left(x^{\gamma} + \sum P(x_{i_1}, x_{i_2}, \ldots, x_{i_m})\right)^{\pi}.$$

We claim that ϕ is well-defined. Indeed, suppose $x + \sum f(x_{i_1}, x_{i_2}, \ldots, x_{i_m}) = 0$, where $x, x_{i_j} \in \mathcal{S}_{\mathcal{C}}$. Then, for $t \in \mathcal{S}_{\mathcal{C}}$, we have

$$0 = \left[x + \sum f(x_{i_1}, x_{i_2}, \ldots, x_{i_m}), t\right]^{\gamma} = [x, t]^{\gamma} + \sum \left[f(x_{i_1}, \ldots, x_{i_m}), t\right]^{\gamma}$$
$$\equiv [x^{\gamma} + \sum P(x_{i_1}, x_{i_2}, \ldots, x_{i_m}), t^{\gamma}].$$

Setting $q = x^\gamma + \sum P(x_{i_1}, x_{i_2}, \ldots, x_{i_m})$, we see that $[q, \mathcal{S}_\mathcal{C}^\gamma] \equiv 0$. Since $\mathcal{S}_\mathcal{C}^\gamma$ is in particular 2-free in $\mathcal{Q}$ we conclude that $q \in \mathcal{C}$ (see e.g., Observation 5 following Definition 3.1). Thus $q^\pi = 0$ and the claim is verified.

Now we wish to show that ϕ is a weak Lie homomorphism. In the proof we will tacitly use the facts that for $u, v \in \mathcal{Q}$ we have $u \equiv u^\pi$ and $[u, v] = [u, v^\pi] = [u^\pi, v^\pi] \equiv [u, v]^\pi$. There are three cases to consider.

First consider two elements x and y in $\mathcal{S}_\mathcal{C}$. Then

$$[x^\phi, y^\phi] = [x^{\gamma\pi}, y^{\gamma\pi}] = [x^\gamma, y^\gamma] \equiv [x, y]^\gamma \equiv [x, y]^{\gamma\pi} = [x, y]^\phi.$$

Next consider $f(x_1, \ldots, x_m)$ with $x_i \in \mathcal{S}_\mathcal{C}$, and an element t in $\mathcal{S}_\mathcal{C}$. Since $\mathcal{S}_\mathcal{C}$ is a Lie ideal of $\mathcal{L}_\mathcal{C}$ we have $[f(x_1, \ldots, x_m), t] \in \mathcal{S}_\mathcal{C}$. Thus

$$[f(x_1, \ldots, x_m), t]^\phi = [f(x_1, \ldots, x_m), t]^{\gamma\pi} \equiv [f(x_1, \ldots, x_m), t]^\gamma$$
$$\equiv [P(x_1, \ldots, x_m), t^\gamma] = [P(x_1, \ldots, x_m)^\pi, t^{\gamma\pi}] = [f(x_1, \ldots, x_m)^\phi, t^\phi].$$

Finally, consider $a = f(x_1, \ldots, x_m)$ and $b = f(y_1, \ldots, y_m)$, where $x_i, y_i \in \mathcal{S}_\mathcal{C}$. Let $t \in \mathcal{S}_\mathcal{C}$. Since $\mathcal{S}_\mathcal{C}$ is a Lie ideal of $\mathcal{L}_\mathcal{C}$ we have $[a, t] \in \mathcal{S}_\mathcal{C}$ and $[b, t] \in \mathcal{S}_\mathcal{C}$. We also remark that $[a, b]$ of course lies in $\mathcal{U}_\mathcal{C}$. We can summarize the above conclusions regarding the first two cases into $[u, x]^\phi \equiv [u^\phi, x^\phi]$ for all $u \in \mathcal{U}_\mathcal{C}$ and $x \in \mathcal{S}_\mathcal{C}$. Making use of this we get

$$[[a, b]^\phi, t^\gamma] = [[a, b]^\phi, t^\phi] \equiv [[a, b], t]^\phi$$
$$= [a, [b, t]]^\phi + [[a, t], b]^\phi \equiv [a^\phi, [b, t]^\phi] + [[a, t]^\phi, b^\phi]$$
$$= [a^\phi, [b^\phi, t^\phi]] + [[a^\phi, t^\phi], b^\phi] = [[a^\phi, b^\phi], t^\phi]$$
$$= [[a^\phi, b^\phi], t^\gamma].$$

Thus $[[a, b]^\phi - [a^\phi, b^\phi], \mathcal{S}_\mathcal{C}^\gamma] \equiv 0$. As we have seen earlier, since $\mathcal{S}_\mathcal{C}$ is 2-free, this implies that $[a, b]^\phi \equiv [a^\phi, b^\phi]$. Thus $\phi : \mathcal{U}_\mathcal{C} \to \mathcal{Q}$ is indeed a weak Lie homomorphism.

Clearly the restriction ϕ' of ϕ to $\mathcal{U}$ is a weak Lie homomorphism extending β. Finally, letting ϕ'' denote the Lie homomorphism of $\mathcal{U}$ into $\overline{\mathcal{Q}}$ given by $x^{\phi''} = \overline{x^{\phi'}}$, we see that $\phi'' : \mathcal{U} \to \overline{\mathcal{Q}}$ is an extension of $\alpha : \mathcal{S} \to \overline{\mathcal{Q}}$. By the maximality of $\alpha : \mathcal{S} \to \overline{\mathcal{Q}}$ we conclude that $\mathcal{U} = \mathcal{S}$, which proves the lemma. $\square$

We remark that if $\mathcal{L}$ admits the operator $\frac{1}{2}$, then $\mathcal{M}$ can be chosen so that it also admits the operator $\frac{1}{2}$. Indeed, when using Zorn's lemma we may confine ourselves only to those Lie ideals that contain $\frac{1}{2}$, and the same proof still works.

Combining Lemma 6.17 with the results from Sections 6.1 and 6.2 we are now ready to establish the main results of this section, and in fact of this chapter.

Theorem 6.18. *Let $\mathcal{S}$ be a Lie ideal of a Lie ring $\mathcal{L}$ of skew elements (of some ring), let $\mathcal{Q}$ be a unital ring with center $\mathcal{C}$, and let $\alpha : \mathcal{S} \to \overline{\mathcal{Q}}$ be a Lie homomorphism. Suppose that both $\mathcal{S}$ and $\mathcal{Q}$ admit the operator $\frac{1}{2}$, suppose that $\mathcal{C}$ is a direct summand of the additive group $\mathcal{Q}$, and suppose there exists a 9-free subset $\mathcal{R}$ of $\mathcal{Q}$ such that $\overline{\mathcal{R}} = \mathcal{S}^\alpha$. Then there exists a homomorphism $\sigma : \langle \mathcal{S} \rangle \to \mathcal{Q}$ such that $x^\alpha = \overline{x^\sigma}$ for all $x \in \mathcal{S}$.*

Proof. We may use Lemma 6.17 for $f = f_2$ (so $m = 3$ and $2m+3 = 9$) to conclude that α can be extended to a Lie homomorphism $\vartheta : \mathcal{M} \to \overline{\mathcal{Q}}$ where $\mathcal{M}$ is a Lie ideal of $\mathcal{L}$ closed under triads. As just pointed out, we may assume that $\mathcal{M}$ also admits the operator $\frac{1}{2}$. This in particular implies that $\mathcal{M} \cap (\mathcal{M} \circ \mathcal{M}) = 0$, and so Lemma 6.14 now tells us that $\mathcal{M}$ is actually equal to the set of all skew elements of the ring $\langle \mathcal{M} \rangle$. Further, it is clear that we may choose $\mathcal{R}_1 \subseteq \mathcal{Q}$ so that $\overline{\mathcal{R}}_1 = \mathcal{M}^\vartheta$ and $\mathcal{R}_1 \supseteq \mathcal{R}$. By Corollary 3.5, $\mathcal{R}_1$ is 9-free. Now we see that $\vartheta : \mathcal{M} \to \overline{\mathcal{Q}}$ satisfies all the conditions of Theorem 6.15. Hence there exists a homomorphism $\sigma : \langle \mathcal{M} \rangle \to \mathcal{Q}$ such that $x^\vartheta = \overline{x^\sigma}$ for all $x \in \mathcal{M}$. In particular, for every $x \in \mathcal{S}$ we have $x^\alpha = \overline{x^\sigma}$. $\square$

Basically the same method is also applicable to the similar, yet simpler problem of describing Lie homomorphisms on Lie ideals of rings.

Theorem 6.19. *Let $\mathcal{S}$ be a Lie ideal of a ring $\mathcal{B}$, let $\mathcal{Q}$ be a unital ring with center $\mathcal{C}$, and let $\alpha : \mathcal{S} \to \overline{\mathcal{Q}}$ be a Lie homomorphism. Suppose that $\mathcal{C}$ is a direct summand of the additive group $\mathcal{Q}$, and suppose there exists a 7-free subset $\mathcal{R}$ of $\mathcal{Q}$ such that $\overline{\mathcal{R}} = \mathcal{S}^\alpha$. Then there exists a direct sum of a homomorphism and the negative of an antihomomorphism $\sigma : \langle \mathcal{S} \rangle \to \mathcal{Q}$ such that $x^\alpha = \overline{x^\sigma}$ for all $x \in \mathcal{S}$.*

Proof. Taking $f = f_1$ (so $m = 2$ and $2m + 3 = 7$), we see by Lemma 6.17 that $\mathcal{S}$ is contained in an f-closed Lie ideal $\mathcal{M}$ of $\mathcal{B}$ such that α can be extended to a Lie homomorphism $\phi : \mathcal{M} \to \overline{\mathcal{Q}}$. Of course, the condition that $\mathcal{M}$ is f-closed simply means that it is a subring of $\mathcal{B}$. Since $\mathcal{S}$ is contained in $\mathcal{M}$ we may assume that $\mathcal{R}$ is contained in $\mathcal{T}$, where $\mathcal{M}^\phi = \overline{\mathcal{T}}$, and so $\mathcal{T}$ is also 7-free by Corollary 3.5. Let $\gamma : \mathcal{M}_\mathcal{C} \to \mathcal{Q}$ be the weak Lie homomorphism associated with ϕ, noting that $\mathcal{M}_\mathcal{C}$ is an associative ring. By Remark 6.2 there exists $\sigma : \mathcal{M}_\mathcal{C} \to \mathcal{Q}$ and $\tau : \mathcal{M}_\mathcal{C} \to \mathcal{C}$ such that $\gamma = \sigma + \tau$, where σ is the direct sum of a homomorphism and the negative of an antihomomorphism. Thus $\overline{x^\sigma} = \overline{x^\gamma} = x^\alpha$ for all $x \in \mathcal{S}$. $\square$

In the rest of this section we will derive a few corollaries to Theorem 6.18. One could derive similar corollaries to Theorem 6.19, but let us confine ourselves only to the (more entangled) context of rings with involution.

Corollary 6.20. *Let $\mathcal{S}$ be a Lie ideal of the Lie ring $\mathcal{L}$ of skew elements (of some ring), let $\mathcal{A}$ be a prime ring with involution, let $\mathcal{C}$ be the extended centroid of $\mathcal{A}$, let $\mathcal{K}$ be the skew elements of $\mathcal{A}$, and let $\mathcal{R}$ be a noncentral Lie ideal of $\mathcal{K}$. Suppose that $\mathcal{S}$ admits the operator $\frac{1}{2}$, and suppose that $\deg(\mathcal{A}) \geq 21$ and $\mathrm{char}(\mathcal{A}) \neq 2$. If α is a Lie homomorphism of $\mathcal{S}$ onto $\overline{\mathcal{R}} = \mathcal{R}/\mathcal{R} \cap \mathcal{C}$; then there exists a homomorphism $\sigma : \langle \mathcal{S} \rangle \to \langle \mathcal{R} \rangle \mathcal{C} + \mathcal{C}$ such that $x^\alpha = \overline{x^\sigma}$ for all $x \in \mathcal{S}$.*

Proof. Let $\mathcal{Q} = \mathcal{Q}_{ml}(\mathcal{A})$. Since $\mathcal{C}$ is a field it is a direct summand of the vector space $\mathcal{Q}$ over $\mathcal{C}$. Further, since $\deg(\mathcal{A}) \geq 21$, Corollary 5.19 implies that $\mathcal{R}$ is 9-free in $\mathcal{Q}$. Thus, Theorem 6.18 can be applied; hence there exists a homomorphism $\sigma : \langle \mathcal{S} \rangle \to \mathcal{Q}$ such that $x^\alpha = \overline{x^\sigma}$ for all $x \in \mathcal{S}$. Since $\overline{x^\sigma} \in \overline{\mathcal{R}}$ if $x \in \mathcal{S}$ it is clear that the image of σ lies in $\langle \mathcal{R} \rangle \mathcal{C} + \mathcal{C}$. $\square$

Both the strength and the weakness of the notion of d-freeness show up in Corollary 6.20: a positive result is obtained for the (presumably more difficult) situation where $\mathcal{A}$ is of sufficiently high degree or is of infinite degree, whereas the situation in which $\mathcal{A}$ is of "low degree" requires other methods. For example, if $\mathcal{A} = M_n(\mathcal{C})$, then Corollary 6.20 is applicable as long as $n \geq 21$. Certain small numbers n really are exceptional. In fact, a detailed analysis of the low degree situation shows that certain cases where $\deg(\mathcal{A}) = 1, 2, 3, 4, 5, 6, 8$ must be excluded. But we shall not include this analysis here. This would lead us too far from the scope of this book.

Let us point out a particular case of Corollary 6.20 which is of special interest. Consider a Lie isomorphism α from $\widehat{\mathcal{L}} = [\mathcal{L}, \mathcal{L}]/\mathcal{Z}(\mathcal{B}) \cap [\mathcal{L}, \mathcal{L}]$ onto $\widehat{\mathcal{K}} = [\mathcal{K}, \mathcal{K}]/\mathcal{Z}(\mathcal{A}) \cap [\mathcal{K}, \mathcal{K}]$, where $\mathcal{L}$ and $\mathcal{K}$ are the skew elements of simple rings $\mathcal{B}$ and $\mathcal{A}$, and $\mathcal{Z}(\mathcal{B})$ and $\mathcal{Z}(\mathcal{A})$ are the centers of these rings. As mentioned at the beginning of this section, $\widehat{\mathcal{L}}$ and $\widehat{\mathcal{K}}$ are (usually) simple Lie rings, and that is why this is especially interesting. Composing α with the canonical projection of $[\mathcal{L}, \mathcal{L}]$ onto $\widehat{\mathcal{L}}$ we thus get a Lie homomorphism from $[\mathcal{L}, \mathcal{L}]$ onto $\widehat{\mathcal{K}}$, for which Corollary 6.20 is applicable. One can show that, under the assumptions of this corollary, there exists an isomorphism σ from $\mathcal{B}$ onto $\mathcal{A}$ such that $\overline{x}^\alpha = \overline{x^\sigma}$ for all $x \in [\mathcal{L}, \mathcal{L}]$ (we omit details of the proof here). So, in particular, the rings $\mathcal{A}$ and $\mathcal{B}$ are isomorphic.

Now we will show how to use Theorem 6.18 to derive an analogous result for Lie derivations. We first note that the notion of a Lie derivation from a Lie subring $\mathcal{R}$ of $\mathcal{Q}$ into $\overline{\mathcal{Q}}$ makes sense; it is of course defined as an additive map $\delta : \mathcal{R} \to \overline{\mathcal{Q}}$ such that $[x, y]^\delta = [x^\delta, \overline{y}] + [\overline{x}, y^\delta]$ for all $x, y \in \mathcal{R}$.

Corollary 6.21. *Let $\mathcal{A}$ be a ring with involution, let $\mathcal{K}$ be the skew elements of $\mathcal{A}$, let $\mathcal{R}$ be a Lie ideal of $\mathcal{K}$, let $\mathcal{Q} \supseteq \mathcal{A}$ be a unital ring with center $\mathcal{C}$, and let $\delta : \mathcal{R} \to \overline{\mathcal{Q}}$ be a Lie derivation. Suppose that both $\mathcal{S}$ and $\mathcal{Q}$ admit the operator $\frac{1}{2}$, suppose that $\mathcal{C}$ is a direct summand of the additive group $\mathcal{Q}$, and suppose that $\mathcal{R}$ is a 9-free subset of $\mathcal{Q}$. Then there exists a derivation $d : \langle \mathcal{R} \rangle \to \mathcal{Q}$ such that $x^\delta = \overline{x^d}$ for all $x \in \mathcal{R}$.*

Proof. With reference to Section 3.3 we recall the ring $\check{\mathcal{Q}} = \mathcal{Q} \times \mathcal{Q}$, with pointwise addition and multiplication given by $(x, y)(z, w) = (xz, xw + yz)$. The center $\check{\mathcal{C}}$ of $\check{\mathcal{Q}}$ is easily seen to be $\mathcal{C} \times \mathcal{C}$, and so $\check{\mathcal{C}}$ is a direct summand of the additive group $\check{\mathcal{Q}}$, and $\frac{1}{2}$ exists in $\check{\mathcal{C}}$. Let $\gamma : \mathcal{R} \to \mathcal{Q}$ be any set-theoretic mapping such that $\overline{x^\gamma} = x^\delta$ for every $x \in \mathcal{R}$ (such always exists). Now define $\alpha : \mathcal{R} \to \widetilde{\mathcal{Q}} = \check{\mathcal{Q}}/\check{\mathcal{C}}$ by $x^\alpha = \overline{(x, x^\gamma)}\ (= (x, x^\gamma) + \check{\mathcal{C}})$. One can easily check that α is a Lie homomorphism. Theorem 3.7 tells us that $\check{\mathcal{R}} = \{(x, x^\gamma) \mid x \in \mathcal{R}\}$ is a 9-free subset of $\check{\mathcal{Q}}$. Since the conditions of Theorem 6.18 are now satisfied we may conclude that there exists a homomorphism $\sigma : \langle \mathcal{R} \rangle \to \check{\mathcal{Q}}$ such that $\overline{x^\sigma} = x^\alpha = \overline{(x, x^\gamma)}$ for all $x \in \mathcal{R}$. Thus $x^\sigma - (x, x^\gamma) \in \check{\mathcal{C}}$ for every $x \in \mathcal{R}$. Write $x^\sigma - (x, x^\gamma) = (\nu(x), \mu(x))$ where $\nu(x), \mu(x) \in \mathcal{C}$. So we have $x^\sigma = (x + \nu(x), x^\gamma + \mu(x))$ for every $x \in \mathcal{R}$. As σ is a homomorphism we have $[xyz + zyx, u]^\sigma = [x^\sigma y^\sigma z^\sigma + z^\sigma y^\sigma x^\sigma, u^\sigma]$ for all

$x, y, z, u \in \mathcal{R}$. Using $[xyz + zyx, u] \in \mathcal{R}$ it follows from this identity that

$$[xyz + zyx, u] + \nu([xyz + zyx, u])$$
$$= [(x + \nu(x))(y + \nu(y))(z + \nu(z)) + (z + \nu(z))(y + \nu(y))(x + \nu(x)), u].$$

Lemma 4.4 allows us to equate coefficients, which forces $\nu = 0$ (just consider the coefficient at yzu, for example). So we now have $x^\sigma = (x, x^\gamma + \mu(x))$ for $x \in \mathcal{R}$. Considering the identity $(x_1 x_2 \ldots x_n)^\sigma = x_1^\sigma x_2^\sigma \ldots x_n^\sigma$ for all $x_i \in \mathcal{R}$ it is now immediate that for every $y \in \langle \mathcal{R} \rangle$ there exists a unique element in $\mathcal{Q}$, which we denote by y^d, such that $y^\sigma = (y, y^d)$. Moreover, for $x \in \mathcal{R}$ we have $x^d = x^\gamma + \mu(x)$ and hence $\overline{x^d} = \overline{x^\gamma} = x^\delta$. Furthermore, since σ is additive, d is also additive, and since σ is multiplicative, it follows from the definition of the multiplication in $\check{\mathcal{Q}}$ that d satisfies the derivation law. Thus d is a derivation. $\square$

Finally we state the prime ring version of Corollary 6.21.

Corollary 6.22. *Let $\mathcal{A}$ be a prime ring with involution, let $\mathcal{C}$ be the extended centroid of $\mathcal{A}$, let $\mathcal{K}$ be the skew elements of $\mathcal{A}$, and let $\mathcal{R}$ be a noncentral Lie ideal of $\mathcal{K}$. Suppose that $\deg(\mathcal{A}) \geq 21$ and $\operatorname{char}(\mathcal{A}) \neq 2$. If δ is Lie derivation of $\mathcal{R}$ into $\overline{\mathcal{R}} = \mathcal{R}/\mathcal{R} \cap \mathcal{C}$, then there exists a derivation $d : \langle \mathcal{R} \rangle \to \langle \mathcal{R} \rangle \mathcal{C} + \mathcal{C}$ such that $x^\delta = \overline{x^d}$ for all $x \in \mathcal{R}$.*

Proof. Since $\mathcal{C}$ is a field, it is a direct summand of $\mathcal{Q} = \mathcal{Q}_{ml}(\mathcal{A})$, and (using Corollary 5.19) we have that $\mathcal{R}$ is 9-free in $\mathcal{Q}$. So we may apply Corollary 6.21 to conclude that there is a derivation $d : \langle \mathcal{R} \rangle \to \mathcal{Q}$ such that $x^\delta = \overline{x^d}$ for all $x \in \mathcal{R}$. However, as δ is assumed to have its image in $\overline{\mathcal{R}}$ it follows that $x^d \in \mathcal{R} + \mathcal{C}$ for every $x \in \mathcal{R}$. Consequently, $y^d \in \langle \mathcal{R} \rangle \mathcal{C} + \mathcal{C}$ for every $y \in \langle \mathcal{R} \rangle$. $\square$

Corollary 6.22 can serve as a sample for demonstrating the applicability of abstract FI theory. Almost all the machinery developed in Part II was used in its proof. Indeed, the main results of Sections 3.2, 3.3, 3.4, 3.5, 4.2, 4.3, 5.1 and 5.2 were at least indirectly applied.

6.4   Jordan Maps

The Lie map problems that were considered in the preceding sections have their Jordan analogues. Assume now that $\mathcal{S}$ is a Jordan subring of an associative ring. Recall that a Jordan homomorphism from $\mathcal{S}$ into a ring $\mathcal{Q}$ is an additive map that preserves the Jordan product, i.e., $(s \circ t)^\alpha = s^\alpha \circ t^\alpha$. Similarly, a Jordan derivation from $\mathcal{S}$ into a ring containing $\mathcal{S}$ is an additive map that acts as a derivation on the Jordan product, i.e., $(s \circ t)^\delta = s^\delta \circ t + s \circ t^\delta$.

Let us start our consideration of Jordan maps by pointing out an important identity

$$[[s, t], u] = s \circ (t \circ u) - t \circ (s \circ u)$$

which links the Jordan product and the Lie product. It implies that every Jordan homomorphism α satisfies the Lie-type identity

$$[[s, t], u]^\alpha = [[s^\alpha, t^\alpha], u^\alpha]. \tag{6.26}$$

This will play a crucial role in our approach.

We begin by treating Jordan homomorphisms that are defined on rings. Obvious examples are homomorphisms and antihomomorphisms, and their direct sums. By a direct sum of a homomorphism and an antihomomorphism we of course mean a map $\alpha : \mathcal{B} \to \mathcal{Q}$ such that there exists a central idempotent ε in $\mathcal{Q}$ with the property that $x \mapsto \varepsilon x^\alpha$ is a homomorphism and $x \mapsto (1 - \varepsilon)x^\alpha$ is an antihomomorphism. Our first theorem is an analogue of Theorem 6.1.

Theorem 6.23. *Let $\mathcal{B}$ be any ring, and let $\mathcal{Q}$ be a unital ring admitting the operator $\frac{1}{2}$. If $\alpha : \mathcal{B} \to \mathcal{Q}$ is a Jordan homomorphism such that $\mathcal{B}^\alpha$ is a 4-free subset of $\mathcal{Q}$, then α is a direct sum of a homomorphism and an antihomomorphism.*

Proof. In Section 1.4 we already saw how to derive an FI for Jordan homomorphisms. The idea is to compute $[[x, y], [z, w]]^\alpha$ in two different ways. By making use of (6.26) this is very easy. On the one hand we have

$$[[x, y], [z, w]]^\alpha = [[x^\alpha, y^\alpha], [z, w]^\alpha],$$

and on the other hand,

$$[[x, y], [z, w]]^\alpha = [[x, y]^\alpha, [z^\alpha, w^\alpha]].$$

Consequently,

$$[[x^\alpha, y^\alpha], B(z, w)] = [B(x, y), [z^\alpha, w^\alpha]], \tag{6.27}$$

where $B(x, y) = [x, y]^\alpha$. Since $\mathcal{B}^\alpha$ is 4-free in $\mathcal{Q}$, Theorem 4.13 implies that B is a quasi-polynomial, i.e.,

$$B(x, y) = \lambda x^\alpha y^\alpha + \lambda' y^\alpha x^\alpha + \mu_1(x)y^\alpha + \mu_2(y)x^\alpha + \nu(x, y)$$

where $\lambda, \lambda' \in \mathcal{C}$, $\mu_1, \mu_2 : \mathcal{B} \to \mathcal{C}$ and $\nu : \mathcal{B}^2 \to \mathcal{C}$. Substituting this form back in (6.27), and then expanding the identity that we get, it follows that

$$(\lambda + \lambda')x^\alpha y^\alpha w^\alpha z^\alpha + (\lambda + \lambda')z^\alpha w^\alpha y^\alpha x^\alpha - (\lambda + \lambda')y^\alpha x^\alpha z^\alpha w^\alpha$$
$$- (\lambda + \lambda')w^\alpha z^\alpha x^\alpha y^\alpha + \mu_1(z)[[x^\alpha, y^\alpha], w^\alpha] + \mu_2(w)[[x^\alpha, y^\alpha], z^\alpha]$$
$$- \mu_1(x)[y^\alpha, [z^\alpha, w^\alpha]] - \mu_2(y)[x^\alpha, [z^\alpha, w^\alpha]] = 0.$$

In view of 4-freeness of $\mathcal{B}^\alpha$, Lemma 4.4 now forces $\lambda + \lambda' = 0$ and $\mu_1 = \mu_2 = 0$. Thus $B(x, y) = [x, y]^\alpha = \lambda[x^\alpha, y^\alpha] + \nu(x, y)$, which together with $(x \circ y)^\alpha = x^\alpha \circ y^\alpha$ yields

$$(xy)^\alpha = \varepsilon x^\alpha y^\alpha + \varepsilon' y^\alpha x^\alpha + \nu(x, y),$$

where $\varepsilon = \frac{1}{2}(1+\lambda)$ and $\varepsilon' = \frac{1}{2}(1-\lambda)$. In the proof of Theorem 6.1 we arrived at a similar, though more complicated identity (6.6). We now proceed as in that proof, that is, we compute $(xyz)^\alpha$ in two different ways. On the one hand we have

$$((xy)z)^\alpha = \varepsilon(xy)^\alpha z^\alpha + \varepsilon' z^\alpha (xy)^\alpha + \nu(xy, z)$$
$$= \varepsilon^2 x^\alpha y^\alpha z^\alpha + \varepsilon\varepsilon' y^\alpha x^\alpha z^\alpha + \varepsilon\nu(x, y)z^\alpha$$
$$+ \varepsilon\varepsilon' z^\alpha x^\alpha y^\alpha + \varepsilon'^2 z^\alpha y^\alpha x^\alpha + \varepsilon'\nu(x, y)z^\alpha + \nu(xy, z),$$

and on the other hand,

$$(x(yz))^\alpha = \varepsilon x^\alpha (yz)^\alpha + \varepsilon'(yz)^\alpha x^\alpha + \nu(x, yz)$$
$$= \varepsilon^2 x^\alpha y^\alpha z^\alpha + \varepsilon\varepsilon' x^\alpha z^\alpha y^\alpha + \varepsilon\nu(y, z)x^\alpha$$
$$+ \varepsilon\varepsilon' y^\alpha z^\alpha x^\alpha + \varepsilon'^2 z^\alpha y^\alpha x^\alpha + \varepsilon'\nu(y, z)x^\alpha + \nu(x, yz).$$

Comparing we get

$$\varepsilon\varepsilon'[y^\alpha, [x^\alpha, z^\alpha]] + \nu(x, y)z^\alpha - \nu(y, z)x^\alpha \in \mathcal{C}.$$

Lemma 4.4 implies that $\varepsilon\varepsilon' = 0$ and $\nu = 0$. So we now have

$$(xy)^\alpha = \varepsilon x^\alpha y^\alpha + \varepsilon' y^\alpha x^\alpha, \qquad (6.28)$$

and hence

$$x^\alpha \circ y^\alpha = (xy)^\alpha + (yx)^\alpha = (\varepsilon + \varepsilon')x^\alpha \circ y^\alpha.$$

Using Lemma 4.4 again we obtain $c \mid c' - 1$. Thus ε is an idempotent, $\varepsilon' = 1 - \varepsilon$, and (6.28) shows that α is a direct sum of a homomorphism and an antihomomorphism. □

Corollary 6.24. *Let $\mathcal{Q}$ be a unital ring admitting the operator $\frac{1}{2}$. If $\mathcal{B}$ is a 4-free subring of $\mathcal{Q}$, then every Jordan derivation $\delta : \mathcal{B} \to \mathcal{Q}$ is a derivation.*

Proof. The proof is based on Theorem 3.7, so the reader should recall the ring $\check{\mathcal{Q}} = \mathcal{Q} \times \mathcal{Q}$ from this theorem. We define $\alpha : \mathcal{B} \to \check{\mathcal{Q}}$ by $x^\alpha = (x, x^\delta)$. One can check that α is a Jordan homomorphism. Since $\mathcal{B}^\alpha$ is a 4-free subset of $\check{\mathcal{Q}}$ by Theorem 3.7, we may apply Theorem 6.23 to conclude that α is a direct sum of a homomorphism and an antihomomorphism. Suppose that ε' is a central idempotent in $\check{\mathcal{Q}}$ such that $x \mapsto \varepsilon' x^\alpha$ is an antihomomorphism. It is easy to see that ε' can, as a central idempotent, only be of the form $\varepsilon' = (\omega, 0)$ where ω is a central idempotent in $\mathcal{Q}$. From $\varepsilon'(xy)^\alpha = \varepsilon' y^\alpha x^\alpha$ we in particular infer that $\omega xy = \omega yx$ for all $x, y \in \mathcal{B}$. Since $\mathcal{B}$ is 4-free (and hence 2-free) it follows that $\omega = 0$, and so also $\varepsilon' = 0$. This shows that α is necessarily a homomorphism. This in turn implies that δ is a derivation. □

We have thereby illustrated once more the applicability of Theorem 3.7. But it should be mentioned that Corollary 6.24 can be proved directly in an even shorter way. All we need is the well-known identity

$$\left((xy)^\delta - x^\delta y - xy^\delta\right)[x, y] = 0, \tag{6.29}$$

which every Jordan derivation satisfies. This can be easily proved; see for example [114, p. 53]. Linearizing (6.29) we then arrive at a situation where we can simply use the definition of a 4-free set to conclude that $(xy)^\delta - x^\delta y - xy^\delta = 0$. So there is nothing deep hidden in Corollary 6.24. In some sense Jordan derivations (on rings) are too "easy" to be treated through FI's. But it is still of some interest to notice that they also fit into the FI context.

We now turn to the more difficult problem concerning Jordan maps of symmetric elements. So, let $\mathcal{B}$ be a ring with involution, let $\mathcal{S}$ be the Jordan subring of symmetric elements in $\mathcal{B}$, and let $\alpha : \mathcal{S} \to \mathcal{Q}$ be a Jordan homomorphism. The problem is: Can α be extended to a homomorphism on $\langle \mathcal{S} \rangle$? The absence of antihomomorphisms can be explained in a similar fashion as in the similar problem on Lie homomorphisms of skew elements (see Section 6.2). Of course one might ask a more general question about Jordan maps on Jordan ideals of symmetric elements. But the motivation for such higher level of generality is not so clear as in the context of skew elements. For example, if $\mathcal{B}$ is a simple ring with char$(\mathcal{B}) \neq 2$, then $\mathcal{S}$ is a simple Jordan ring [113, 114], whereas the Lie ring $\mathcal{K}$ of skew elements of $\mathcal{B}$ may not be simple (specifically, $[\mathcal{K}, \mathcal{K}]$ is sometimes its proper Lie ideal). We shall therefore confine ourselves to the set of all symmetric elements, which is at any rate a rather general and very important example of a Jordan ring (in particular this is evident from Zelmanov's classification of prime nondegenerate Jordan algebras [199]).

Jordan map problems are seemingly simpler than analogous Lie map problems. In particular, there are no central maps in conclusions, so one might expect that their treatment should be easier and more direct. Indeed, as already mentioned earlier, there are other powerful methods available for studying Jordan maps. Concerning FI's, however, it seems that in the Lie context it is more obvious how to derive appropriate identities than in the Jordan context. We shall solve the problem that we just posed by reducing it to one of the Lie map problems that was resolved in the preceding section. The next lemma points out the Lie ring that will do the trick.

Lemma 6.25. *Let $\mathcal{S}$ be the set of symmetric elements of a ring with involution. If $\mathcal{S}$ admits the operator $\frac{1}{2}$, then $\mathcal{S} + [\mathcal{S}, \mathcal{S}]$ is a Lie ideal of the ring $\langle \mathcal{S} \rangle$.*

Proof. From the identity

$$s_1 s_2 \ldots s_n = \left(\frac{s_1}{2} s_2 \ldots s_n + s_n \ldots s_2 \frac{s_1}{2}\right) + \left(\frac{s_1}{2} s_2 \ldots s_n - s_n \ldots s_2 \frac{s_1}{2}\right)$$

we easily infer that every element $x \in \langle S \rangle$ can be written as $x = s + k$ where $s \in S$ and k is a skew element in $\langle S \rangle$. Therefore, for all $t, u \in S$ we have

$$[t, x] = [t, s] + [t, k] \in [S, S] + S,$$
$$[[t, u], x] = [[t, u], s] + [[t, u], k]$$
$$= [[t, u], s] + \big([u, [k, t]] + [t, [u, k]] \big) \in S + [S, S].$$

This proves the lemma. $\square$

Theorem 6.26. *Let S be the set of symmetric elements of a ring with involution, let Q be a unital ring with center C, and let $\alpha : S \to Q$ be a Jordan homomorphism. Suppose that both S and Q admit the operator $\frac{1}{2}$, suppose that C is a direct summand of the additive group Q, and suppose that S^α is a 7-free subset of Q. Then α can be extended to a homomorphism $\sigma : \langle S \rangle \to Q$.*

Proof. Let us define $\beta : S + [S, S] \to \overline{Q} = Q/C$ according to

$$\left(s + \sum_i [s_i, t_i] \right)^\beta = \overline{s^\alpha + \sum_i [s_i^\alpha, t_i^\alpha]}.$$

To show that β is well-defined, we assume that $\sum_i [s_i, t_i] = 0$ for some $s_i, t_i \in S$. Then we have $\sum_i [[s_i, t_i], u] = 0$, which in view of (6.26) yields $\sum_i [[s_i^\alpha, t_i^\alpha], u^\alpha] = 0$. Thus $a = \sum_i [s_i^\alpha, t_i^\alpha]$ commutes with every element from a 7-free (and hence 2-free) set S^α, which implies $a \in C$ and so $\bar{a} = 0$. Since $S \cap [S, S] = 0$, this proves that β is well-defined. We claim that β is a Lie homomorphism, i.e., $[x, y]^\beta = [x^\beta, y^\beta]$ for all $x, y \in S + [S, S]$. We will consider three cases.

The first case is when both $x, y \in S$, and it is trivial. Indeed, we have $[x, y]^\beta = \overline{[x^\alpha, y^\alpha]} = [\overline{x^\alpha}, \overline{y^\alpha}] = [x^\beta, y^\beta]$.

The second case is when $x \in S$ and $y \in [S, S]$. Without loss of generality we may take $y = [s, t]$, $s, t \in S$. Again using (6.26) we have

$$[x, [s, t]]^\beta = \overline{[x, [s, t]]^\alpha} = \overline{[x^\alpha, [s^\alpha, t^\alpha]]} = [\overline{x^\alpha}, [\overline{s^\alpha}, \overline{t^\alpha}]] = [x^\beta, [s^\beta, t^\beta]] = [x^\beta, [s, t]^\beta].$$

Finally, assume that $x = [u, v]$ and $y = [s, t]$ with $u, v, s, t \in S$. Using $[[S, S], S] \subseteq S$ and the conclusions of the first two cases we have

$$[[u, v], [s, t]]^\beta = [v, [[s, t], u]]^\beta + [u, [v, [s, t]]]^\beta$$
$$= [v^\beta, [[s, t], u]^\beta] + [u^\beta, [v, [s, t]]^\beta]$$
$$= [v^\beta, [[s, t]^\beta, u^\beta]] + [u^\beta, [v^\beta, [s, t]^\beta]]$$
$$= [[u^\beta, v^\beta], [s, t]^\beta] = [[u, v]^\beta, [s, t]^\beta].$$

Thus β is indeed a Lie homomorphism.

Since $\overline{S^\alpha} \subseteq S^\beta$, we may choose $R \subseteq Q$ so that $\overline{R} = S^\beta$ and $S^\alpha \subseteq R$. Therefore β satisfies all the conditions of Theorem 6.19; indeed, by Lemma 6.25

it is defined on a Lie ideal of a ring, and by Corollary 3.5 $\mathcal{R}$ is 7-free in $\mathcal{Q}$. Therefore there exists a direct sum of a homomorphism and the negative of an antihomomorphism $\sigma : \langle \mathcal{S} \rangle \to \mathcal{Q}$ such that $x^\beta = \overline{x^\sigma}$ for all $x \in \mathcal{S} + [\mathcal{S}, \mathcal{S}]$. In particular, $\overline{s^\alpha} = s^\beta = \overline{s^\sigma}$ for every $s \in \mathcal{S}$, so that $\mu(s) = s^\sigma - s^\alpha \in \mathcal{C}$ for all $s \in \mathcal{S}$.

Let $\varepsilon \in \mathcal{C}$ be an idempotent such that $x \mapsto \varepsilon x^\sigma$ is a homomorphism and $x \mapsto (1 - \varepsilon)x^\sigma$ is the negative of an antihomomorphism. For all $s, t \in \mathcal{S}$ we have

$$
\begin{aligned}
s^\alpha \circ t^\alpha = (s \circ t)^\alpha &= (s \circ t)^\sigma - \mu(s \circ t) \\
&= \varepsilon(s \circ t)^\sigma + (1 - \varepsilon)(s \circ t)^\sigma - \mu(s \circ t) \\
&= \varepsilon s^\sigma \circ t^\sigma - (1 - \varepsilon)s^\sigma \circ t^\sigma - \mu(s \circ t) \\
&= (2\varepsilon - 1)(s^\alpha + \mu(s)) \circ (t^\alpha + \mu(t)) - \mu(s \circ t) \\
&= (2\varepsilon - 1)s^\alpha \circ t^\alpha + (4\varepsilon - 2)\mu(s)t^\alpha + (4\varepsilon - 2)\mu(t)s^\alpha \\
&\quad + (4\varepsilon - 2)\mu(s)\mu(t) - \mu(s \circ t),
\end{aligned}
$$

and so

$$
2(1 - \varepsilon)s^\alpha \circ t^\alpha - (4\varepsilon - 2)\mu(s)t^\alpha - (4\varepsilon - 2)\mu(t)s^\alpha - (4\varepsilon - 2)\mu(s)\mu(t) + \mu(s \circ t) = 0.
$$

Since $\mathcal{S}^\alpha$ is in particular 3-free, a standard application of Lemma 4.4 yields $2(1 - \varepsilon) = 0$ and $(4\varepsilon - 2)\mu(s) = 0$. Thus $\varepsilon = 1$, i.e., σ is a homomorphism, and $\mu(s) = 0$, i.e., σ is an extension of α. This proves the theorem. □

Corollary 6.27. *Let $\mathcal{S}$ be the set of symmetric elements of a ring with involution, let $\mathcal{Q} \supseteq \langle \mathcal{S} \rangle$ be a unital ring with center $\mathcal{C}$, and let $\delta : \mathcal{S} \to \mathcal{Q}$ be a Jordan derivation. Suppose that both $\mathcal{S}$ and $\mathcal{Q}$ admit the operator $\frac{1}{2}$, suppose that $\mathcal{C}$ is a direct summand of the additive group $\mathcal{Q}$, and suppose that $\mathcal{S}$ is a 7-free subset of $\mathcal{Q}$. Then δ can be extended to a derivation $d : \langle \mathcal{S} \rangle \to \mathcal{Q}$.*

Proof. As in the proof of Corollary 6.24 we define $\alpha : \mathcal{S} \to \check{\mathcal{Q}}$ by $x^\alpha = (x, x^\delta)$. Then α is a Jordan homomorphism and $\mathcal{S}^\alpha$ is 7-free in $\check{\mathcal{Q}}$ by Theorem 3.7. Noting that all the conditions of Theorem 6.26 are met, it follows that α can be extended to a homomorphism $\sigma : \langle \mathcal{S} \rangle \to \check{\mathcal{Q}}$. It is clear from the definition of α that for every $y \in \langle \mathcal{S} \rangle$ there exists a unique element $y^d \in \mathcal{Q}$ such that $y^\sigma = (y, y^d)$. Since σ is a homomorphism, d must be a derivation. It is also clear that d extends δ. □

We could easily state corollaries to our results that concern the classical context of simple or prime rings. However, as we already mentioned, Jordan maps can also be treated by other methods. It turns out that in the classical situation these methods yield slightly better results. For example, as a corollary to Theorem 6.23 we have that a Jordan homomorphism from any ring onto a prime ring $\mathcal{A}$ with char$(\mathcal{A}) \neq 2$ is either a homomorphism or an antihomomorphism, provided that deg$(\mathcal{A}) \geq 4$. But this degree assumption is absolutely superfluous here, as can be shown by elementary means [111, 114] (see also the proof of Corollary 7.10 below). Thus the "error" that the FI approach usually causes is somehow more disturbing in the Jordan context than in the Lie one (maybe just because in the Lie case there are, to the best of our knowledge, no other methods available). Treating Jordan homomorphisms on symmetric elements is much more complicated than treating Jordan homomorphisms on rings; as far as we know the former cannot be handled by elementary means. But the advanced methods based on Zelmanov's approach [199] yield definitive results in the prime ring situation (see for example [132, 160]), while Theorem 6.26 leaves a gap. So it does not make much sense to stress applications of our results to classical cases. But d-freeness assumption makes sense also in rings that are not so nice from the structural point of view. So the results of this section are still potentially applicable in concrete situations.

6.5 f-Homomorphisms and f-Derivations

Lie homomorphisms are additive maps preserving the polynomial $x_1 x_2 - x_2 x_1$, and Jordan homomorphisms are additive maps preserving the polynomial $x_1 x_2 + x_2 x_1$. Our aim now is to consider additive maps that preserve an arbitrary multilinear polynomial $f = f(x_1, \ldots, x_m) \in \mathbb{Z}\langle x_1, x_2 \ldots \rangle$ of degree $m \geq 2$. Only one mild assumption will be imposed on f, namely, we assume that at least one of its coefficients is equal to 1. We fix such a polynomial f.

Let us now specify the problem we are about to tackle. Let $\mathcal{B}$ and $\mathcal{Q}$ be rings, and let $\mathcal{S}$ be an f-closed additive subgroup of $\mathcal{B}$ (meaning that $f(\bar{x}_m) \in \mathcal{S}$ for all $\bar{x}_m \in \mathcal{S}^m$). An additive map $\alpha : \mathcal{S} \to \mathcal{Q}$ will be a called an f-*homomorphism* if

$$f(x_1, x_2, \ldots, x_m)^\alpha = f(x_1^\alpha, x_2^\alpha, \ldots, x_m^\alpha) \quad \text{for all } x_1, \ldots, x_m \in \mathcal{S}$$

(sometimes we will use an abbreviated notation and write this as $f(\bar{x}_m)^\alpha = f(\bar{x}_m^\alpha)$). The goal is, of course, to describe the form of α.

In this generality there is little hope that anything interesting can be said. We will first confine ourselves to the case where $\mathcal{S}$ is a Lie subring of $\mathcal{B}$ and study the action of α on the Lie product, and later, in the main result, to the case when $\mathcal{S} = \mathcal{B}$ is a ring.

We remark that the case when f is a polynomial identity for both $\mathcal{S}$ and $\mathcal{Q}$ must certainly be excluded; namely, in such a case every additive map from $\mathcal{S}$ into $\mathcal{Q}$ is an f-homomorphism. However, the d-freeness assumption which we shall impose on $\mathcal{S}^\alpha$ will prevent this type of situation from occurring.

The condition that f is multilinear is not indispensable, it is included only for convenience. Indeed, if f was not multilinear, then we could linearize $f(\bar{x}_m)^\alpha = f(\bar{x}_m^\alpha)$ and thereby arrive at an $\tilde{f}$-homomorphism where $\tilde{f}$ is multilinear and of the same degree as f.

The basic problem is to "conjure up" an appropriate FI involving an f-homomorphism. Let us reveal the main idea of our approach. The crucial computations will be made in the free algebra $\mathbb{Z}\langle X \rangle$. First we recall (from the second paragraph of the proof of Lemma 6.17) the following easily obtained but very useful identity:

$$[f(\overline{x}_m), y] = \sum_{i=1}^{m} f(x_1, \ldots, x_{i-1}, [x_i, y], x_{i+1}, \ldots, x_m). \qquad (6.30)$$

In particular (6.30) gives

$$[f(\overline{x}_m), f(\overline{y}_m)] = \sum_{i=1}^{m} f(x_1, \ldots, x_{i-1}, [x_i, f(\overline{y}_m)], x_{i+1}, \ldots, x_m).$$

Since

$$[x_i, f(\overline{y}_m)] = \sum_{j=1}^{m} f(y_1, \ldots, y_{j-1}, [x_i, y_j], y_{j+1}, \ldots, y_m),$$

it follows that

$$[f(\overline{x}_m), f(\overline{y}_m)] = \sum_{i=1}^{m} \sum_{j=1}^{m} f(x_1, \ldots, x_{i-1}, g_{ij}, x_{i+1}, \ldots, x_m), \qquad (6.31)$$

where

$$g_{ij} = f(y_1, \ldots, y_{j-1}, [x_i, y_j], y_{j+1}, \ldots, y_m).$$

On the other hand, since $[f(\overline{x}_m), f(\overline{y}_m)] = -[f(\overline{y}_m), f(\overline{x}_m)]$, we can replace the roles of x_i's and y_i's on the right-hand side of (6.31) and change this sign. Comparing the identity, so obtained, with (6.31), we get

$$\sum_{i=1}^{m} \sum_{j=1}^{m} f(x_1, \ldots, x_{i-1}, g_{ij}, x_{i+1}, \ldots, x_m)$$

$$+ \sum_{i=1}^{m} \sum_{j=1}^{m} f(y_1, \ldots, y_{i-1}, h_{ij}, y_{i+1}, \ldots, y_m) = 0,$$

where

$$h_{ij} = f(x_1, \ldots, x_{j-1}, [y_i, x_j], x_{j+1}, \ldots, x_m).$$

This is the key formula from which an FI for *f*-homomorphisms can be derived. This FI will eventually lead to the following lemma.

Lemma 6.28. *Let S be an f-closed Lie subring of a ring $\mathcal{B}$, let $\mathcal{Q}$ be a unital ring with center $\mathcal{C}$, and let $\alpha : S \to \mathcal{Q}$ be an f-homomorphism. If S^α is a (2m)-free subset of $\mathcal{Q}$, then there exists $\eta \in \mathcal{C}$ such that $[x, y]^\alpha - \eta[x^\alpha, y^\alpha] \in \mathcal{C}$ for all $x, y \in S$.*

Proof. Define $B : \mathcal{S}^2 \to \mathcal{Q}$ by $B(x,y) = [x,y]^\alpha$, and note that the identity preceding the statement of the lemma immediately yields

$$\sum_{i=1}^{m}\sum_{j=1}^{m} f(x_1^\alpha, \ldots, x_{i-1}^\alpha, g_{ij}^\alpha, x_{i+1}^\alpha, \ldots, x_m^\alpha)$$

$$+ \sum_{i=1}^{m}\sum_{j=1}^{m} f(y_1^\alpha, \ldots, y_{i-1}^\alpha, h_{ij}^\alpha, y_{i+1}^\alpha, \ldots, y_m^\alpha) = 0,$$

where

$$g_{ij}^\alpha = f(y_1^\alpha, \ldots, y_{j-1}^\alpha, B(x_i, y_j), y_{j+1}^\alpha, \ldots, y_m^\alpha),$$
$$h_{ij}^\alpha = f(x_1^\alpha, \ldots, x_{j-1}^\alpha, B(y_i, x_j), x_{j+1}^\alpha, \ldots, x_m^\alpha).$$

This identity might appear complicated, but actually this is exactly a core FI of the type for which Corollary 4.14 (for $P = 0$) is applicable. Indeed, expanding this identity (in one's mind) and using $B(x,y) = -B(y,x)$ (so that the order of variables does not cause problems), we see that (referring to the notation in Corollary 4.14) each of the summands is of the form $c_{M,N} MBN$. Without loss of generality we may assume that f involves the term $x_1 x_2 \ldots x_m$. Then one of the terms in the expanded form of our FI is equal to

$$-B(x_1, y_1) x_2^\alpha \ldots x_m^\alpha y_2^\alpha \ldots y_m^\alpha. \tag{6.32}$$

Indeed, the term involving $B(x_1, y_1) x_2^\alpha \ldots x_m^\alpha y_2^\alpha \ldots y_m^\alpha$ can only appear in the second summation with $i = j = 1$, and using $B(x_1, y_1) = -B(y_1, x_1)$ it clearly follows that it appears multiplied by -1. For our purposes it is only important to note that -1 is invertible in $\mathcal{C}$ and that (6.32) is a term in our FI with B appearing as a leftmost middle function. Since $\mathcal{S}^\alpha$ is $(2m)$-free in $\mathcal{Q}$, all conditions of Corollary 4.14 are met, and we may conclude that B is a quasi-polynomial. Since $B(x,y) = -B(y,x)$ we can use Lemma 4.4 in a standard fashion to conclude that

$$[x,y]^\alpha = \eta[x^\alpha, y^\alpha] + \mu(x)y^\alpha - \mu(y)x^\alpha + \nu(x,y)$$

where $\eta \in \mathcal{C}$, $\mu : \mathcal{S} \to \mathcal{C}$ is an additive map, and $\nu : \mathcal{S}^2 \to \mathcal{C}$ is a biadditive map. We now return to (6.30). Applying α to this identity it follows that

$$\eta[f(x_1^\alpha, x_2^\alpha, \ldots, x_m^\alpha), y^\alpha] + \mu(f(\overline{x}_m))y^\alpha - \mu(y)f(\overline{x}_m^\alpha) + \nu(f(\overline{x}_m), y)$$

$$= \sum_{i=1}^{m} f(x_1^\alpha, \ldots, x_{i-1}^\alpha, \eta[x_i^\alpha, y^\alpha] + \mu(x_i)y^\alpha - \mu(y)x_i^\alpha + \nu(x_i, y), x_{i+1}^\alpha, \ldots, x_m^\alpha).$$

This identity could be simplified, but for our purposes it is only important to note that again this is an identity which can be interpreted as that a quasi-polynomial (this time of degree $\leq m+1$) is zero. Since $m+1 \leq 2m$, we are again in a position to apply Lemma 4.4. It is easy to see that the coefficient at $y^\alpha x_2^\alpha \ldots x_m^\alpha$ is equal to $\mu(x_1)$, which implies that $\mu = 0$. So we have $[x,y]^\alpha - \eta[x^\alpha, y^\alpha] \in \mathcal{C}$. $\square$

Remark 6.29. The conclusion of Lemma 6.28 still holds if we replace the assumption that α is an f-homomorphism by a slightly weaker assumption that f satisfies the condition

$$f(x_1, x_2, \ldots, x_m)^\alpha = \lambda f(x_1^\alpha, x_2^\alpha, \ldots, x_m^\alpha) \quad \text{for all } x_1, \ldots, x_m \in \mathcal{S},$$

where λ is a fixed invertible element in $\mathcal{C}$. The proof is basically the same, the necessary changes are obvious. The reason for mentioning this is that such a situation will actually appear below, in the proof of Theorem 7.17.

Lemma 6.28 gives useful information about the action of an f-homomorphism on the Lie product. To obtain a more definitive conclusion, i.e., expressing α through associative (anti)homomorphisms, we have to impose some further restrictions. The presence of the element η causes some problems. Note that the lemma implies that the map $x^\beta = \eta x^\alpha$ is a weak Lie homomorphism. We already saw how to handle weak Lie homomorphisms in some important Lie subrings, so apparently we are close to the final goal. The problem is, however, that there is no reason to believe that the image of β satisfies any d-free condition. In the worse case η can be 0 and the approach based on β clearly fails. Anyway, in various situations this problem can be settled. We will present one of them.

Theorem 6.30. *Let $\mathcal{B}$ be a ring, let $\mathcal{Q}$ be a unital ring such that its center $\mathcal{C}$ is a field, and let $\alpha : \mathcal{B} \to \mathcal{Q}$ be an f-homomorphism. If $\mathcal{B}^\alpha$ is a $(2m)$-free subset of $\mathcal{Q}$, then $x^\alpha = \lambda x^\sigma + x^\mu$ for all $x \in \mathcal{B}$, where $\lambda \in \mathcal{C}$, $\sigma : \mathcal{B} \to \mathcal{Q}$ is a homomorphism or an antihomomorphism, and $\mu : \mathcal{B} \to \mathcal{C}$ is an additive map.*

Proof. We proceed from Lemma 6.28. So, we know that there is $\eta \in \mathcal{C}$ such that $[x, y]^\alpha - \eta[x^\alpha, y^\alpha] \in \mathcal{C}$ for all $x, y \in \mathcal{B}$.

Suppose first that $\eta = 0$, that is, $\nu(x, y) = [x, y]^\alpha \in \mathcal{C}$ for all $x, y \in \mathcal{B}$. Let f_i be the partial derivative of f at x_i (i.e., f_i is an element in $\mathbb{Z}\langle X \rangle$ which we get by formally replacing x_i in $f(x_1, \ldots, x_m)$ by 1). Applying α to (6.30) we get

$$\nu(f(\overline{x}_m), y) = \sum_{i=1}^{m} f(x_1^\alpha, \ldots, x_{i-1}^\alpha, \nu(x_i, y), x_{i+1}^\alpha, \ldots, x_m^\alpha)$$

$$= \sum_{i=1}^{m} \nu(x_i, y) f_i(x_1^\alpha, \ldots, x_{i-1}^\alpha, x_{i+1}^\alpha, \ldots, x_m^\alpha).$$

Again invoking Lemma 4.4 it follows that the product of $\nu(x_i, y)$ with every coefficient of f_i is 0. Since $\mathcal{C}$ is assumed to be a field this yields a sharp conclusion about ν and f_i, but let us state only what we need, that is

$$\nu(x_i, y) f_i(x_1^\alpha, \ldots, x_{i-1}^\alpha, x_{i+1}^\alpha, \ldots, x_m^\alpha) = 0.$$

In other notation we can write this as

$$[x, y]^\alpha f_i(\overline{z}_{m-1}^\alpha) = 0 \tag{6.33}$$

for all $x, y \in \mathcal{B}$, $\bar{z}_{m-1} \in \mathcal{B}^{m-1}$, and $i = 1, \ldots, m$.

We claim that $f(\bar{x}_m)$ can be written as

$$f(\bar{x}_m) = x_m p(\bar{x}_{m-1}) + q(\bar{x}_m)$$

where $p(\bar{x}_{m-1}) \in \mathbb{Z}\langle X \rangle$ and $q(\bar{x}_m) \in [\mathbb{Z}\langle X \rangle, \mathbb{Z}\langle X \rangle]$. Indeed, this follows immediately from $g_1 x_m g_2 = x_m g_2 g_1 + [g_1, x_m g_2]$. On the other hand, we can write $f(\bar{x}_m) = p(\bar{x}_{m-1}) x_m + [x_m, p(\bar{x}_{m-1})] + q(\bar{x}_m)$ and so

$$f(\bar{x}_m) = p(\bar{x}_{m-1}) x_m + r(\bar{x}_m)$$

where $r(\bar{x}_m)$ also belongs to $[\mathbb{Z}\langle X \rangle, \mathbb{Z}\langle X \rangle]$. Since $q(\bar{y}_m), r(\bar{y}_m) \in [\mathcal{B}, \mathcal{B}]$ for all $\bar{y}_m \in \mathcal{B}^m$, we have

$$\omega(\bar{y}_m) = q(\bar{y}_m)^\alpha \in [\mathcal{B}, \mathcal{B}]^\alpha \subseteq \mathcal{C} \quad \text{and} \quad \rho(\bar{y}_m) = r(\bar{y}_m)^\alpha \in [\mathcal{B}, \mathcal{B}]^\alpha \subseteq \mathcal{C}.$$

In view of (6.33) we in particular have

$$\omega(\bar{y}_m) f_m(\bar{z}_{m-1}^\alpha) = 0 \quad \text{and} \quad \rho(\bar{y}_m) f_m(\bar{z}_{m-1}^\alpha) = 0 \tag{6.34}$$

for all $\bar{y}_m \in \mathcal{B}^m$, $\bar{z}_{m-1} \in \mathcal{B}^{m-1}$. Further, we have

$$f(\bar{y}_m)^\alpha = \big(y_m p(\bar{y}_{m-1})\big)^\alpha + \omega(\bar{y}_m) = \big(p(\bar{y}_{m-1}) y_m\big)^\alpha + \rho(\bar{y}_m). \tag{6.35}$$

Using (6.34) and (6.35) it now follows that for all $\bar{y}_{m-1}, \bar{z}_{m-1} \in \mathcal{B}^{m-1}$ and $w \in \mathcal{B}$ we have

$$f\Big(\bar{y}_{m-1}^\alpha, f(\bar{z}_{m-1}^\alpha, w^\alpha)\Big) = f\Big(\bar{y}_{m-1}^\alpha, f(\bar{z}_{m-1}, w)^\alpha\Big)$$

$$= f\Big(\bar{y}_{m-1}^\alpha, \big(w p(\bar{z}_{m-1})\big)^\alpha + \omega(\bar{z}_{m-1}, w)\Big)$$

$$= f\Big(\bar{y}_{m-1}, w p(\bar{z}_{m-1})\Big)^\alpha + \omega(\bar{z}_{m-1}, w) f_m(\bar{y}_{m-1}^\alpha)$$

$$= \Big(p(\bar{y}_{m-1}) w p(\bar{z}_{m-1})\Big)^\alpha + \rho\big(\bar{y}_{m-1}, w p(\bar{z}_{m-1})\big).$$

On the other hand,

$$f\Big(\bar{z}_{m-1}^\alpha, f(\bar{y}_{m-1}^\alpha, w^\alpha)\Big) = f\Big(\bar{z}_{m-1}^\alpha, f(\bar{y}_{m-1}, w)^\alpha\Big)$$

$$= f\Big(\bar{z}_{m-1}^\alpha, \big(p(\bar{y}_{m-1}) w\big)^\alpha + \rho(\bar{y}_{m-1}, w)\Big)$$

$$= f\Big(\bar{z}_{m-1}, p(\bar{y}_{m-1}) w\Big)^\alpha + \rho(\bar{y}_{m-1}, w) f_m(\bar{z}_{m-1}^\alpha)$$

$$= \Big(p(\bar{y}_{m-1}) w p(\bar{z}_{m-1})\Big)^\alpha + \omega\big(\bar{z}_{m-1}, p(\bar{y}_{m-1}) w\big).$$

So in each case the result is the same modulo $\mathcal{C}$. Therefore

$$f\Big(\bar{y}_{m-1}^\alpha, f(\bar{z}_{m-1}^\alpha, w^\alpha)\Big) - f\Big(\bar{z}_{m-1}^\alpha, f(\bar{y}_{m-1}^\alpha, w^\alpha)\Big) \in \mathcal{C} \tag{6.36}$$

for all $\overline{y}_{m-1}, \overline{z}_{m-1} \in \mathcal{B}^{m-1}$ and $w \in \mathcal{B}$. Since $\mathcal{B}$ is $(2m)$-free we are again in a position to apply Lemma 4.4. Consequently, all coefficients of terms in the expansion of (6.36) are zero. However, assuming that f involves the term $x_1 x_2 \dots x_m$, it follows that (6.36) involves the term $y_1^\alpha \dots y_{m-1}^\alpha z_1^\alpha \dots z_{m-1}^\alpha w^\alpha$, a contradiction.

Thus $\eta \neq 0$, and hence, since $\mathcal{C}$ is assumed to be a field, η is invertible. Now, $x^\beta = \eta x^\alpha$ defines a weak Lie homomorphism from $\mathcal{B}$ into $\mathcal{Q}$, and $\mathcal{B}^\beta = \eta \mathcal{B}^\alpha$ is $(2m)$-free in $\mathcal{Q}$ (see Lemma 3.3 (v)). But actually all we need now is that the range of β is 4-free. Namely, Remark 6.2 tells us that in this case we have $\beta = \theta + \tau$ where $\tau : \mathcal{B} \to \mathcal{C}$ is an additive map, and $\theta : \mathcal{B} \to \mathcal{Q}$ is a direct sum of a homomorphism and the negative of an antihomomorphism. However, as a field $\mathcal{C}$ does not contain nontrivial idempotents. Therefore θ is either a homomorphism or the negative of an antihomomorphism. Now we define μ by $x^\mu = \eta^{-1} x^\tau$, we set $\sigma = \theta$ if θ is a homomorphism and $\sigma = -\theta$ if θ is an antihomomorphism, and similarly, we set $\lambda = \eta^{-1}$ if θ is a homomorphism and $\lambda = -\eta^{-1}$ if θ is an antihomomorphism. In any case we have $x^\alpha = \lambda x^\sigma + x^\mu$, and the theorem is proved. $\square$

One can derive some further conclusions about λ and μ, namely, sometimes (depending on f!) μ is 0, and $\lambda^{m-1} = \pm 1$ (just think of Lie and Jordan homomorphisms). But we shall not go into detail with regard to this matter.

Corollary 6.31. *Let $\mathcal{B}$ be any ring, let $\mathcal{A}$ be a prime ring with extended centroid $\mathcal{C}$, and let α be an f-homomorphism from $\mathcal{B}$ onto $\mathcal{A}$. If $\deg(\mathcal{A}) \geq 2m$, then $x^\alpha = \lambda x^\sigma + x^\mu$ for all $x \in \mathcal{B}$, where $\lambda \in \mathcal{C}$, $\sigma : \mathcal{B} \to \mathcal{AC} + \mathcal{C}$ is a homomorphism or an antihomomorphism, and $\mu : \mathcal{B} \to \mathcal{C}$ is an additive map.*

Proof. Apply Theorem 6.30 and Corollary 5.12; note that in the present situation the image of σ necessarily lies in $\mathcal{AC} + \mathcal{C}$. $\square$

Obtaining analogous results for derivations is easier. We define an f-derivation to be an additive map $\delta : \mathcal{A} \to \mathcal{Q}$ such that

$$f(x_1, \dots, x_m)^\delta = \sum_{i=1}^{m} f(x_1, \dots, x_{i-1}, x_i^\delta, x_{i+1}, \dots, x_m) \qquad (6.37)$$

for all $x_1, \dots, x_m \in \mathcal{A}$; here $\mathcal{A}$ can be in principle any f-closed additive subgroup of $\mathcal{Q}$. But we shall examine only the situation where $\mathcal{A}$ is a ring.

One would expect, of course, that f-derivations can be expressed through derivations and central maps. Note, however, that if $\mathrm{char}(\mathcal{A}) = m - 1$, then the identity map is always an f-derivation. So we have to take this "degenerate" example into account.

Theorem 6.32. *Let $\mathcal{A}$ be a ring, let $\mathcal{Q} \supseteq \mathcal{A}$ be a unital ring with center $\mathcal{C}$, and let $\delta : \mathcal{A} \to \mathcal{Q}$ be an f-derivation. If $\mathcal{A}^\alpha$ is an $(2m)$-free subset of $\mathcal{Q}$, then $x^\delta = \lambda x + x^d + x^\mu$ for all $x \in \mathcal{A}$, where $\lambda \in \mathcal{C}$ is such that $(m-1)\lambda = 0$, $d : \mathcal{A} \to \mathcal{Q}$ is a derivation, and $\mu : \mathcal{B} \to \mathcal{C}$ is an additive map.*

Proof. Let $\check{\mathcal{Q}}$ and $\check{\mathcal{C}}$ have their usual meaning (see Section 3.3), and define $\alpha :$ $\mathcal{A} \to \check{\mathcal{Q}}$ according to $x^\alpha = (x, x^\delta)$. One can check that α is an f-homomorphism. Theorem 3.7 makes it possible for us to use the conclusion of Lemma 6.28. Therefore there exists $\eta \in \check{\mathcal{C}}$ such that $[x, y]^\alpha - \eta[x^\alpha, y^\alpha] \in \check{\mathcal{C}}$ for all $x, y \in \mathcal{A}$. Let us write $\eta = (\eta_1, \eta_2)$, where $\eta_i \in \mathcal{C}$, and so we have $[x, y] - \eta_1[x, y] \in \mathcal{C}$ and $[x, y]^\delta - \eta_1[x^\delta, y] - \eta_1[x, y^\delta] - \eta_2[x, y] \in \mathcal{C}$. Since $\mathcal{A}$ is in particular 3-free in $\mathcal{Q}$, the first relation immediately yields $\eta_1 = 1$. The second relation therefore implies that $x^\Delta = x^\delta + \eta_2 x$ is a weak Lie derivation. Therefore there exists a derivation $d : \mathcal{A} \to \mathcal{Q}$ and an additive map $\mu : \mathcal{A} \to \mathcal{C}$ such that $\Delta = d + \mu$ (see Remark 6.8). Accordingly, $x^\delta = \lambda x + x^d + x^\mu$ where $\lambda = -\eta_2$. Using this form of δ in (6.37) it is easy to show, by a standard application of Lemma 4.4, that $(m-1)\lambda = 0$. $\square$

Corollary 6.33. *Let $\mathcal{A}$ be a prime ring with extended centroid $\mathcal{C}$, and let $\delta : \mathcal{A} \to \mathcal{A}$ be an f-derivation. If $\deg(\mathcal{A}) \geq 2m$, then $x^\delta = \lambda x + x^d + x^\mu$ for all $x \in \mathcal{A}$, where $\lambda \in \mathcal{C}$, $d : \mathcal{A} \to \mathcal{A}\mathcal{C} + \mathcal{C}$ is a derivation, and $\mu : \mathcal{A} \to \mathcal{C}$ is an additive map. Moreover, if $\mathrm{char}(\mathcal{A})$ does not divide $m-1$, then $\lambda = 0$.*

Proof. Apply Theorem 6.32 and Corollary 5.12. $\square$

Literature and Comments. Let us first make some historic remarks about the development of the Lie map topic before the FI methods were discovered. The description of Lie homomorphisms on important Lie subalgebras of $M_n(\mathbb{F})$ has been well-known for a long time (see for example [119, Chapter 10]). In 1951 Hua characterized Lie automorphisms of a simple Artinian ring $M_n(\mathcal{D})$, $n \geq 3$ [118] (and Kaplansky studied Lie derivations on $M_n(\mathcal{B})$, $n \geq 3$ and $\mathcal{B}$ any unital ring; however, he did not publish this). Motivated by these results and his own work on Lie structures in associative rings, Herstein in his 1961 "AMS Hour Talk" [113] posed various conjectures on Lie maps. Specifically, he conjectured the description of Lie maps on (a) $\mathcal{A}$, (b) $[\mathcal{A}, \mathcal{A}]$ and $[\mathcal{A}, \mathcal{A}]/\mathcal{Z} \cap [\mathcal{A}, \mathcal{A}]$, (c) $\mathcal{K}$, and (d) $[\mathcal{K}, \mathcal{K}]$ and $[\mathcal{K}, \mathcal{K}]/\mathcal{Z} \cap [\mathcal{K}, \mathcal{K}]$; here $\mathcal{A}$ was a simple ring (although he also mentioned the possibility of tackling the more general prime ring case), $\mathcal{Z}$ is its center, and (in case $\mathcal{A}$ has an involution) $\mathcal{K}$ is the set of skew elements of $\mathcal{A}$. In the 1960s and 1970s, many of these conjectures were settled by Martindale and some of his students, however, under the assumption that the rings in question contain nontrivial idempotents [116, 147, 148, 150, 151, 153, 154, 155, 185]. Incidentally we mention that it was the Lie homomorphism problem that motivated Martindale to introduce the concept of the extended centroid. Over many years Lie map problems were also considered in operator algebras and the techniques there also rest heavily on the presence of idempotents [8, 9, 10, 11, 110, 161, 162, 163, 164, 165, 166].

The question whether Lie maps can be described in rings without idempotents (say, even in division rings) was open for a long time. The first breakthrough in this regard was made by Brešar [58] in 1993 who characterized Lie isomorphisms and Lie derivations on arbitrary prime rings with $\deg(\mathcal{A}) \geq 3$. The proof was based on commuting traces of biadditive maps. This was also the first paper in which the applicability of FI's was noticed. Herstein's conjectures concerning the case (a) were thereby settled, however, with a small exception: the case when $\deg(\mathcal{A}) = 2$ was handled later by Blau [53] using classical PI methods, and after that also by Brešar and Šemrl [84] using commuting maps (this approach was presented in Section 6.1). Further, Banning and Mathieu [12] extended the results from [58] to semiprime rings, and Chebotar considered the char($\mathcal{A}$) = 2 case [87].

After [58], the next important step in settling Herstein's conjectures was made in 1994 by Beidar, Martindale and Mikhalev [39] who characterized Lie isomorphisms on the skew elements $\mathcal{K}$ of a prime ring with involution (of the first kind). In their proof they combined FI's (more precisely, commuting traces of triadditive maps) with GPI techniques. Only later, after the advanced theory of FI's was created, did it turn out that GPI's can be avoided; this was shown in 1999 by Chebotar [90] who also slightly improved the result from [39]. Using the techniques invented in [39], in 1996 Swain [192] described Lie derivations on $\mathcal{K}$. Herstein's conjectures concerning the case (c) were thereby (neglecting some technical questions) basically settled.

Even after all these results were obtained it was still not clear at all how to approach the cases (b) and (d). Commuting maps, upon which everything so far was based, are obviously not applicable in case the Lie subring in question is not closed under certain powers of its elements ($\mathcal{A}$ is closed under squares, $\mathcal{K}$ is closed under cubes, while $[\mathcal{A}, \mathcal{A}]$ and $[\mathcal{K}, \mathcal{K}]$ are in general not closed under any powers). The breakthrough in this regard was made in the 2001 paper [31] by Beidar and Chebotar. They noticed that one can apply FI's in extending Lie maps from Lie ideals to "associative-friendly" structures. Furthermore, making use of their theory of d-free sets [29, 30] which at that time already existed, they were able to get rid of the assumption that Lie homomorphisms must be injective which had always been used before. In particular [31], together with the subsequent paper [32] on Lie derivations, settled Herstein's conjectures concerning the case (b).

The final step in settling Herstein's conjectures was made in the trilogy [23, 24, 25] by Beidar, Brešar, Chebotar and Martindale. Most importantly, these papers settle the case (d). Besides that, they systematically treat all natural questions related to Herstein's conjectures that were often omitted in preceding papers. In particular, the last paper in the series, [25], studies Lie maps in "low degree" rings; this topic is more or less omitted in our treatise. Otherwise, in Sections 6.1–6.3 we surveyed the most important ideas and results from almost all papers that were listed. Still, among them [23] is the closest one to our exposition. But unlike in [23, 24, 25], in this book we do not consider all of Herstein's conjectures systematically. We believe that what we have presented is enough for the reader to get the picture, especially since we have considered in detail the extreme cases from the point of view of difficulty. Roughly speaking, Corollary 6.5 settles the simplest one among Herstein's conjectures (see the second problem in [113, p. 528]), and Corollary 6.20 settles (modulo the low degree cases) the most difficult one among them (see the fifth problem in [113, p. 529]).

Appropriate modifications of the techniques that were presented also work in the context of Banach algebras. On the one hand, structure theorems for Lie maps on certain Banach algebras were obtained (e.g., [79] describes Lie isomorphisms on W^*-algebras), and on the other hand FI methods were also applied to the topic which is by nature entirely analytic, namely, the automatic continuity theory. For example, combining FI's with analytic tools Berenguer and Villena [49] proved that the separating space (which somehow measures the discontinuity of operators on Banach algebras) of a Lie derivation on a semisimple Banach algebra $\mathcal{A}$ is always contained in the center of $\mathcal{A}$; see also related papers [50, 51, 70]. Using various techniques, including FI's, Villena described in [194] Lie derivations on Banach algebras through their action on primitive ideals. Another example where algebraic FI methods were successfully "glued" with analytic methods is the paper [1] by Alaminos, Brešar and Villena. This paper studies the question of whether all Lie and Jordan derivations from a C^*-algebra into its Banach bimodule are of standard forms. Johnson showed that this is true under the assumption of continuity [124]. The problem whether it is also true without this assumption is to the best of our knowledge still open. Among other results, [1] solves this problem for W^*-algebras. The proof nicely illustrates both the power and the imperfection of the FI approach. For those algebras that are "fairly noncommutative" this approach works, while for commutative ones, and for those that are "close" to commutative ones, entirely different analytic methods are used.

An account on Lie maps in the framework of C^*-algebras is given in the monograph by Ara and Mathieu [7]. FI methods play an important role there as well.

Let us also mention the papers [146, 170, 200] in which Lie map results from Section 6.1 are applied to different Lie algebra topics.

Jordan maps have a very long history. Already in the 1940s and 1950s they were studied by Ancochea [4, 5], Kaplansky [126], Hua [117], Jacobson and Rickart [121, 122] and Herstein [111, 112], to mention just a few. In 1967 Martindale [149] showed that Jordan homomorphisms on the symmetric elements of a ring $\mathcal{A}$ with involution can be extended to homomorphisms on $\mathcal{A}$, provided that $\mathcal{A}$ contains nontrivial idempotents satisfying certain technical assumptions. Similarly as in the Lie map case, in the Jordan context it was also an open question for a long time whether the involvement of idempotents is really necessary. In his celebrated paper [199] in which he classified prime nondegenerate Jordan algebras, Zelmanov also obtained the first idempotent-free result on Jordan homomorphisms. The ideas from [199] were later systematically developed in papers devoted only to Jordan maps [132, 157, 160]. Section 3.4 introduces an alternative approach, avoiding the application of the so-called Zelmanov polynomial. The main result, Theorem 6.26, is implicitly contained in the paper [26] which actually studies a more general problem of describing f-homomorphisms on certain Jordan algebras. This book does not consider f-homomorphisms at this level of generality. In Section 6.5 we have basically confined ourselves to f-homomorphisms and f-derivations defined on rings. A version of the main result, Theorem 6.30, is at least indirectly contained in the paper [34] by Beidar and Fong, which is the first work devoted to f-homomorphisms with f being an arbitrary polynomial (for example, an earlier paper [78] studies the case of a polynomial x^n which was considered already by Herstein [113, 114]). In other papers where more general problems on f-homomorphisms and f-derivations were considered [23, 24, 26], the proofs rely heavily on the ideas from [34]. We conjecture that these ideas could turn out to be useful in various problems involving arbitrary polynomials (one illustration of this is given in [47]).

Let us mention at the end that although the results that were presented in Chapter 6 are not essentially new (i.e., in some forms they appeared in papers that we listed), many of them differ in various details from those in the literature. Preparing a survey of these topics also somehow forced us to examine carefully technical assumptions in all results, and as a consequence we were able to improve at least slightly quite a few of them.

Chapter 7

Linear Preserver Problems

"Linear Preserver Problems" are a very popular research area especially in Linear Algebra, and also in Operator Theory and Functional Analysis. These problems deal with linear maps between algebras that, roughly speaking, preserve certain properties; the goal is to find the form of these maps. This is indeed a rather vague description, and certainly one could explain what is a linear preserver problem in a more precise and systematic manner. But let us instead give a couple of illustrative examples.

An invertibility preserving map is a map that sends invertible elements into invertible elements — finding all surjective invertibility preserving linear maps between semisimple unital Banach algebras is an intriguing open problem (the conjecture is that, up to a multiple by an invertible element, they all are Jordan homomorphisms). An idempotent preserving map is a map that sends idempotents into idempotents — clearly Jordan homomorphisms preserve idempotents, and in algebras having "enough" idempotents it often turns out that these are the only idempotent preserving linear maps. A commutativity preserving map is a map that sends commuting pairs of elements into commuting pairs — homomorphisms and antihomomorphisms are obvious examples, but so are maps having a commutative range.

As these examples indicate, linear preserver problems sometimes make sense only in some special classes of algebras (say, having "enough but not too many" invertible elements, idempotents, etc.), and sometimes one has to impose some additional restrictions (like surjectivity) on maps in question. Further, it is natural to expect that (Jordan) homomorphisms often appear in solutions of these problems.

Linear preserver problems have a rich history (Frobenius' result [109] on determinant preserving maps on matrix algebras dates back to the 19th century!), and the literature in this area is really vast. We have absolutely no intention to treat these problems systematically; we shall only refer to a few survey articles [83, 143, 144, 178] that give a more complete account. Our sole aim is to point out a couple of instances where FI's can be effectively applied.

Perhaps the most interesting linear preserver problem that can be solved using FI techniques is the one on commutativity preservers. In Section 7.1 we shall give a rather detailed analysis of this problem, carefully examining the necessity of the conditions that will be imposed in our results. In other Sections 7.2-7.4 we will treat certain other problems a bit more superficially, concentrating just on the main features. Some of the results will be presented only in terms of d-free sets, without stating corollaries to concrete classes of rings. An interested reader can then combine these results with those from Parts I and II.

The title of this chapter is actually slightly misleading. Mostly we will consider rings and not algebras, and our maps will be therefore only additive and not linear. Anyhow, "Linear Preserver Problems" is a common name for the circle of ideas to which this chapter certainly belongs.

7.1 Commutativity Preserving Maps

Let $\mathcal{B}$ and $\mathcal{Q}$ be rings, and let $\mathcal{J}$ be a subset of $\mathcal{B}$. As already mentioned, a map $\alpha : \mathcal{J} \to \mathcal{Q}$ is said to be *commutativity preserving* if for all $x, y \in \mathcal{J}$, $[x, y] = 0$ implies $[x^\alpha, y^\alpha] = 0$. Note that Lie homomorphisms are examples of commutativity preserving maps. So the problem that we are now facing, that is to describe commutativity preservers, is in principle more difficult than the one to describe Lie homomorphisms. But, of course, we will confine ourselves to some less general situations than in the Lie map case. First of all, we will assume that $\mathcal{J}$ is a Jordan subring of $\mathcal{B}$. Thus, for every $x \in \mathcal{J}$ also $x^2 \in \mathcal{J}$. Since x and x^2 certainly commute, every commutativity preserving map $\alpha : \mathcal{J} \to \mathcal{Q}$ satisfies

$$[(x^2)^\alpha, x^\alpha] = 0 \quad \text{for all } x \in \mathcal{J}. \tag{7.1}$$

This is the identity we have met before, in particular when dealing with Lie homomorphisms between rings. If α is additive (and of course we shall assume throughout this chapter that our maps are additive), then we can linearize (7.1) and hence obtain

$$[(x \circ y)^\alpha, z^\alpha] + [(z \circ x)^\alpha, y^\alpha] + [(y \circ z)^\alpha, x^\alpha] = 0 \quad \text{for all } x, y, z \in \mathcal{J}. \tag{7.2}$$

This is an FI we can handle. So already now it is not surprising that something can be done using the FI approach. We shall see, however, that the problem is more subtle than one might expect at first glance.

The title of this section is "misleading" in a similar sense as the title of the chapter: we shall not really consider commutativity preserving maps, but only maps satisfying the weaker condition (7.1). The main reason for this is that (7.1) is all we need; in our proofs we simply do not need the stronger condition that α preserves commutativity. On the other hand, the condition (7.1) is of interest in its own right. Indeed it originally appeared only because of technical reasons, but later it has turned to be more useful than first expected; see the comments at the end of this chapter.

A standard example of a map satisfying (7.1) is the one of the form

$$x^\alpha = \lambda x^\sigma + x^\gamma \quad \text{for all } x \in \mathcal{J}, \tag{7.3}$$

where σ is a Jordan homomorphism from $\mathcal{J}$ into $\mathcal{Q}$, λ is an element in the center $\mathcal{C}$ of $\mathcal{Q}$, and γ is a map from $\mathcal{J}$ into $\mathcal{C}$. We remark that these maps "usually" also preserve commutativity. For example, if σ is a homomorphism or an antihomomorphism, or more generally a direct sum of a homomorphism and an antihomomorphism, then this is certainly true.

Of course there are other examples. For example, every map with a commutative range trivially preserves commutativity. But we will only consider maps that have in some sense a "large" range, so such examples will be excluded. Our goal will be to show that under appropriate assumptions a map satisfying (7.1) must be of the form (7.3). So what are these assumptions? One of them can be easily guessed. In order to resolve (7.2) we will assume that $\mathcal{J}^\alpha$ is a 3-free subset of $\mathcal{Q}$. Further, we will assume that $\mathcal{C}$ is a field with $\text{char}(\mathcal{C}) \neq 2$. This is not absolutely necessary, but it will simplify our arguing, and in the main application, where $\mathcal{A}$ is a prime ring with $\text{char}(\mathcal{A}) \neq 2$ and $\mathcal{Q} = \mathcal{Q}_{ml}(\mathcal{A})$, this condition is fulfilled.

The next examples show that these assumptions are not sufficient. The first one indicates that some conditions must be imposed also on $\mathcal{J}$, not only on $\mathcal{J}^\alpha$, in order to obtain the desired conclusion.

Example 7.1. Suppose that $\mathcal{B}$ is an algebra over a field $\mathbb{F}$, and suppose that $\mathcal{J}$ is a Jordan subalgebra of $\mathcal{B}$ such that every element in $\mathcal{J}$ is algebraic of degree ≤ 2 over $\mathbb{F}$. Assuming also that $\mathcal{B}$ is unital and that $1 \in \mathcal{J}$, it clearly follows that every linear map $\alpha : \mathcal{J} \to \mathcal{Q}$ that sends 1 into the center of $\mathcal{Q}$ automatically satisfies (7.1). But of course there is no reason to believe that α is of the form (7.3). A typical example of this situation is when $\mathcal{B} = \mathcal{J} = M_2(\mathbb{F})$. It is an easy exercise to show that two matrices in $\mathcal{B}$ commute if and only if one of them is a linear combination of the other one and the identity matrix I. Therefore, every linear map $\alpha : \mathcal{B} \to \mathcal{Q}$ that sends I into a central element not only satisfies (7.1), but even preserves commutativity. It is easy to find concrete examples of such maps that are not of the form (7.3).

Recall the useful identity

$$[[x, y], z] = x \circ (y \circ z) - y \circ (x \circ z), \tag{7.4}$$

which in particular implies that every Jordan subring $\mathcal{J}$ satisfies $[[\mathcal{J}, \mathcal{J}], \mathcal{J}] \subseteq \mathcal{J}$. Let $\mathcal{J}$ be a Jordan subalgebra from Example 7.1, and assume also that $\text{char}(\mathbb{F}) \neq 2$. Then, as is evident from the proof of Lemma 3.3 (vii), there exists a linear functional τ on $\mathcal{J}$ such that $\tau(1) = 2$ and $x^2 - \tau(x)x \in \mathbb{F}1$ for all $x \in \mathcal{J}$. Linearizing this relation we get

$$x \circ y - \tau(x)y - \tau(y)x \in \mathbb{F}1 \quad \text{for all } x, y \in \mathcal{J}. \tag{7.5}$$

As $\tau\left(x - \frac{\tau(x)}{2}1\right) = 0$, we see that every $x \in \mathcal{J}$ can be written as $x = x' + \omega 1$ where $\tau(x') = 0$ and $\omega \in \mathbb{F}$. Accordingly, for any $x, y, z \in \mathcal{J}$ the element $[[x,y],z]$ can be written as $[[x',y'],z']$ where $\tau(x') = \tau(y') = \tau(z') = 0$. From (7.5) we see that $y' \circ z', x' \circ z' \in \mathbb{F}1$, hence (7.4) yields $[[x,y],z] = [[x',y'],z'] \in \mathbb{F}x' + \mathbb{F}y'$, and this finally implies $\tau([[x,y],z]) = 0$. From (7.5) we thus see that

$$[[\mathcal{J},\mathcal{J}],\mathcal{J}] \circ [[\mathcal{J},\mathcal{J}],\mathcal{J}] \subseteq \mathbb{F}1,$$

and so, in particular,

$$\left[[\mathcal{J},\mathcal{J}],[[\mathcal{J},\mathcal{J}],\mathcal{J}] \circ [[\mathcal{J},\mathcal{J}],\mathcal{J}]\right] = 0. \tag{7.6}$$

So, Jordan subrings from Example 7.1 satisfy this special identity. The reason for pointing this out will soon become clear.

   The next example is more sophisticated, and shows that we have to reconcile ourselves with further restrictions.

Example 7.2. Let $\mathbb{F}$ be a field, let X be a countably infinite set and let $\mathcal{B} = \mathbb{F}\langle X \rangle$ be the (unital) free algebra on X over $\mathbb{F}$. Let $\mathcal{B}_1$ be the linear span of all elements from X, and let $\mathcal{B}_2$ be the linear span of all monomials of degree ≥ 2. Clearly, $\mathcal{B} = \mathbb{F}1 \oplus \mathcal{B}_1 \oplus \mathcal{B}_2$ (the vector space direct sum). Next, let $\mathbb{E}$ be a countably infinite dimensional field extension of $\mathbb{F}$, and let $\mathcal{A} = \mathbb{E}\langle X \rangle$. Of course, $\mathcal{A} = \mathbb{E}1 \oplus \mathcal{A}_1 \oplus \mathcal{A}_2$ where $\mathcal{A}_1$ and $\mathcal{A}_2$ are defined analogously as $\mathcal{B}_1$ and $\mathcal{B}_2$. Note that $\dim_\mathbb{F}(\mathcal{B}_1) = \dim_\mathbb{F}(\mathcal{A}_1 \oplus \mathcal{A}_2)$ and $\dim_\mathbb{F}(\mathbb{F}1 \oplus \mathcal{B}_2) = \dim_\mathbb{F}(\mathbb{E}1)$. Therefore there exists a bijective $\mathbb{F}$-linear map $\alpha : \mathcal{B} \to \mathcal{A}$ such that $\mathcal{B}_1^\alpha = \mathcal{A}_1 \oplus \mathcal{A}_2$ and $(\mathbb{F}1 \oplus \mathcal{B}_2)^\alpha = \mathbb{E}1$. Pick $u, v \in \mathcal{B}$ such that $[u,v] = 0$. Let us write

$$u = \lambda 1 + u_1 + u_2, \quad v = \mu 1 + v_1 + v_2,$$

where $\lambda, \mu, \nu \in \mathbb{F}$, $u_1, v_1 \in \mathcal{B}_1$, and $u_2, v_2 \in \mathcal{B}_2$. Now, $[u,v] = 0$ can be rewritten as $[u_1,v_1] + [u_1,v_2] + [u_2,v_1] + [u_2,v_2] = 0$, which clearly implies $[u_1,v_1] = 0$. Write $u_1 = \sum \lambda_i x_i$ and $v_1 = \sum_j \mu_j x_j$. Then $0 = [u_1,v_1] = \sum_{i<j}(\lambda_i\mu_j - \lambda_j\mu_i)[x_i,x_j]$, whence $\lambda_i\mu_j = \lambda_j\mu_i$ for each pair $i < j$. Applying α we have $u^\alpha = (\lambda 1 + u_2)^\alpha + u_1^\alpha = \epsilon 1 + \sum_i \lambda_i x_i^\alpha$ where $\epsilon \in \mathbb{E}$, and similarly $v^\alpha = \gamma 1 + \sum_j \mu_j x_j^\alpha$ with $\gamma \in \mathbb{E}$. Thus $[u^\alpha,v^\alpha] = [\sum_i \lambda_i x_i^\alpha, \sum_j \mu_j x_j^\alpha] = \sum_{i<j}(\lambda_i\mu_j - \lambda_j\mu_i)[x_i^\alpha,x_j^\alpha] = 0$. This proves that α preserves commutativity (incidentally, the proof could be slightly shortened by using Bergman's centralizer theorem [52]). However, it is easy to see that α is not of the form (7.3).

   Algebras $\mathcal{B}$ and $\mathcal{A}$ from this example are prime, even primitive, and $\mathcal{A}$ is a d-free subset of $\mathcal{Q}_{ml}(\mathcal{A})$ for every $d \in \mathbb{N}$ (Corollary 5.13). But nevertheless there are nonstandard examples of commutativity preserving bijective linear maps between $\mathcal{B}$ and $\mathcal{A}$. Now it appears that we also have to impose some condition on α (not only on its domain and its range) if we can hope for the desired solution.

   Note that α from Example 7.2 sends a "large piece" of $\mathcal{B}$ into the center. In the following theorem we will show that maps $\alpha : \mathcal{J} \to \mathcal{Q}$ satisfying (7.1) are of the form (7.3) provided that their range is 3-free and they do not map a certain Jordan ideal $\mathcal{L}$ of $\mathcal{J}$ into $\mathcal{C}$. From the definition of $\mathcal{L}$ (see below) and from (7.6) we see that $\mathcal{L} = 0$ if $\mathcal{J}$ is as in Example 7.1. Thus, although Examples 7.1 and 7.2 are of a very different nature, both of them justify the presence of the condition $\mathcal{L}^\alpha \not\subseteq \mathcal{C}$.

Theorem 7.3. *Let $\mathcal{J}$ be a Jordan subring (of some ring), let $\mathcal{K}$ be the Jordan ideal of $\mathcal{J}$ generated by $\Big[[\mathcal{J},\mathcal{J}],[[\mathcal{J},\mathcal{J}],\mathcal{J}] \circ [[\mathcal{J},\mathcal{J}],\mathcal{J}]\Big]$, and let $\mathcal{L}$ be the Jordan ideal of $\mathcal{J}$ generated by $\mathcal{K}\circ\mathcal{K}$. Let $\mathcal{Q}$ be a unital ring such that its center $\mathcal{C}$ is a field with $\mathrm{char}(\mathcal{C}) \neq 2$. Further, let $\alpha : \mathcal{J} \to \mathcal{Q}$ be an additive map satisfying $[(x^2)^\alpha, x^\alpha] = 0$ for all $x \in \mathcal{J}$. Suppose that $\mathcal{J}^\alpha$ is a 3-free subset of $\mathcal{Q}$, and suppose that $\mathcal{L}^\alpha \not\subseteq \mathcal{C}$. Then there exist a Jordan homomorphism $\sigma : \mathcal{J} \to \mathcal{Q}$, an element $\lambda \in \mathcal{C}$ and an additive map $\gamma : \mathcal{J} \to \mathcal{C}$ such that $x^\alpha = \lambda x^\sigma + x^\gamma$ for all $x \in \mathcal{J}$.*

Proof. The first part of the proof is standard. We know that (7.1) implies (7.2), and hence, since $\mathcal{J}^\alpha$ is 3-free, Corollary 4.14 can be applied. Thus $(x \circ y)^\alpha$ is a quasi-polynomial, and its coefficients are multiadditive by Lemma 4.6. Since the function $(x \circ y)^\alpha$ is symmetric (i.e., $(x \circ y)^\alpha = (y \circ x)^\alpha$), a standard application of Lemma 4.4 gives

$$(x \circ y)^\alpha = \eta x^\alpha \circ y^\alpha + \mu(x)y^\alpha + \mu(y)x^\alpha + \nu(x,y), \qquad (7.7)$$

where $\eta \in \mathcal{C}$, $\mu : \mathcal{J} \to \mathcal{C}$ is an additive map, and $\nu : \mathcal{J}^2 \to \mathcal{C}$ is a symmetric biadditive map.

For $u, v \in \mathcal{Q}$ we shall write $u \equiv v$ if $u - v \in \mathcal{C}$. Applying (7.7) we get

$$
\begin{aligned}
((x^2 \circ y) \circ x)^\alpha \equiv{}& \eta(x^2 \circ y)^\alpha \circ x^\alpha + \mu(x^2 \circ y)x^\alpha + \mu(x)(x^2 \circ y)^\alpha \\
\equiv{}& \eta^2((x^2)^\alpha \circ y^\alpha) \circ x^\alpha + \eta\mu(x^2)y^\alpha \circ x^\alpha + \eta\mu(y)(x^2)^\alpha \circ x^\alpha \\
&+ \{2\eta\nu(x^2,y) + \mu(x^2 \circ y)\}x^\alpha + \eta\mu(x)(x^2)^\alpha \circ y^\alpha \\
&+ \mu(x)\mu(x^2)y^\alpha + \mu(x)\mu(y)(x^2)^\alpha.
\end{aligned}
$$

Similarly we obtain

$$
\begin{aligned}
(x^2 \circ (y \circ x))^\alpha \equiv{}& \eta(x^2)^\alpha \circ (y \circ x)^\alpha + \mu(x^2)(y \circ x)^\alpha + \mu(y \circ x)(x^2)^\alpha \\
\equiv{}& \eta^2(x^2)^\alpha \circ (y^\alpha \circ x^\alpha) + \eta\mu(y)(x^2)^\alpha \circ x^\alpha + \eta\mu(x)(x^2)^\alpha \circ y^\alpha \\
&+ \{2\eta\nu(x,y) + \mu(y \circ x)\}(x^2)^\alpha + \eta\mu(x^2)y^\alpha \circ x^\alpha \\
&+ \mu(x^2)\mu(x)y^\alpha + \mu(x^2)\mu(y)x^\alpha.
\end{aligned}
$$

However, $(x^2 \circ y) \circ x = x^2 \circ (y \circ x)$, and so $((x^2 \circ y) \circ x)^\alpha = (x^2 \circ (y \circ x))^\alpha$. Let us now compare the right-hand sides of the two above relations. Noting that the first terms, $\eta^2((x^2)^\alpha \circ y^\alpha) \circ x^\alpha$ and $\eta^2(x^2)^\alpha \circ (y^\alpha \circ x^\alpha)$, are equal because $(x^2)^\alpha$ and x^α commute, it follows that

$$\{\mu(x)\mu(y) - 2\eta\nu(x,y) - \mu(y \circ x)\}(x^2)^\alpha \equiv \{\mu(x^2)\mu(y) - 2\eta\nu(x^2,y) - \mu(x^2 \circ y)\}x^\alpha.$$

As $\mathrm{char}(\mathcal{C}) \neq 2$, setting $y = x$ in (7.7) we get

$$(x^2)^\alpha = \eta(x^\alpha)^2 + \mu(x)x^\alpha + \frac{1}{2}\nu(x,x), \tag{7.8}$$

and so, by substituting (7.8) in the preceding identity, it follows that

$$\varepsilon(x,y)(x^\alpha)^2 + \zeta(x,x,y)x^\alpha \equiv 0 \tag{7.9}$$

for all $x, y \in \mathcal{J}$, where $\varepsilon : \mathcal{J}^2 \to \mathcal{C}$ is a symmetric biadditive map defined by

$$\varepsilon(x,y) = \eta\{\mu(x)\mu(y) - 2\eta\nu(x,y) - \mu(y \circ x)\}, \tag{7.10}$$

and $\zeta : \mathcal{J}^3 \to \mathcal{C}$ is a triadditive map defined by

$$\zeta(x,x',y) = \eta\{\nu(x \circ x',y) - 2\nu(x,y)\mu(x')\}$$
$$+ \mu(x)\mu(x')\mu(y) + \frac{1}{2}\mu((x \circ x') \circ y) - \mu(y \circ x)\mu(x') - \frac{1}{2}\mu(x \circ x')\mu(y). \tag{7.11}$$

Let us show that ε is 0 (incidentally, if $\mathcal{J}^\alpha$ was 4-free, then (by fixing y in 7.9) this would follow immediately from Corollary 4.21, but we are only assuming 3-freeness). The complete linearization of (7.9) gives

$$\varepsilon(x,y)u^\alpha \circ v^\alpha + \varepsilon(u,y)x^\alpha \circ v^\alpha + \varepsilon(v,y)x^\alpha \circ u^\alpha + (\zeta(x,u,y) + \zeta(u,x,y))v^\alpha$$
$$+ (\zeta(x,v,y) + \zeta(v,x,y))u^\alpha + (\zeta(v,u,y) + \zeta(u,v,y))x^\alpha \equiv 0 \tag{7.12}$$

for all $x, y, u, v \in \mathcal{J}$. By Lemma 3.3 (vii) there exists $x_1 \in \mathcal{J}$ such that x_1^α is not algebraic of degree ≤ 2 over $\mathcal{C}$. Therefore $\varepsilon(x_1,y) = \zeta(x_1,x_1,y) = 0$ for every $y \in \mathcal{J}$ by (7.9). Setting $x = u = x_1$ and $v = y$ in (7.12) we thus get

$$2\varepsilon(y,y)(x_1^\alpha)^2 \in \mathcal{C}x_1^\alpha + \mathcal{C},$$

which implies $\varepsilon(y,y) = 0$ for every $y \in \mathcal{J}$. As ε is symmetric it follows that $\varepsilon(x,y) = 0$ for all $x, y \in \mathcal{J}$.

Suppose that $\eta \neq 0$. Since $\varepsilon = 0$, it now follows from (7.10) that

$$\mu(x)\mu(y) - 2\eta\nu(x,y) - \mu(y \circ x) = 0$$

for all $x, y \in \mathcal{J}$, and so (7.7) can be rewritten as

$$(x \circ y)^\alpha = \eta x^\alpha \circ y^\alpha + \mu(x)y^\alpha + \mu(y)x^\alpha + \frac{1}{2}\eta^{-1}\mu(x)\mu(y) - \frac{1}{2}\eta^{-1}\mu(y \circ x).$$

Note that this implies that the map $\sigma : \mathcal{J} \to \mathcal{Q}$ defined according to

$$x^\sigma = \eta x^\alpha + \frac{1}{2}\mu(x)$$

is a Jordan homomorphism. So we just set $\lambda = \eta^{-1}$ and $x^\gamma = -\frac{1}{2}\eta^{-1}\mu(x)$ and the proof is complete.

Thus, the only problem that remains is to show that $\eta \neq 0$. Suppose, on the contrary, that $\eta = 0$. The relation (7.9) now reduces to $\zeta(x, x, y)x^\alpha \equiv 0$ for all $x, y \in \mathcal{J}$. Therefore, for every $x \in \mathcal{J}$ we have either $x^\alpha \equiv 0$ or $\zeta(x, x, y) = 0$ for all $y \in \mathcal{J}$. We claim that the latter condition actually holds for every $x \in \mathcal{J}$. If this was not true, there would be $x_0, y_0 \in \mathcal{J}$ such that $\zeta(x_0, x_0, y_0) \neq 0$ (and so $x_0^\alpha \equiv 0$). Of course there exists $x_1 \in \mathcal{J}$ such that $x_1^\alpha \neq 0$, and so also $(x_0 + x_1)^\alpha \neq 0$ and $(x_0 - x_1)^\alpha \neq 0$. Consequently,

$$\zeta(x_1, x_1, y_0) = 0, \quad \zeta(x_0 + x_1, x_0 + x_1, y_0) = 0, \quad \text{and} \quad \zeta(x_0 - x_1, x_0 - x_1, y_0) = 0.$$

Since ζ is additive in each argument, these three identities readily imply that $2\zeta(x_0, x_0, y_0) = 0$. But since $\text{char}(\mathcal{C}) \neq 2$, this is a contradiction. Thus $\zeta(x, x, y) = 0$ for all $x, y \in \mathcal{J}$, and hence $\zeta(x, z, y) + \zeta(z, x, y) = 0$ for all $x, y, z \in \mathcal{J}$. In view of (7.11) this implies

$$\mu((x \circ z) \circ y)$$
$$= \mu(y \circ x)\mu(z) + \mu(z \circ y)\mu(x) + \mu(x \circ z)\mu(y) - 2\mu(x)\mu(z)\mu(y) \quad (7.13)$$

for all $x, y, z \in \mathcal{J}$. Since y and z appear symmetrically on the right-hand side of (7.13), the left-hand side remains the same after changing the roles of y and z. That is, we have $\mu((x \circ z) \circ y) = \mu((x \circ y) \circ z)$, or equivalently $\mu([[y, z], x]) = 0$. Denoting by $\mathcal{M}$ the kernel of μ, we thus have

$$[[\mathcal{J}, \mathcal{J}], \mathcal{J}] \subseteq \mathcal{M}. \quad (7.14)$$

Consequently, for $x, y \in \mathcal{J}$ and $u_1, u_2 \in \mathcal{M}$ we have

$$[[x, y], u_1 \circ u_2] = [[x, y], u_1] \circ u_2 + [[x, y], u_2] \circ u_1 \in \mathcal{M} \cap (\mathcal{M} \circ \mathcal{M}).$$

That is,

$$[[\mathcal{J}, \mathcal{J}], \mathcal{M} \circ \mathcal{M}] \subseteq \mathcal{M} \cap (\mathcal{M} \circ \mathcal{M}),$$

and so in particular

$$\left[[\mathcal{J}, \mathcal{J}], [[\mathcal{J}, \mathcal{J}], \mathcal{J}] \circ [[\mathcal{J}, \mathcal{J}], \mathcal{J}]\right] \subseteq \mathcal{M} \cap (\mathcal{M} \circ \mathcal{M}). \quad (7.15)$$

We claim that the Jordan ideal of $\mathcal{J}$ generated by $\mathcal{M} \cap (\mathcal{M} \circ \mathcal{M})$ is contained in $\mathcal{M}$. Pick $u \in \mathcal{M} \cap (\mathcal{M} \circ \mathcal{M})$. We can write $u = \sum_i x_i \circ z_i$ where $x_i, z_i \in \mathcal{M}$. From (7.13) we see that $u \circ y \in \mathcal{M}$ for every $y \in \mathcal{J}$. Consequently, by induction on n it follows, by using (7.14), that

$$(((\ldots((u \circ y_1) \circ y_2)\ldots) \circ y_{n-2}) \circ y_{n-1}) \circ y_n$$
$$= \left[y_n, (\ldots((u \circ y_1) \circ y_2)\ldots) \circ y_{n-2}], y_{n-1}\right]$$
$$+ \left((\ldots((u \circ y_1) \circ y_2)\ldots) \circ y_{n-2}\right) \circ (y_{n-1} \circ y_n) \in [[\mathcal{J}, \mathcal{J}], \mathcal{J}] + \mathcal{M} \subseteq \mathcal{M}$$

$u \circ (u \circ x) = 0$ for every $x \in \mathcal{B}$. These two identities together yield $2uxu = 0$. But this forces $u = 0$. $\qquad\qquad\qquad\qquad\qquad\qquad\qquad\qquad\qquad\qquad\qquad\qquad\square$

In the situation which we are about to consider, it will be possible to show that a Jordan homomorphism σ from (7.3) is either a homomorphism or an anti-homomorphism. Unfortunately, the results from Section 6.4 are not directly applicable for proving this, and neither is the classical theorem by Herstein on Jordan homomorphisms onto prime rings [111]. But by inspecting (one of) its proof(s) one can see that only small changes must be made to get the desired conclusion. Although we could say, as we did in Remark 7.7, that this proof is not connected with the theme of this book, we shall nevertheless include it here. In particular it nicely illustrates the advantages of the elementary approach to Jordan homomorphisms (cf. the discussion at the end of Section 6.4). The bulk of the proof is the following elementary lemma.

Lemma 7.9. *Let σ be a Jordan homomorphism from a ring $\mathcal{B}$ into a 2-torsion free ring $\mathcal{Q}$. Then for all $x, y, z \in \mathcal{B}$ we have*

$$\big((xy)^\sigma - x^\sigma y^\sigma\big) z^\sigma \big((xy)^\sigma - y^\sigma x^\sigma\big) + \big((xy)^\sigma - y^\sigma x^\sigma\big) z^\sigma \big((xy)^\sigma - x^\sigma y^\sigma\big) = 0,$$
$$\big((xy)^\sigma - x^\sigma y^\sigma\big)\big((xy)^\sigma - y^\sigma x^\sigma\big) = \big((xy)^\sigma - y^\sigma x^\sigma\big)\big((xy)^\sigma - x^\sigma y^\sigma\big) = 0.$$

Proof. Since $\mathcal{Q}$ is 2-torsion free, σ clearly satisfies $(x^2)^\sigma = (x^\sigma)^2$ for all $x \in \mathcal{B}$. Therefore, using $2xzx = (x \circ z) \circ x - x^2 \circ z$ it follows that $(xzx)^\sigma = x^\sigma z^\sigma x^\sigma$ for all $x, z \in \mathcal{B}$. Accordingly, $(x(yzy)x)^\sigma = x^\sigma y^\sigma z^\sigma y^\sigma x^\sigma$, and so

$$(xyzyx + yxzxy)^\sigma = x^\sigma y^\sigma z^\sigma y^\sigma x^\sigma + y^\sigma x^\sigma z^\sigma x^\sigma y^\sigma \qquad\qquad (7.19)$$

for all $x, y, z \in \mathcal{B}$. On the other hand, linearizing $(xyx)^\sigma = x^\sigma y^\sigma x^\sigma$ we see that $(x_1 y x_2 + x_2 y x_1)^\sigma = x_1^\sigma y^\sigma x_2^\sigma + x_2^\sigma y^\sigma x_1^\sigma$, and so writing $xyzyx + yxzxy$ as $(xy)z(yx) + (yx)z(xy)$ we get

$$(xyzyx + yxzxy)^\sigma = (xy)^\sigma z^\sigma (yx)^\sigma + (yx)^\sigma z^\sigma (xy)^\sigma \qquad\qquad (7.20)$$

for all $x, y, z \in \mathcal{B}$. Comparing (7.19) and (7.20), and also using $(yx)^\sigma = x^\sigma y^\sigma + y^\sigma x^\sigma - (xy)^\sigma$, we obtain the first identity which we wished to prove.

The second identity can be derived by computing $(xyxy + xy^2 x)^\sigma$ in two different ways: on the one hand we have $(xy \cdot xy + xy^2 x)^\sigma = (xy)^\sigma \cdot (xy)^\sigma + x^\sigma(y^\sigma)^2 x^\sigma$, and on the other hand $(xy(xy) + (xy)yx)^\sigma = x^\sigma y^\sigma (xy)^\sigma + (xy)^\sigma y^\sigma x^\sigma$. Comparing, we get $\big((xy)^\sigma - x^\sigma y^\sigma\big)\big((xy)^\sigma - y^\sigma x^\sigma\big) = 0$. In a similar fashion, by computing $(xyxy + yx^2 y)^\sigma$ in two different ways, we get the last identity. $\qquad\square$

Corollary 7.10. *Let $\mathcal{A}$ and $\mathcal{B}$ be prime rings such that $\deg(\mathcal{A}) \geq 3$, $\deg(\mathcal{B}) \geq 3$, $\mathrm{char}(\mathcal{A}) \neq 2$, and $\mathrm{char}(\mathcal{B}) \neq 2$. Let $\alpha : \mathcal{B} \to \mathcal{A}$ be a surjective additive map satisfying $[(x^2)^\alpha, x^\alpha] = 0$ for all $x \in \mathcal{B}$. Suppose that α does not map any nonzero ideal of $\mathcal{B}$ into the center of $\mathcal{A}$. Then α is of the form $x^\alpha = \lambda x^\sigma + x^\gamma$ for all $x \in \mathcal{B}$, where $\lambda \in \mathcal{C}$, the extended centroid of $\mathcal{A}$, $\sigma : \mathcal{B} \to \mathcal{A}\mathcal{C} + \mathcal{C}$ is either a homomorphism or an antihomomorphism, and $\gamma : \mathcal{B} \to \mathcal{C}$ is an additive map.*

Proof. We claim that $\left[[x_1, x_2], [[x_3, x_4], x_5] \circ [[x_6, x_7], x_8]\right]$ is not a polynomial identity of $\mathcal{B}$. If it was, then by Theorem C.2 it would also be a polynomial identity of $M_n(\mathbb{F})$ for some field $\mathbb{F}$ and some $n \geq 3$ (as $\deg(\mathcal{B}) \geq 3$). However, from (7.17) we see that this is not true, and our claim is thus proved.

In other words, we have found out that the set $\left[[\mathcal{B}, \mathcal{B}], [[\mathcal{B}, \mathcal{B}], \mathcal{B}] \circ [[\mathcal{B}, \mathcal{B}], \mathcal{B}]\right]$ is nonzero. Therefore the Jordan ideal $\mathcal{K}$ of $\mathcal{B}$ generated by this set is also nonzero. Furthermore, $\mathcal{K} \circ \mathcal{K}$ is also nonzero by the second assertion in Lemma 7.8, and so $\mathcal{L}$, the Jordan ideal of $\mathcal{B}$ generated by $\mathcal{K} \circ \mathcal{K}$, is nonzero as well. But then it follows from (both assertions of) Lemma 7.8 that $\mathcal{L}$ contains a nonzero ideal of $\mathcal{B}$. According to our assumption we thus have that $\mathcal{L}^\alpha$ is not contained in the center of $\mathcal{A}$, and hence it also is not contained in $\mathcal{C}$.

One of the conditions of Theorem 7.3 is therefore fulfilled. Further, $\mathcal{A} = \mathcal{B}^\alpha$ is a 3-free subset of $\mathcal{Q} = \mathcal{Q}_{ml}(\mathcal{A})$ by Corollary 5.12, and of course the center $\mathcal{C}$ of $\mathcal{Q}$ is a field with $\mathrm{char}(\mathcal{C}) \neq 2$. Theorem 7.3 therefore tells us that there exist a Jordan homomorphism $\sigma : \mathcal{B} \to \mathcal{Q}$, an element $\lambda \in \mathcal{C}$ and an additive map $\gamma : \mathcal{B} \to \mathcal{C}$ such that $x^\alpha = \lambda x^\sigma + x^\gamma$ for all $x \in \mathcal{B}$. Of course, $\lambda \neq 0$, and so σ actually maps into $\mathcal{AC} + \mathcal{C}$.

It remains to prove that σ is either a homomorphism or an antihomomorphism. Pick $x, y \in \mathcal{B}$ and set $a = (xy)^\sigma - x^\sigma y^\sigma$ and $b = (xy)^\sigma - y^\sigma x^\sigma$. Note that Lemma 7.9 implies that $aub + bua = 0$ for all $u \in \mathcal{CB}^\sigma + \mathcal{C}$. Therefore, in particular this holds for all $u \in \mathcal{B}^\alpha = \mathcal{A}$, which in turn implies that it holds for all $u \in \mathcal{AC} + \mathcal{C}$. Of course, $\mathcal{AC} + \mathcal{C}$ is also a prime ring and clearly $a, b \in \mathcal{AC} + \mathcal{C}$. In view of Theorem A.7, the identity $aub + bua = 0$ implies that a and b are linearly dependent over the extended centroid of $\mathcal{AC} + \mathcal{C}$ (which is, incidentally, equal to $\mathcal{C}$). But then $aub + bua = 0$, together with the assumption that $\mathrm{char}(\mathcal{A}) \neq 2$, clearly implies that either $a = 0$ or $b = 0$.

So we now know that for each pair $x, y \in \mathcal{B}$ we have either $(xy)^\sigma = x^\sigma y^\sigma$ or $(xy)^\sigma = y^\sigma x^\sigma$. We claim that one of these possibilities must actually hold for all $x, y \in \mathcal{B}$. Proving this is easy. Given any $x \in \mathcal{B}$, we set $\mathcal{G}_x = \{y \in \mathcal{B} \,|\, (xy)^\sigma = x^\sigma y^\sigma\}$ and $\mathcal{H}_x = \{y \in \mathcal{B} \,|\, (xy)^\sigma = y^\sigma x^\sigma\}$. Clearly $\mathcal{G}_x$ and $\mathcal{H}_x$ are additive subgroups of $\mathcal{B}$ and their union is $\mathcal{B}$. But a group can not be the union of two proper subgroups, so we have $\mathcal{G}_x = \mathcal{B}$ or $\mathcal{H}_x = \mathcal{B}$. Thus, $\mathcal{B}$ is the union of its subsets $\mathcal{G} = \{x \in \mathcal{B} \,|\, \mathcal{G}_x = \mathcal{B}\}$ and $\mathcal{H} = \{x \in \mathcal{B} \,|\, \mathcal{H}_x = \mathcal{B}\}$. Since $\mathcal{G}$ and $\mathcal{H}$ are also additive subgroups of $\mathcal{B}$, we have either $\mathcal{G} = \mathcal{B}$ (i.e., σ is a homomorphism) or $\mathcal{H} = \mathcal{B}$ (i.e., σ is an antihomomorphism). $\qquad\square$

Let us point out once again that Example 7.1 justifies the degree assumption in Corollary 7.10, and that Example 7.2 justifies the assumption that α does not map nonzero ideals in the center (note that $\mathcal{B}_2$ in this example is an ideal!).

Remark 7.11. The assumption (in Corollary 7.10) that α does not map nonzero ideals in the center is redundant if one of the following two conditions is fulfilled:

(a) $\mathcal{B}$ is a simple ring,

(b) α preserves commutativity in both directions (i.e., $[x, y] = 0$ if and only if $[x^\alpha, y^\alpha] = 0$).

In case (a) this is obvious, and in case (b) this follows from the (well-known and easily established) fact that nonzero ideals of noncommutative prime rings can not be commutative.

There is another situation where this assumption is redundant, and moreover, such that the conclusion can be given in a particularly nice form. Before stating this last result in this section, we mention that the free algebra $\mathbb{F}\langle X \rangle$ is centrally closed over $\mathbb{F}$ [156]. So, the algebra $\mathcal{A}$ from Example 7.2 is centrally closed over $\mathbb{F}$, while the algebra $\mathcal{B}$ is centrally closed over $\mathbb{E}$ and not over $\mathbb{F}$.

Corollary 7.12. *Let $\mathcal{A}$ and $\mathcal{B}$ be unital centrally closed prime algebras over a field $\mathbb{F}$ with $\mathrm{char}(\mathbb{F}) \neq 2$, and let $\alpha : \mathcal{B} \to \mathcal{A}$ be a bijective $\mathbb{F}$-linear map such that $[(x^2)^\alpha, x^\alpha] = 0$ for all $x \in \mathcal{B}$. If $\deg(\mathcal{B}) \geq 3$, then α is of the form $x^\alpha = \lambda x^\sigma + x^\gamma 1$ for all $x \in \mathcal{B}$, where $\lambda \in \mathbb{F}$, σ is either an isomorphism or an antiisomorphism from $\mathcal{B}$ onto $\mathcal{A}$, and $\gamma : \mathcal{B} \to \mathbb{F}$ is a linear functional.*

Proof. Substituting $x + 1$ for x in $[(x^2)^\alpha, x^\alpha] = 0$ we get $[(x^2 + x)^\alpha, 1^\alpha] = 0$. Now, substituting $-x$ for x in this identity clearly yields $2[x^\alpha, 1^\alpha] = 0$, and so $[x^\alpha, 1^\alpha] = 0$. Since $\mathcal{A} = \mathcal{B}^\alpha$ is centrally closed it follows that $1^\alpha \in \mathbb{F}1$. Since α is linear and bijective this means that $(\mathbb{F}1)^\alpha = \mathbb{F}1$, and moreover, α maps only elements from $\mathbb{F}1$ into $\mathbb{F}1$. In particular, α cannot map a nonzero ideal of $\mathcal{B}$ into the center $\mathbb{F}1$ of $\mathcal{A}$.

Suppose that $\deg(\mathcal{A}) \leq 2$. Since $\mathcal{A}$ is centrally closed, Theorem C.2 implies that $\dim_\mathbb{F} \mathcal{A} \leq 4$. As α is bijective, we then also have $\dim_\mathbb{F} \mathcal{B} \leq 4$. But this, again in view of Theorem C.2, contradicts the assumption that $\deg(\mathcal{B}) \geq 3$. Thus, $\deg(\mathcal{A}) \geq 3$.

Note that all the conditions of Corollary 7.10 are fulfilled. So we have that $x^\alpha = \lambda x^\sigma + x^\gamma$, where $\lambda \in \mathbb{F}$, $\sigma : \mathcal{B} \to \mathcal{A}$ is either a homomorphism or an antihomomorphism, and $\gamma : \mathcal{B} \to \mathbb{F}1$. Moreover, clearly $\lambda \neq 0$, and σ and γ are linear (not only additive). The latter can be shown by noticing at the beginning of the proof of Theorem 7.3 that the map μ from (7.7) must be linear (see Remark 4.7).

We still have to show that σ is bijective. The kernel $\mathcal{I}$ of σ is an ideal of $\mathcal{B}$, and $\mathcal{I}^\alpha \subseteq \mathbb{F}1$. As noted above, this implies that $\mathcal{I} = 0$. Thus σ is injective. Since $\lambda \neq 0$ and since $1^\alpha \in \mathbb{F}1$ it follows that $1^\sigma \in \mathbb{F}1$ — but then 1^σ can be only equal to 1. Accordingly, $x^\alpha = (\lambda x + x^\gamma 1)^\sigma$, showing that σ is surjective. $\quad\square$

Let us mention that there is a short cut to the proof of Corollary 7.12. One can avoid dealing with the Jordan ideals $\mathcal{K}$ and $\mathcal{L}$ from Theorem 7.3, and instead use the observation from Remark 7.4. Thus one has to show that the situation that there is a map $\mu : \mathcal{B} \to \mathbb{F}$ such that $(x^2 - \mu(x)x)^\alpha \in \mathbb{F}1$ for every $x \in \mathcal{B}$ can not occur. But this is easy to see. Namely, since $\mathbb{F}1 = (\mathbb{F}1)^\alpha$, $(x^2 - \mu(x)x)^\alpha \in \mathbb{F}1$ implies that $x^2 - \mu(x)x \in \mathbb{F}1$, which is impossible since $\deg(\mathcal{B}) > 2$.

7.2   Normality Preserving Maps

Let $\mathcal{A}$ be an algebra with involution $*$. We say that an element $x \in \mathcal{A}$ is *normal* if x commutes with x^*. By a *normality preserving map* we mean a map α between algebras with involution that sends normal elements into normal elements; that is to say, $[x, x^*] = 0$ implies $[x^\alpha, (x^\alpha)^*] = 0$. Obvious examples are $*$-*homomorphisms* and $*$-*antihomomorphisms* - by these we mean algebra homomorphisms and anti-homomorphisms which are also $*$-*linear* (that is, they are linear and preserve $*$ in the sense that $(x^*)^\alpha = (x^\alpha)^*$ for every x). One can therefore expect that *Jordan* $*$-*homomorphisms* (i.e., $*$-linear Jordan homomorphisms) also "often" preserve normality. Another example is a map whose range consists of normal elements; for example, any map sending the first ring into the center of the second ring. The problem which we are about to consider is to find conditions under which a normality preserving map can be expressed through these basic examples. This problem is similar to the one concerning commutativity preserving maps, but technically it turns out to be more complicated. We shall therefore search for a solution in a less general framework than in the preceding section. One restriction has already been indicated: although the problem obviously makes sense in the context of rings, we will only consider algebras over a field $\mathbb{F}$. We will also assume that $\text{char}(\mathbb{F}) \neq 2$, and, more importantly, we will consider only central $\mathbb{F}$-algebras. We say that an $\mathbb{F}$-algebra is *central* if its center is equal to $\mathbb{F}1$ (so in particular it is unital). For simplicity we will consider only maps from an algebra into itself. It is clear from the proofs that this restriction is not really necessary; we could easily state the results for maps between different algebras. But the formulations would then become a bit lengthy and the notation more complicated, so we decided to avoid this.

So, let $\mathcal{A}$ be a central $\mathbb{F}$-algebra with involution $*$, and let $\mathcal{S}$ (resp. $\mathcal{K}$) be the set of its symmetric (resp. skew) elements. Every element $x \in \mathcal{A}$ can be written as $x = s + k$ where $s \in \mathcal{S}$ and $k \in \mathcal{K}$; indeed, just take $s = \frac{1}{2}(x + x^*)$ and $k = \frac{1}{2}(x - x^*)$. Moreover, this decomposition is unique. Note that x is normal if and only if s and k commute. It is clear that $(\mathbb{F}1)^* = \mathbb{F}1$. We shall say that $*$ is of the *first kind* if it acts as the identity on $\mathbb{F}1$, i.e., $\mathbb{F}1 \subseteq \mathcal{S}$; equivalently, $*$ is an $\mathbb{F}$-linear map. Otherwise $*$ is said to be of the *second kind*. (We remark that this is the classical definition of "involution of the first (second) kind" and suffices for our purposes here; more recently this definition has been modified in order to accommodate problems with prime rings, with the extended centroid playing the role of the center).

If $*$ is of the second kind, we obviously have $\mathcal{K} \cap \mathbb{F}1 \neq 0$. Thus there exists a nonzero $\varepsilon \in \mathbb{F}$ such that $(\varepsilon 1)^* = -\varepsilon 1$. Of course, $(\varepsilon^{-1} 1)^* = -\varepsilon^{-1} 1$, and so from $k = \varepsilon(\varepsilon^{-1} k)$ we see that $\mathcal{K} = \varepsilon \mathcal{S}$. Because of this simple connection between symmetric and skew elements, the case when the involution is of the second kind is often easier to handle. Let us begin with this simpler case.

Lemma 7.13. *Let* $\mathbb{F}$ *be a field with* char$(\mathbb{F}) \neq 2$, *and let* $\mathcal{A}$ *be a central* $\mathbb{F}$-*algebra with involution of the second kind. Then every normality preserving linear map* $\alpha : \mathcal{A} \to \mathcal{A}$ *satisfies* $[(x^2)^\alpha, ((x^*)^\alpha)^*] = 0$ *and* $[x^\alpha, ((x^*)^\alpha)^*] = 0$ *for every* $x \in \mathcal{A}$. *Moreover, if* α *is surjective, then* $1^\alpha \in \mathbb{F}1$.

Proof. Pick $s \in \mathcal{S}$. Then $s, s^2, s^2 + s$ and $s^2 + \varepsilon s$ are all normal, and hence their images are normal. Thus,

$$[s^\alpha, (s^\alpha)^*] = 0, \quad [(s^2)^\alpha, ((s^2)^\alpha)^*] = 0, \quad [(s^2 + s)^\alpha, ((s^2 + s)^\alpha)^*] = 0,$$
$$[(s^2 + \varepsilon s)^\alpha, ((s^2 + \varepsilon s)^\alpha)^*] = 0.$$

It is easy to see that these identities imply that

$$[(s^2)^\alpha, (s^\alpha)^*] = 0. \tag{7.21}$$

Linearizing (7.21) we get

$$[(s \circ t)^\alpha, (u^\alpha)^*] + [(u \circ s)^\alpha, (t^\alpha)^*] + [(t \circ u)^\alpha, (s^\alpha)^*] = 0$$

for all $s, t, u \in \mathcal{S}$, and so in particular by setting $u = s$ we get

$$[(s \circ t)^\alpha, (s^\alpha)^*] + [(s^2)^\alpha, (t^\alpha)^*] = 0 \tag{7.22}$$

for all $s, t \in \mathcal{S}$. As every $x \in \mathcal{A}$ can be written as $s + \varepsilon t$ with $s, t \in \mathcal{S}$, one easily deduces from (7.21) and (7.22) that $[(x^2)^\alpha, ((x^*)^\alpha)^*] = 0$.

To establish the second identity, we begin by observing that for every $s \in \mathcal{S}$ the elements $s^\alpha, 1^\alpha, s^\alpha + 1^\alpha$ and $s^\alpha + \varepsilon 1^\alpha$ are all normal since $s, 1, s+1$ and $s + \varepsilon 1$ are normal. It easily follows that $[1^\alpha, (s^\alpha)^*] = 0$ for every $s \in \mathcal{S}$. Again writing $x \in \mathcal{A}$ as $s + \varepsilon t$ we then see that $[1^\alpha, (x^\alpha)^*] = 0$ for every $x \in \mathcal{A}$. Therefore also $[(1^\alpha)^*, x^\alpha] = 0$. Consequently, replacing x by $x + 1$ in $[(x^2)^\alpha, ((x^*)^\alpha)^*] = 0$ it follows that $[x^\alpha, ((x^*)^\alpha)^*] = 0$. Of course, if α is surjective, then $[1^\alpha, (x^\alpha)^*] = 0$ implies that 1^α lies in the center $\mathbb{F}1$ of $\mathcal{A}$. $\square$

If α in Lemma 7.13 was $*$-linear, then the first identity from the conclusion of the lemma becomes the familiar one $[(x^2)^\alpha, x^\alpha] = 0$. Thus, in this case, the problem has been reduced to the one treated thoroughly in Section 7.1. On the other hand, the second identity from the conclusion of the lemma, which can be also written as $[(x^\alpha)^*, (x^*)^\alpha] = 0$, indicates that it is quite likely that α must necessarily be $*$-linear, or at least "almost" $*$-linear. These ideas lead to the following theorem.

Theorem 7.14. *Let* $\mathbb{F}$ *be a field with* char$(\mathbb{F}) \neq 2$, *and let* $\mathcal{A}$ *be a central* $\mathbb{F}$-*algebra with involution of the second kind. Let* $\alpha : \mathcal{A} \to \mathcal{A}$ *be a bijective normality preserving linear map. Suppose there exists a central* $\mathbb{F}$-*algebra* $\mathcal{Q} \supseteq \mathcal{A}$ *such that* $\mathcal{A}$ *is a 3-free subset of* $\mathcal{Q}$. *Then* α *is of the form* $x^\alpha = \lambda x^\sigma + x^\gamma 1$ *for all* $x \in \mathcal{A}$, *where* $\lambda \in \mathbb{F}$, σ *is a Jordan* $*$-*automorphism of* $\mathcal{A}$, *and* $\gamma : \mathcal{A} \to \mathbb{F}$ *is a linear functional.*

Proof. By Lemma 7.13 we have $[x^\alpha, ((x^*)^\alpha)^*] = 0$ for every $x \in \mathcal{A}$. As $\mathcal{A}$ is in particular 2-free, Corollary 4.15 shows that there exists $\omega \in \mathbb{F}$ such that $((x^*)^\alpha)^* - \omega x^\alpha \in \mathbb{F}1$ for every $x \in \mathcal{A}$. Moreover, $\omega \neq 0$ since otherwise $\mathcal{A}$ would be commutative (which is impossible since it is 2-free). But then $[(x^2)^\alpha, ((x^*)^\alpha)^*] = 0$ implies $[(x^2)^\alpha, x^\alpha] = 0$. Corollary 4.15 shows that there are $\eta \in \mathbb{F}$ and $\mu : \mathcal{A} \to \mathbb{F}$ such that $(x^2)^\alpha - \eta(x^\alpha)^2 - \mu(x)x^\alpha \in \mathbb{F}1$. By Lemma 7.13 we know that $1^\alpha \in \mathbb{F}1$ and so $(\mathbb{F}1)^\alpha = \mathbb{F}1$. Accordingly, $\eta = 0$ implies $x^2 - \mu(x)x \in \mathbb{F}1$, a contradiction. Therefore $\eta \neq 0$.

Theorem 7.3 together with Remark 7.4 now shows that α is of the form $x^\alpha = \lambda x^\sigma + x^\gamma 1$ for all $x \in \mathcal{A}$, where $\lambda \in \mathbb{F}$, $\sigma : \mathcal{A} \to \mathcal{A}$ is a Jordan homomorphism, and $\gamma : \mathcal{A} \to \mathbb{F}$ is a linear functional (the fact that σ and γ are linear (not only additive) can be easily proved (cf. the proof of Corollary 7.12)). It is clear that $\lambda \neq 0$.

Let us show that σ is bijective. The proof is just slightly different from the proof that σ in Corollary 7.12 is bijective. Suppose that $x^\sigma = 0$ for some $x \in \mathcal{A}$. Then $x^\alpha \in \mathbb{F}1$, and so $x \in \mathbb{F}1$. If x was not 0, then it would follow that $1^\sigma = 0$, whence $x^\sigma = \frac{1}{2}(x \circ 1)^\sigma = \frac{1}{2}(x^\sigma \circ 1^\sigma) = 0$ for every $x \in \mathcal{A}$, which implies $\mathcal{A} = \mathcal{A}^\alpha \subseteq \mathbb{F}1$ — a contradiction. Therefore σ is injective. As $\lambda \neq 0$ and $1^\alpha \in \mathbb{F}1$, we also have $1^\sigma \in \mathbb{F}1$. Since σ is an injective Jordan homomorphism, the only possibility is that $1^\sigma = 1$. But then we have $x^\alpha = (\lambda x + x^\gamma 1)^\sigma$, proving the surjectivity of σ.

It remains to show that $(x^*)^\sigma = (x^\sigma)^*$ for every $x \in \mathcal{A}$. From $((x^*)^\alpha)^* - \omega x^\alpha \in \mathbb{F}1$ and $x^\alpha - \lambda x^\sigma \in \mathbb{F}1$ one easily infers that there is $\varepsilon \in \mathbb{F}$ such that $((x^*)^\sigma)^* - \varepsilon x^\sigma \in \mathbb{F}1$ for every $x \in \mathcal{A}$. Replacing x by x^2 in this relation it follows that $(((x^*)^\sigma)^*)^2 - \varepsilon(x^\sigma)^2 \in \mathbb{F}1$. Writing $((x^*)^\sigma)^*$ as $\varepsilon x^\sigma + \xi(x)$, where $\xi(x) \in \mathbb{F}1$, it follows that

$$(\varepsilon^2 - \varepsilon)(x^\sigma)^2 + 2\varepsilon\xi(x)x^\sigma \in \mathbb{F}1.$$

Since σ is onto it follows from Corollary 4.21 that $\varepsilon^2 = \varepsilon$ and $\varepsilon\xi(x) = 0$ for every $x \in \mathcal{A}$. It is clear that ε cannot be 0, so we must have $\varepsilon = 1$ and $\xi = 0$. Accordingly, $((x^*)^\sigma)^* = x^\sigma$ and so $(x^*)^\sigma = (x^\sigma)^*$. □

By Corollary 5.12, a unital centrally closed prime algebra with $\deg(\mathcal{A}) \geq 3$ satisfies the conditions of Theorem 7.14. Moreover, in this situation σ is either an automorphism or an antiautomorphism (see Corollary 7.12 and its proof). So we have the following corollary.

Corollary 7.15. *Let $\mathbb{F}$ be a field with $\mathrm{char}(\mathbb{F}) \neq 2$, and let $\mathcal{A}$ be a unital centrally closed prime $\mathbb{F}$-algebra with involution of the second kind. Let $\alpha : \mathcal{A} \to \mathcal{A}$ be a bijective normality preserving linear map. If $\deg(\mathcal{A}) \geq 3$, then α is of the form $x^\alpha = \lambda x^\sigma + x^\gamma 1$ for all $x \in \mathcal{A}$, where $\lambda \in \mathbb{F}$, σ is either a $*$-automorphism or a $*$-antiautomorphism of $\mathcal{A}$, and $\gamma : \mathcal{A} \to \mathbb{F}$ is a linear functional.*

We now direct our attention to the more difficult case where the involution is of the first kind. In this situation we will assume that the map α in question is $*$-linear. Thus, $\mathcal{S}^\alpha \subseteq \mathcal{S}$ and $\mathcal{K}^\alpha \subseteq \mathcal{K}$, and the condition that α is normality where $\lambda \in \mathbb{F}$, $\mu : \mathcal{K} \to \mathbb{F}$, $\nu : \mathcal{K}^2 \to \mathbb{F}$, and $\zeta : \mathcal{K}^3 \to \mathbb{F}$. However, since $\mu(k)k^2$ and $\zeta(k, k, k)1$ lie in $\mathcal{S}$, while all other terms belong to $\mathcal{K}$, it follows that

$$(k^3)^\alpha = \lambda(k^\alpha)^3 + \nu(k, k)k^\alpha \quad \text{for all } k \in \mathcal{K}. \tag{7.29}$$

Note also that ν is bilinear (cf. Remark 4.7), and without loss of generality we may assume that it is symmetric (otherwise we can replace it by the map $(k, l) \mapsto \frac{1}{2}(\nu(k, l) + \nu(l, k))$).

If $\lambda = 0$, then $(k^3 - \nu(k, k)k)^\alpha = 0$. Thus $k^3 - \nu(k, k)k = 0$, a contradiction in view of Corollary 4.21. Thus $\lambda \neq 0$.

Linearizing (7.29), and also using char($\mathbb{F}$) $\neq 2$, we get

$$(k^2l + klk + lk^2)^\alpha = \lambda\Big((k^\alpha)^2l^\alpha + k^\alpha l^\alpha k^\alpha + l^\alpha(k^\alpha)^2\Big)$$
$$+ \nu(k, k)l^\alpha + 2\nu(k, l)k^\alpha \quad \text{for all } k, l \in \mathcal{K}. \tag{7.30}$$

The identities that were derived so far make it possible for us to compute $(k^2lk + klk^2)^\alpha$ in two different ways. First, according to (7.27) we have

$$2(k^2lk + klk^2)^\alpha = (k \circ (k^2l + klk + lk^2))^\alpha - (k^3 \circ l)^\alpha$$
$$= k^\alpha \circ (k^2l + klk + lk^2)^\alpha + \varepsilon(k, k^2l + klk + lk^2)1 - (k^3)^\alpha \circ l^\alpha - \varepsilon(k^3, l)1.$$

Thus,

$$2(k^2lk + klk^2)^\alpha - k^\alpha \circ (k^2l + klk + lk^2)^\alpha + (k^3)^\alpha \circ l^\alpha \in \mathbb{F}1.$$

Therefore it follows from (7.29) and (7.30) that

$$2(k^2lk + klk^2)^\alpha$$
$$- k^\alpha\Big(\lambda(k^\alpha)^2l^\alpha + \lambda k^\alpha l^\alpha k^\alpha + \lambda l^\alpha(k^\alpha)^2 + \nu(k, k)l^\alpha + 2\nu(k, l)k^\alpha\Big)$$
$$- \Big(\lambda(k^\alpha)^2l^\alpha + \lambda k^\alpha l^\alpha k^\alpha + \lambda l^\alpha(k^\alpha)^2 + \nu(k, k)l^\alpha + 2\nu(k, l)k^\alpha\Big)k^\alpha$$
$$+ \Big(\lambda(k^\alpha)^3 + \nu(k, k)k^\alpha\Big)l^\alpha + l^\alpha\Big(\lambda(k^\alpha)^3 + \nu(k, k)k^\alpha\Big) \in \mathbb{F}1,$$

and hence

$$(k^2lk + klk^2)^\alpha - \lambda\big((k^\alpha)^2l^\alpha k^\alpha + k^\alpha l^\alpha(k^\alpha)^2\big) - 2\nu(k, l)(k^\alpha)^2 \in \mathbb{F}1.$$

On the other hand, (7.27) gives

$$(k^2lk + klk^2)^\alpha = (k \circ klk)^\alpha = k^\alpha \circ (klk)^\alpha + \varepsilon(k, klk)1.$$

Comparing the last two relations we obtain

$$k^\alpha\Big((klk)^\alpha - \lambda k^\alpha l^\alpha k^\alpha - \nu(k, l)k^\alpha\Big) + \Big((klk)^\alpha - \lambda k^\alpha l^\alpha k^\alpha - \nu(k, l)k^\alpha\Big)k^\alpha \in \mathbb{F}1$$

for all $k, l \in \mathcal{K}$. Regarding this relation for any fixed $l \in \mathcal{K}$ we see that again we are in a position to apply Corollary 4.17, more precisely, its modified version explained in Remark 4.18. Hence it follows that

$$(klk)^\alpha = \lambda k^\alpha l^\alpha k^\alpha + \nu(k, l)k^\alpha \quad \text{for all } k, l \in \mathcal{K}. \tag{7.31}$$

Linearizing this identity we get

$$(k_1 l k_2 + k_2 l k_1)^\alpha = \lambda k_1^\alpha l^\alpha k_2^\alpha + \lambda k_2^\alpha l^\alpha k_1^\alpha + \nu(k_1, l)k_2^\alpha + \nu(k_2, l)k_1^\alpha \tag{7.32}$$

for all $k_1, k_2, l \in \mathcal{K}$. Our next goal is to show that $\nu(k, l) = 0$ for all $k, l \in \mathcal{K}$. The proof is based on computing $(k l_1 k l_2 k + k l_2 k l_1 k)^\alpha$. Using (7.32) and (7.31) we obtain

$$\begin{aligned}
(k l_1 k l_2 k + k l_2 k l_1 k)^\alpha &= ((k l_1 k) l_2 k + k l_2 (k l_1 k))^\alpha \\
&= \lambda (k l_1 k)^\alpha l_2^\alpha k^\alpha + \lambda k^\alpha l_2^\alpha (k l_1 k)^\alpha + \nu(k l_1 k, l_2)k^\alpha + \nu(k, l_2)(k l_1 k)^\alpha \\
&= \lambda^2 k^\alpha l_1^\alpha k^\alpha l_2^\alpha k^\alpha + \lambda^2 k^\alpha l_2^\alpha k^\alpha l_1^\alpha k^\alpha + 2\lambda \nu(k, l_1)k^\alpha l_2^\alpha k^\alpha \\
&\quad + \lambda \nu(k, l_2)k^\alpha l_1^\alpha k^\alpha + \nu(k l_1 k, l_2)k^\alpha + \nu(k, l_2)\nu(k, l_1)k^\alpha.
\end{aligned}$$

Note that l_1 and l_2 appear symmetrically on the left-hand side of this identity. Therefore the left-hand side will remain the same if we change the roles of l_1 and l_2. Thus,

$$\begin{aligned}
&\lambda^2 k^\alpha l_1^\alpha k^\alpha l_2^\alpha k^\alpha + \lambda^2 k^\alpha l_2^\alpha k^\alpha l_1^\alpha k^\alpha + 2\lambda \nu(k, l_1)k^\alpha l_2^\alpha k^\alpha \\
&\quad + \lambda \nu(k, l_2)k^\alpha l_1^\alpha k^\alpha + \nu(k l_1 k, l_2)k^\alpha + \nu(k, l_2)\nu(k, l_1)k^\alpha \\
&= \lambda^2 k^\alpha l_2^\alpha k^\alpha l_1^\alpha k^\alpha + \lambda^2 k^\alpha l_1^\alpha k^\alpha l_2^\alpha k^\alpha + 2\lambda \nu(k, l_2)k^\alpha l_1^\alpha k^\alpha \\
&\quad + \lambda \nu(k, l_1)k^\alpha l_2^\alpha k^\alpha + \nu(k l_2 k, l_1)k^\alpha + \nu(k, l_1)\nu(k, l_2)k^\alpha.
\end{aligned}$$

Since $\lambda \neq 0$, this clearly yields $\nu(k, l) = 0$ by comparing coefficients at the $k^\alpha l_2^\alpha k^\alpha$ term and using Lemma 4.20.

Therefore, (7.32) reduces to

$$(k_1 l k_2 + k_2 l k_1)^\alpha = \lambda \left(k_1^\alpha l^\alpha k_2^\alpha + k_2^\alpha l^\alpha k_1^\alpha \right) \quad \text{for all } k_1, k_2, l \in \mathcal{K}. \tag{7.33}$$

Thus, α "almost" preserves the polynomial $f = x_1 x_2 x_3 + x_3 x_2 x_1$, just the factor λ makes some perturbation. However, as noted in Remark 6.29, in order to use the result of Lemma 6.28, this is not a problem. Using the assumption that $\mathcal{K}$ is 6-free (incidentally, this is the only place where 6-freeness is used) it now follows that there exists $\eta \in \mathbb{F}$ such that $[k, l]^\alpha - \eta[k^\alpha, l^\alpha] \in \mathbb{F}1$ for all $k, l \in \mathcal{K}$. However, since $[k, l]^\alpha \in \mathcal{K}$ and $\eta[k^\alpha, l^\alpha] \in \mathcal{K}$, we actually have

$$[k, l]^\alpha = \eta[k^\alpha, l^\alpha] \quad \text{for all } k, l \in \mathcal{K}. \tag{7.34}$$

Accordingly,

$$[[k, l], m]^\alpha = \eta[[k, l]^\alpha, m^\alpha] = \eta^2[[k^\alpha, l^\alpha], m^\alpha]$$

for all $k, l, m \in \mathcal{K}$. On the other hand,

$$[[k, l], m]^\alpha = (klm + mlk)^\alpha - (lkm + mkl)^\alpha$$
$$= \lambda(k^\alpha l^\alpha m^\alpha + m^\alpha l^\alpha k^\alpha) - \lambda(l^\alpha k^\alpha m^\alpha + m^\alpha k^\alpha l^\alpha) = \lambda[[k^\alpha, l^\alpha], m^\alpha].$$

Therefore $(\eta^2 - \lambda)[[\mathcal{K}, \mathcal{K}], \mathcal{K}] = 0$. As $\mathcal{K}$ is 3-free, a standard argument shows that $[[\mathcal{K}, \mathcal{K}], \mathcal{K}] \neq 0$. Therefore $\eta^2 = \lambda$.

We recall that $\langle \mathcal{K} \rangle = \mathcal{K} + \mathcal{K} \circ \mathcal{K}$ by Lemma 6.13. It is clear that the restriction of $*$ to $\langle \mathcal{K} \rangle$ is an involution on $\langle \mathcal{K} \rangle$, and that $\mathcal{K} \circ \mathcal{K}$ and $\mathcal{K}$ are the sets of symmetric and skew elements of $\langle \mathcal{K} \rangle$, respectively. We now define $\sigma : \langle \mathcal{K} \rangle \to \langle \mathcal{K} \rangle$ by

$$\Big(k + \sum_i l_i \circ m_i\Big)^\sigma = \eta k^\alpha + \lambda \sum_i l_i^\alpha \circ m_i^\alpha.$$

To show that α is well-defined, assume that $\sum_i l_i \circ m_i = 0$ for some $l_i, m_i \in \mathcal{K}$. Then also

$$\sum_i \Big((l_i m_i p + p m_i l_i) + (m_i l_i p + p l_i m_i)\Big) = \Big(\sum_i l_i \circ m_i\Big) \circ p = 0$$

for every $p \in \mathcal{K}$, and hence (7.33) implies that

$$\sum_i \Big((l_i^\alpha m_i^\alpha p^\alpha + p^\alpha m_i^\alpha l_i^\alpha) + (m_i^\alpha l_i^\alpha p^\alpha + p^\alpha l_i^\alpha m_i^\alpha)\Big) = 0.$$

This can be rewritten as

$$\Big(\sum_i l_i^\alpha \circ m_i^\alpha\Big) \circ p^\alpha = 0.$$

Since p^α is an arbitrary element in $\mathcal{K}$, a standard d-freeness argument shows that $\sum_i l_i^\alpha \circ m_i^\alpha = 0$. This proves that σ is well-defined.

It is clear that σ is $*$-linear and surjective. In order to prove that it is injective, it is enough to check that $\sum_i l_i^\alpha \circ m_i^\alpha = 0$ implies $\sum_i l_i \circ m_i = 0$. This can be done in a similar way as establishing that σ is well-defined. From $\sum_i l_i^\alpha \circ m_i^\alpha = 0$ it follows that

$$\lambda\Big(\sum_i l_i^\alpha \circ m_i^\alpha\Big) \circ p^\alpha = 0$$

for every $p \in \mathcal{K}$. Expanding this relation and applying (7.33) we see that it can be rewritten as $\big((\sum_i l_i \circ m_i) \circ p\big)^\alpha = 0$. Hence $(\sum_i l_i \circ m_i) \circ p = 0$, which in turn implies $\sum_i l_i \circ m_i = 0$.

Next, for arbitrary $k, l \in \mathcal{K}$ we have

$$(kl)^\sigma = \frac{1}{2}\Big([k, l] + k \circ l\Big)^\sigma = \frac{1}{2}\Big(\eta[k, l]^\alpha + \lambda k^\alpha \circ l^\alpha\Big).$$

Using (7.34) and $\eta^2 = \lambda$ it follows immediately that

$$(kl)^\sigma = k^\sigma l^\sigma \quad \text{for all } k, l \in \mathcal{K}. \tag{7.35}$$

Further, using (7.33) and (7.34) we get

$$
\begin{aligned}
(kl^2)^\sigma &= \frac{1}{2}\left(l^2 k + kl^2 + l \circ [k,l]\right)^\sigma = \frac{1}{2}\left(\eta(l^2 k + kl^2)^\alpha + \lambda l^\alpha \circ [k,l]^\alpha\right) \\
&= \frac{1}{2}\left(\eta^3\left((l^\alpha)^2 k^\alpha + k^\alpha (l^\alpha)^2\right) + \eta^3 l^\alpha \circ [k^\alpha, l^\alpha]\right) \\
&= k^\sigma (l^\sigma)^2.
\end{aligned}
$$

Since $(l^\sigma)^2 = (l^2)^\sigma$ by (7.35), we also have $(kl^2)^\sigma = k^\sigma (l^2)^\sigma$. Consequently,

$$
(k(l \circ m))^\sigma = k^\sigma (l \circ m)^\sigma \quad \text{for all } k,l,m \in \mathcal{K}. \tag{7.36}
$$

Note that (7.35) and (7.36) together show that $(ky)^\sigma = k^\sigma y^\sigma$ for all $k \in \mathcal{K}$ and $y \in \langle \mathcal{K} \rangle$. Since $\langle \mathcal{K} \rangle$ is generated by $\mathcal{K}$, this implies that $(xy)^\sigma = x^\sigma y^\sigma$ for all $x, y \in \langle \mathcal{K} \rangle$. Thus σ is a $*$-automorphism.

Since $(k \circ l)^\sigma = k^\sigma \circ l^\sigma = \lambda k^\alpha \circ l^\alpha$ it follows from (7.27) that $\lambda s^\alpha - s^\sigma \in \mathbb{F}1$ for every $s \in \mathcal{K} \circ \mathcal{K}$. Therefore there exists a linear functional ξ on $\mathcal{K} \circ \mathcal{K}$ such that

$$
\lambda s^\alpha - s^\sigma = s^\xi 1 \quad \text{for all } s \in \mathcal{K} \circ \mathcal{K}.
$$

Accordingly, for $x \in \langle \mathcal{K} \rangle$ we have

$$
\begin{aligned}
x^\alpha &= \frac{1}{2}(x + x^*)^\alpha + \frac{1}{2}(x - x^*)^\alpha \\
&= \frac{1}{2}\left(\lambda^{-1}(x + x^*)^\sigma + \lambda^{-1}(x + x^*)^\xi 1\right) + \frac{1}{2}\eta^{-1}(x - x^*)^\sigma \\
&= \left(\frac{1}{2}(\lambda^{-1} + \eta^{-1})x + \frac{1}{2}(\lambda^{-1} - \eta^{-1})x^*\right)^\sigma + \frac{1}{2}\lambda^{-1}(x + x^*)^\xi 1.
\end{aligned}
$$

We now set

$$
\lambda_1 = \frac{1}{2}(\lambda^{-1} + \eta^{-1}), \quad \lambda_2 = \frac{1}{2}(\lambda^{-1} - \eta^{-1}), \quad x^\gamma = \frac{1}{2}\lambda^{-1}(x + x^*)^\xi,
$$

and the proof is complete. $\qquad\square$

An inspection of the proof of Theorem 7.17 shows that we can replace the assumption that α preserves normality by a weaker assumption that α satisfies the relations (7.24), (7.25), and (7.28).

Combining Theorem 7.17 with Corollary 5.18 we get the following result.

Corollary 7.18. *Let $\mathbb{F}$ be a field with $\operatorname{char}(\mathbb{F}) \neq 2,3$, and let $\mathcal{A}$ be a centrally closed prime $\mathbb{F}$-algebra with involution of the first kind. Let $\alpha : \mathcal{A} \to \mathcal{A}$ be a bijective normality preserving $*$-linear map. If $\deg(\mathcal{A}) \geq 13$, then there exist $\lambda_1, \lambda_2 \in \mathbb{F}$, $\lambda_1 \neq \pm\lambda_2$, a linear functional $\gamma : \langle \mathcal{K} \rangle \to \mathbb{F}$ such that $\mathcal{K}^\gamma = 0$, and a $*$-automorphism σ of $\langle \mathcal{K} \rangle$ such that $x^\alpha = (\lambda_1 x + \lambda_2 x^*)^\sigma + x^\gamma 1$ for all $x \in \langle \mathcal{K} \rangle$.*

Let us finally remark that if the algebra $\mathcal{A}$ in Corollary 7.18 is simple, then $\langle \mathcal{K} \rangle = \mathcal{A}$. This is a well-known result of Herstein (see for example [114, Theorem 2.2]). So, in this case we can get a complete description of α.

7.3   Zero Jordan Product Preserving Maps

Let $\mathcal{B}$ and $\mathcal{Q}$ be rings. We say that a map $\alpha : \mathcal{B} \to \mathcal{Q}$ is *zero Jordan product preserving* if $x^\alpha \circ y^\alpha = 0$ whenever $x, y \in \mathcal{B}$ satisfy $x \circ y = 0$. We remark that a commutativity preserving map can be described as a map that preserves zero Lie product. Thus, the problem of determining the structure of zero Jordan product preserving maps is the Jordan-type version of the Lie-type problem treated in Section 7.1.

An obvious example of a zero Jordan product preserving map is a Jordan homomorphism multiplied by a central element in $\mathcal{Q}$. Our goal is to show that under appropriate assumptions this is the only possible example. Of course, in the course of the proof we shall use FI's at some point; however, unlike in the case of zero Lie product (i.e., commutativity) preserving maps it does not seem possible to derive an FI directly from the condition that a map preserves zero Jordan product. Therefore we shall have to confine ourselves to a certain special class of rings, in which applying FI's will be possible. Specifically, we will consider the case when $\mathcal{B} = M_n(\mathcal{R})$ with $n \geq 3$ and $\mathcal{R}$ an arbitrary unital ring. So our intention is to derive an analogue of Corollary 7.5 (there will be some differences in technical assumptions, though). Let us point out that $\mathcal{B}$ is additively generated by the matrices ae_{ij}, $a \in \mathcal{R}$ and $1 \leq i, j \leq n$. We shall denote ae_{ij} by a_{ij} for simplicity. Clearly, an additive map $\alpha : \mathcal{B} \to \mathcal{Q}$ is completely determined by the values $(a_{ij})^\alpha$.

Our proof consists of two steps. The first lemma is elementary and does not depend on FI's. It states that under very mild assumptions α satisfies a much stronger condition than preserving zero Jordan products. In particular, it preserves *equal* Jordan products. By this we mean that $x^\alpha \circ y^\alpha = u^\alpha \circ v^\alpha$ whenever $x, y, u, v \in \mathcal{B}$ are such that $x \circ y = u \circ v$. This will make it possible for us to apply FI's and thereby obtain the desired conclusion.

Lemma 7.19. *Let $\mathcal{R}$ and $\mathcal{Q}$ be unital rings with $\frac{1}{2}$, let $\mathcal{B} = M_n(\mathcal{R})$ with $n \geq 3$, and let $\alpha : \mathcal{B} \to \mathcal{Q}$ be a zero Jordan product preserving additive map. Then for all $x_i, y_i \in \mathcal{B}$, $\sum_{i=1}^{m} x_i \circ y_i = 0$ implies $\sum_{i=1}^{m} x_i^\alpha \circ y_i^\alpha = 0$. In particular, α preserves equal Jordan products.*

Proof. We first remark that $\mathcal{B}$ is, of course, also unital and contains $\frac{1}{2}$.

Let $a, b \in \mathcal{R}$ and $1 \leq i, j, k, l \leq n$. We claim that

$$(a_{ij})^\alpha \circ (b_{kl})^\alpha = 0 \quad \text{if } i \neq l \text{ and } j \neq k. \tag{7.37}$$

Indeed, since $i \neq l$ and $j \neq k$ we have $a_{ij} \circ b_{kl} = 0$, which according to our assumption implies $(a_{ij})^\alpha \circ (b_{kl})^\alpha = 0$.

Assume now that $i \neq k$. Then $(ab)_{ik} \circ (e_{ii} - e_{kk}) = 0$, and hence

$$\big((ab)_{ik}\big)^\alpha \circ (e_{kk})^\alpha = \big((ab)_{ik}\big)^\alpha \circ (e_{ii})^\alpha.$$

Now consider two cases, when $j \neq k$ and when $j = k$. In the first case, since also $i \neq k$, we have $(a_{ij} + (ab)_{ik}) \circ (b_{jk} - e_{kk}) = 0$. Consequently,

$$((a_{ij})^\alpha + ((ab)_{ik})^\alpha) \circ ((b_{jk})^\alpha - (e_{kk})^\alpha) = 0.$$

Since $(a_{ij})^\alpha \circ (e_{kk})^\alpha = 0$ and $((ab)_{ik})^\alpha \circ (b_{jk})^\alpha = 0$ by (7.37), this further yields $(a_{ij})^\alpha \circ (b_{jk})^\alpha = ((ab)_{ik})^\alpha \circ (e_{kk})^\alpha$. In the second case, when $j = k$, we have $(a_{ik} - e_{ii}) \circ ((ab)_{ik} + b_{kk}) = 0$, from which one can derive $(a_{ik})^\alpha \circ (b_{kk})^\alpha = ((ab)_{ik})^\alpha \circ (e_{ii})^\alpha$ by a similar argument. To summarize, we have

$$(a_{ij})^\alpha \circ (b_{jk})^\alpha = ((ab)_{ik})^\alpha \circ (e_{kk})^\alpha = ((ab)_{ik})^\alpha \circ (e_{ii})^\alpha \quad \text{if } i \neq k. \tag{7.38}$$

Now assume $i \neq j$. Note that then

$$\left(\frac{1}{2}(ab)_{ii} + a_{ij} - \frac{1}{2}(ba)_{jj}\right) \circ \left(-e_{ii} + b_{ji} + e_{jj}\right) = 0$$

and so

$$\left(\frac{1}{2}((ab)_{ii})^\alpha + (a_{ij})^\alpha - \frac{1}{2}((ba)_{jj})^\alpha\right) \circ \left(-(e_{ii})^\alpha + (b_{ji})^\alpha + (e_{jj})^\alpha\right) = 0.$$

This identity together with $((ab)_{ii})^\alpha \circ (e_{jj})^\alpha = 0$ (by (7.37)), $((ba)_{jj})^\alpha \circ (e_{ii})^\alpha = 0$ (by (7.37)), $((ab)_{ii})^\alpha \circ (b_{ji})^\alpha = ((bab)_{ji})^\alpha \circ (e_{ii})^\alpha = ((ba)_{jj})^\alpha \circ (b_{ji})^\alpha$ (by (7.38)), and $(a_{ij})^\alpha \circ (e_{ii})^\alpha = (a_{ij})^\alpha \circ (e_{jj})^\alpha$ (by (7.38)), implies that

$$(a_{ij})^\alpha \circ (b_{ji})^\alpha = \frac{1}{2}((ab)_{ii})^\alpha \circ (e_{ii})^\alpha + \frac{1}{2}((ba)_{jj})^\alpha \circ (e_{jj})^\alpha. \tag{7.39}$$

We claim that (7.39) holds even when $i = j$. That is, we are going to show that

$$(a_{ii})^\alpha \circ (b_{ii})^\alpha = \frac{1}{2}((a \circ b)_{ii})^\alpha \circ (e_{ii})^\alpha.$$

Note that for every $k \neq i$ we have

$$(a_{ii} - b_{ik} + b_{ki} - a_{kk})^\alpha \circ (b_{ii} - a_{ik} + a_{ki} - b_{kk})^\alpha = 0.$$

Using also $(a_{ii})^\alpha \circ (b_{kk})^\alpha = 0$, $(a_{kk})^\alpha \circ (b_{ii})^\alpha = 0$, $(a_{ii} - a_{kk})^\alpha \circ (-a_{ik} + a_{ki})^\alpha = 0$, $(b_{ik})^\alpha \circ (a_{ik})^\alpha = 0$, $(b_{ki})^\alpha \circ (a_{ki})^\alpha = 0$ and $(-b_{ik} + b_{ki})^\alpha \circ (b_{ii} - b_{kk})^\alpha = 0$, it follows from (7.39) that

$$(a_{ii})^\alpha \circ (b_{ii})^\alpha + (a_{kk})^\alpha \circ (b_{kk})^\alpha = \frac{1}{2}\left(((a \circ b)_{ii})^\alpha \circ (e_{ii})^\alpha + ((a \circ b)_{kk})^\alpha \circ (e_{kk})^\alpha\right). \tag{7.40}$$

Since $n \geq 3$, there is l such that $l \notin \{i, k\}$. Then, by (7.40), we have

$$2(a_{ii})^\alpha \circ (b_{ii})^\alpha + \left((a_{kk})^\alpha \circ (b_{kk})^\alpha + (a_{ll})^\alpha \circ (b_{ll})^\alpha\right)$$
$$= \left((a_{ii})^\alpha \circ (b_{ii})^\alpha + (a_{kk})^\alpha \circ (b_{kk})^\alpha\right) + \left((a_{ii})^\alpha \circ (b_{ii})^\alpha + (a_{ll})^\alpha \circ (b_{ll})^\alpha\right)$$
$$= \frac{1}{2}\left(\left((a \circ b)_{ii}\right)^\alpha \circ (e_{ii})^\alpha + \left((a \circ b)_{kk}\right)^\alpha \circ (e_{kk})^\alpha\right)$$
$$\quad + \frac{1}{2}\left(\left((a \circ b)_{ii}\right)^\alpha \circ (e_{ii})^\alpha + \left((a \circ b)_{ll}\right)^\alpha \circ (e_{ll})^\alpha\right)$$
$$= \left((a \circ b)_{ii}\right)^\alpha \circ (e_{ii})^\alpha + \frac{1}{2}\left(\left((a \circ b)_{kk}\right)^\alpha \circ (e_{kk})^\alpha + \left((a \circ b)_{ll}\right)^\alpha \circ (e_{ll})^\alpha\right)$$
$$= \left((a \circ b)_{ii}\right)^\alpha \circ (e_{ii})^\alpha + \left((a_{kk})^\alpha \circ (b_{kk})^\alpha + (a_{ll})^\alpha \circ (b_{ll})^\alpha\right).$$

Thus $2(a_{ii})^\alpha \circ (b_{ii})^\alpha = \left((a \circ b)_{ii}\right)^\alpha \circ (e_{ii})^\alpha$, which proves our claim.

Having (7.37), (7.38) and (7.39) at hand it is now easy to prove the lemma. Let $\mathcal{W}$ be the set of all elements a_{ij}, $a \in \mathcal{R}$, $1 \leq i, j \leq n$. Let $x_1, y_1, \ldots, x_m, y_m \in \mathcal{B}$ be such that $x_1 \circ y_1 + \ldots + x_m \circ y_m = 0$, and let us show that then $x_1^\alpha \circ y_1^\alpha + \ldots + x_m^\alpha \circ y_m^\alpha = 0$. Since every element in $\mathcal{B}$ is a sum of elements in $\mathcal{W}$, there is no loss of generality in assuming that all x_i, y_i lie in $\mathcal{W}$. Note that, for $x, y \in \mathcal{W}$, the element $x \circ y$ can be written in one of the following forms:

$$c_{ij}, \quad c_{ii}, \quad c_{ii} + c'_{jj},$$

where $j \neq i$ and $c, c' \in \mathcal{R}$.

Since $x_1 \circ y_1 + \ldots + x_m \circ y_m = 0$, for any fixed i and j with $i \neq j$, the sum of terms of the form c_{ij} is 0. These terms arise from terms of the form $a_{ik} \circ b_{kj}$ with $c = ab$. The corresponding terms $(a_{ik})^\alpha \circ (b_{kj})^\alpha$ can be written as $(c_{ij})^\alpha \circ (e_{ii})^\alpha$ with $c = ab$ by (7.38) and hence their sum is 0.

Similarly, for any i, the sum of terms of the form c_{ii} is 0. These terms arise from terms of the form $a_{ij} \circ b_{ji} = c_{ii} + c'_{jj}$ with $j \neq i$, $c = ab$ and $c' = ba$, or $a_{ii} \circ b_{ii} = c_{ii}$ with $c = a \circ b$. By (7.39), the corresponding terms $(a_{ij})^\alpha \circ (b_{ji})^\alpha$ can be written as $\frac{1}{2}\left((c_{ii})^\alpha \circ (e_{ii})^\alpha\right) + \frac{1}{2}\left((c'_{jj})^\alpha \circ (e_{jj})^\alpha\right)$ with $c = ab$ and $c' = ba$, and $(a_{ii})^\alpha \circ (b_{ii})^\alpha$ can be written as $\frac{1}{2}\left((c_{ii})^\alpha \circ (e_{ii})^\alpha\right)$ with $c = a \circ b$. Thus the sum of terms of the form $(c_{ii})^\alpha \circ (e_{ii})^\alpha$ is 0. Therefore we have $x_1^\alpha \circ y_1^\alpha + \ldots + x_m^\alpha \circ y_m^\alpha = 0$. $\square$

Knowing that α preserves equal Jordan products, it follows from the identity $\frac{1}{2} \circ (x \circ y) = x \circ y$ that α satisfies

$$\frac{1}{2} \cdot 1^\alpha \circ (x \circ y)^\alpha = x^\alpha \circ y^\alpha \quad \text{for all } x, y \in \mathcal{B}.$$

Thus, α is "close" to a Jordan homomorphism; just the presence of the element 1^α creates some problems. In order to show that this element is central and invertible we will apply FI's.

Theorem 7.20. *Let $\mathcal{R}$ and $\mathcal{Q}$ be unital rings with $\frac{1}{2}$, let $\mathcal{B} = M_n(\mathcal{R})$ with $n \geq 3$, and let $\alpha : \mathcal{B} \to \mathcal{Q}$ be a zero Jordan product preserving additive map. Suppose that $\mathcal{B}^\alpha$ is a 4-free subset of $\mathcal{Q}$. Then there exist $\lambda \in \mathcal{C}$, the center of $\mathcal{Q}$, and a Jordan homomorphism $\sigma : \mathcal{B} \to \mathcal{Q}$ such that $x^\alpha = \lambda x^\sigma$ for all $x \in \mathcal{B}$.*

Proof. By Lemma 7.19 we have

$$(xyx)^\alpha \circ y^\alpha = (yxy)^\alpha \circ x^\alpha \tag{7.41}$$

for all $x, y \in \mathcal{B}$. The linearization of (7.41) yields

$$(xyz)^\alpha \circ u^\alpha + (xuz)^\alpha \circ y^\alpha + (zyx)^\alpha \circ u^\alpha + (zux)^\alpha \circ y^\alpha$$
$$= (yxu)^\alpha \circ z^\alpha + (uxy)^\alpha \circ z^\alpha + (yzu)^\alpha \circ x^\alpha + (uzy)^\alpha \circ x^\alpha,$$

or, equivalently,

$$\big((yzu)^\alpha + (uzy)^\alpha\big) \circ x^\alpha - \big((xuz)^\alpha + (zux)^\alpha\big) \circ y^\alpha$$
$$+ \big((yxu)^\alpha + (uxy)^\alpha\big) \circ z^\alpha - \big((xyz)^\alpha + (zyx)^\alpha\big) \circ u^\alpha = 0$$

for all $x, y, z, u \in \mathcal{B}$. Since $\mathcal{B}^\alpha$ is a 4-free subset of $\mathcal{Q}$, applying Theorem 4.13 (together with Lemma 4.6) we get

$$(xyz)^\alpha + (zyx)^\alpha = \lambda'_1 x^\alpha y^\alpha z^\alpha + \lambda'_2 x^\alpha z^\alpha y^\alpha + \lambda'_3 y^\alpha x^\alpha z^\alpha$$
$$+ \lambda'_4 y^\alpha z^\alpha x^\alpha + \lambda'_5 z^\alpha x^\alpha y^\alpha + \lambda'_6 z^\alpha y^\alpha x^\alpha$$
$$+ \mu'_1(x) y^\alpha z^\alpha + \mu'_2(x) z^\alpha y^\alpha + \mu'_3(y) x^\alpha z^\alpha$$
$$+ \mu'_4(y) z^\alpha x^\alpha + \mu'_5(z) x^\alpha y^\alpha + \mu'_6(z) y^\alpha x^\alpha$$
$$+ \nu'_1(x, y) z^\alpha + \nu'_2(x, z) y^\alpha + \nu'_3(y, z) x^\alpha + \tau(x, y, z)$$

for all $x, y, z \in \mathcal{B}$, where $\lambda'_i \in \mathcal{C}$, $\mu'_i : \mathcal{B} \to \mathcal{C}$ are additive maps, $\nu'_i : \mathcal{B}^2 \to \mathcal{C}$ are biadditive maps and $\tau : \mathcal{B}^3 \to \mathcal{C}$ is a triadditive map. In particular, we have

$$(xyx)^\alpha = \lambda_1 (x^\alpha)^2 y^\alpha + \lambda_2 x^\alpha y^\alpha x^\alpha + \lambda_3 y^\alpha (x^\alpha)^2$$
$$+ \mu_1(x) x^\alpha y^\alpha + \mu_2(x) y^\alpha x^\alpha + \mu_3(y)(x^\alpha)^2 \tag{7.42}$$
$$+ \nu_1(x, y) x^\alpha + \nu_2(x, x) y^\alpha + \tau(x, y, x)$$

for all $x, y \in \mathcal{B}$, where $\lambda_i \in \mathcal{C}$, $\mu_i : \mathcal{B} \to \mathcal{C}$ are additive maps and $\nu_i : \mathcal{B}^2 \to \mathcal{C}$ are biadditive maps. Substituting (7.42) into (7.41), we have, on the left-hand side,

$$(xyx)^\alpha \circ y^\alpha$$
$$= \lambda_1\big((x^\alpha)^2(y^\alpha)^2 + y^\alpha(x^\alpha)^2 y^\alpha\big) + \lambda_2\big((x^\alpha y^\alpha)^2 + (y^\alpha x^\alpha)^2\big)$$
$$+ \lambda_3\big(y^\alpha(x^\alpha)^2 y^\alpha + (y^\alpha)^2(x^\alpha)^2\big) + \mu_1(x)\big(x^\alpha(y^\alpha)^2 + y^\alpha x^\alpha y^\alpha\big) \tag{7.43}$$
$$+ \mu_2(x)\big(y^\alpha x^\alpha y^\alpha + (y^\alpha)^2 x^\alpha\big) + \mu_3(y)\big((x^\alpha)^2 y^\alpha + y^\alpha(x^\alpha)^2\big)$$
$$+ \nu_1(x, y)\big(x^\alpha y^\alpha + y^\alpha x^\alpha\big) + 2\nu_2(x, x)(y^\alpha)^2 + 2\tau(x, y, x)y^\alpha,$$

and, on the right-hand side,

$$(yxy)^\alpha \circ x^\alpha$$
$$= \lambda_1\big((y^\alpha)^2(x^\alpha)^2 + x^\alpha(y^\alpha)^2 x^\alpha\big) + \lambda_2\big((y^\alpha x^\alpha)^2 + (x^\alpha y^\alpha)^2\big)$$
$$+ \lambda_3\big(x^\alpha(y^\alpha)^2 x^\alpha + (x^\alpha)^2(y^\alpha)^2\big) + \mu_1(y)\big(y^\alpha(x^\alpha)^2 + x^\alpha y^\alpha x^\alpha\big) \tag{7.44}$$
$$+ \mu_2(y)\big(x^\alpha y^\alpha x^\alpha + (x^\alpha)^2 y^\alpha\big) + \mu_3(x)\big((y^\alpha)^2 x^\alpha + x^\alpha(y^\alpha)^2\big)$$
$$+ \nu_1(y, x)\big(y^\alpha x^\alpha + x^\alpha y^\alpha\big) + 2\nu_2(y, y)(x^\alpha)^2 + 2\tau(y, x, y)x^\alpha.$$

for all $x, y \in \mathcal{B}$. Comparing (7.43) and (7.44) and applying Lemma 4.20, we see that $\lambda_1 = \lambda_3 = 0$, $\mu_1 = \mu_2 = \mu_3 = 0$, $\nu_1(x, y) = \nu_1(y, x)$, $\nu_2(x, x) = 0$ and $\tau(x, y, x) = 0$ for all $x, y \in \mathcal{B}$. Setting $\gamma = \lambda_2$ and $\nu = \nu_1$ in (7.42) we obtain

$$(xyx)^\alpha = \gamma x^\alpha y^\alpha x^\alpha + \nu(x, y)x^\alpha \tag{7.45}$$

for all $x, y \in \mathcal{B}$. When $x = 1$, (7.45) reduces to

$$y^\alpha = \gamma 1^\alpha y^\alpha 1^\alpha + \nu(1, y)1^\alpha \tag{7.46}$$

for all $y \in \mathcal{B}$. Setting $y = 1$ in (7.46) we obtain

$$(1 - \nu(1, 1))1^\alpha = \gamma(1^\alpha)^3. \tag{7.47}$$

Commuting (7.46) with $(1^\alpha)^2$ and taking into account (7.47) we see that $(1^\alpha)^2 \in \mathcal{C}$ — namely, since $\mathcal{B}^\alpha$ is in particular 2-free, the centralizer of $\mathcal{B}^\alpha$ in $\mathcal{Q}$ is just $\mathcal{C}$. Right multiplying (7.46) by 1^α we obtain

$$y^\alpha 1^\alpha - \gamma(1^\alpha)^3 y^\alpha = \nu(1, y)(1^\alpha)^2 \in \mathcal{C}$$

for all $y \in \mathcal{B}$. Since $\mathcal{B}^\alpha$ is 2-free, a standard argument shows that $\gamma(1^\alpha)^3 = 1^\alpha \in \mathcal{C}$ and $\nu(1, y)(1^\alpha)^2 = 0$ for all $y \in \mathcal{B}$. Thus

$$\nu(1, y)1^\alpha = \nu(1, y)\gamma(1^\alpha)^3 = 0$$

and we can rewrite (7.46) as

$$y^\alpha - \gamma(1^\alpha)^2 y^\alpha$$

for all $y \in \mathcal{B}$. By Lemma 4.4, $\gamma(1^\alpha)^2 = 1$. Thus $\lambda = 1^\alpha$ is invertible and $\lambda^{-1} = \gamma 1^\alpha$. Further, we have $\nu(x, 1) = \nu(1, x) = \nu(1, x)\gamma(1^\alpha)^2 = 0$ for all $x \in \mathcal{B}$. Setting $y = 1$ in (7.45) we get $(x^2)^\alpha = \lambda^{-1}(x^\alpha)^2$. Accordingly, $\sigma : \mathcal{B} \to \mathcal{Q}$ defined by $x^\sigma = \lambda^{-1}x^\alpha$, is a Jordan homomorphism. $\square$

In light of Corollary 2.22, $\mathcal{Q} = M_n(\mathcal{R})$ is a 4-free subset of itself for any unital ring $\mathcal{R}$ and $n \geq 4$. Thus setting $\mathcal{Q} = \mathcal{B} = M_n(\mathcal{R})$ for $n \geq 4$ in Theorem 7.20, we obtain the following corollary.

Corollary 7.21. *Let $\mathcal{R}$ be a unital ring with $\frac{1}{2}$, let $\mathcal{B} = M_n(\mathcal{R})$ with $n \geq 4$, and let $\alpha : \mathcal{B} \to \mathcal{B}$ be a surjective zero Jordan product preserving additive map. Then there exist $\lambda \in \mathcal{C}$, the center of $\mathcal{B}$, and a Jordan homomorphism $\sigma : \mathcal{B} \to \mathcal{B}$ such that $x^\alpha = \lambda x^\sigma$ for all $x \in \mathcal{B}$.*

Let us recall that σ from Theorem 7.20 (and hence also from Corollary 7.21) is in fact a sum of a homomorphism and an antihomomorphism (see Remark 7.7).

7.4 Equal Product Preserving Maps

Let $\mathcal{B}$ and $\mathcal{Q}$ be rings, and let $\mathcal{S}$ be a subset of $\mathcal{B}$. We say that a map $\alpha : \mathcal{S} \to \mathcal{Q}$ is *equal product preserving* if $x^\alpha y^\alpha = u^\alpha v^\alpha$ whenever $x, y, u, v \in \mathcal{S}$ satisfy $xy = uv$.

Equal product preserving maps automatically preserve commutativity and zero Jordan products; moreover, if $\mathcal{B}$ and $\mathcal{Q}$ are rings with involution and α preserves $*$ and equal products, then it also preserves normality. Thus, the condition that a map preserves equal products is a rather strong one, certainly much stronger than those considered in previous sections. The challenge now is to describe such maps under rather mild assumptions on $\mathcal{S}$ and $\mathcal{S}^\alpha$.

We will first treat an easy case when $\mathcal{S} = \mathcal{B}$. The problem is trivial if $\mathcal{B}$ is unital, $(\mathcal{B}^\alpha)^2 \neq 0$, and the centralizer $\mathcal{F}$ of $\mathcal{B}^\alpha$ in $\mathcal{Q}$ is a field. Indeed, we have

$$1^\alpha (xy)^\alpha = (xy)^\alpha 1^\alpha = x^\alpha y^\alpha \quad \text{for all } x, y \in \mathcal{B}.$$

Setting $y = 1$ we get $1^\alpha \in \mathcal{F}$. If $1^\alpha = 0$, then $(\mathcal{B}^\alpha)^2 = 0$, a contradiction. Hence, 1^α is invertible and

$$(xy)^\alpha = (1^\alpha)^{-1} x^\alpha y^\alpha \quad \text{for all } x, y \in \mathcal{B}.$$

This means that $x^\sigma = (1^\alpha)^{-1} x^\alpha$ is a homomorphism. Therefore we will not assume that $\mathcal{B}$ is unital.

Proposition 7.22. *Let $\mathcal{B}$ be any ring, and let $\mathcal{Q}$ be a unital ring such that its center $\mathcal{C}$ is a field. Let $\alpha : \mathcal{B} \to \mathcal{Q}$ be an equal product preserving additive map. If $(\mathcal{B}^2)^\alpha \neq 0$ and $\mathcal{B}^\alpha$ is a 3-free subset of $\mathcal{Q}$, then there exist $\lambda \in \mathcal{C}$ and a homomorphism $\sigma : \mathcal{B} \to \mathcal{Q}$ such that $x^\alpha = \lambda x^\sigma$ for all $x \in \mathcal{B}$.*

Proof. We first note that

$$(xy)^\alpha z^\alpha = x^\alpha (yz)^\alpha \quad \text{for all } x, y, z \in \mathcal{B}. \tag{7.48}$$

Applying Theorem 4.13 (and Lemma 4.6) we obtain that $(xy)^\alpha$ is a quasi-polynomial with multiadditive coefficients, that is

$$(xy)^\alpha = \lambda_1 x^\alpha y^\alpha + \lambda_2 y^\alpha x^\alpha + \mu_1(x) y^\alpha + \mu_2(y) x^\alpha + \nu(x, y) \tag{7.49}$$

for some elements $\lambda_1, \lambda_2 \in \mathcal{C}$, additive maps $\mu_1, \mu_2 : \mathcal{B} \to \mathcal{C}$, and a biadditive map $\nu : \mathcal{B}^2 \to \mathcal{C}$. Substituting (7.49) into (7.48) we obtain

$$\lambda_2 x^\alpha z^\alpha y^\alpha - \lambda_2 y^\alpha x^\alpha z^\alpha + \mu_2(z) x^\alpha y^\alpha + [\mu_1(y) - \mu_2(y)] x^\alpha z^\alpha$$
$$- \mu_1(x) y^\alpha z^\alpha + \nu(y, z) x^\alpha - \nu(x, y) z^\alpha = 0$$

for all $x, y, z \in \mathcal{B}$. Since $\mathcal{B}^\alpha$ is a 3-free subset of $\mathcal{Q}$, Lemma 4.4 implies that $\lambda_2 = \mu_1 = \mu_2 = \nu = 0$. Thus $(xy)^\alpha = \lambda_1 x^\alpha y^\alpha$ for all $x, y \in \mathcal{B}$ and so $x^\sigma = \lambda_1 x^\alpha$ is a homomorphism. Since $(\mathcal{B}^2)^\alpha \neq 0$, it follows that $\lambda_1 \neq 0$. Therefore $x^\alpha = \lambda x^\sigma$ where $\lambda = \lambda_1^{-1}$. $\qquad\square$

In our main result in this section we consider the case when $\mathcal{S} = \mathcal{K}$ is the set of skew elements in a ring with involution.

Theorem 7.23. *Let Q be a unital ring such that its center $\mathcal{C}$ is a field with $\mathrm{char}(\mathcal{C}) \neq 2$, let $\mathcal{K}$ be the set of all skew elements (of some ring with involution), and let $\alpha : \mathcal{K} \to Q$ be an additive map. Suppose that $[aba, a]^\alpha \neq 0$ for some $a, b \in \mathcal{K}$, $\mathcal{K}$ admits the operator $\frac{1}{2}$, and $\mathcal{K}^\alpha$ is an 8-free subset of Q. Then α is equal product preserving if and only if there exist $\gamma \in \mathcal{C}$ and a homomorphism $\sigma : \langle \mathcal{K} \rangle \to Q$ such that $x^\alpha = \gamma x^\sigma$ for all $x \in \mathcal{K}$.*

Proof. The "if" part is trivial. So assume that α is equal product preserving. Since for all $x, y \in \mathcal{K}$, xyx and yxy are again elements of $\mathcal{K}$, we see that

$$(xyx)^\alpha y^\alpha = x^\alpha (yxy)^\alpha \quad \text{for all } x, y \in \mathcal{K}. \tag{7.50}$$

Linearizing (7.50) on x and y we get

$$[(xyu)^\alpha + (uyx)^\alpha]v^\alpha + [(xvu)^\alpha + (uvx)^\alpha]y^\alpha$$
$$= x^\alpha[(yuv)^\alpha + (vuy)^\alpha] + u^\alpha[(yxv)^\alpha + (vxy)^\alpha]$$

for all $x, y, u, v \in \mathcal{K}$. Since $\mathcal{K}^\alpha$ is, in particular, 4-free, Theorem 4.13 now tells us that $P(x, y, u) = (xyu)^\alpha + (uyx)^\alpha$ is a quasi-polynomial. Accordingly, $(xyx)^\alpha = \frac{1}{2}P(x, y, x)$ can be written as

$$(xyx)^\alpha = \lambda_1 x^\alpha y^\alpha x^\alpha + \lambda_2 (x^\alpha)^2 y^\alpha + \lambda_3 y^\alpha (x^\alpha)^2$$
$$+ \mu_1(x) y^\alpha x^\alpha + \mu_2(x) x^\alpha y^\alpha + \mu_3(y)(x^\alpha)^2 \tag{7.51}$$
$$+ \nu_1(x, y) x^\alpha + \nu_2(x, x) y^\alpha + \omega(x, y, x),$$

where, by Lemma 4.6, $\lambda_1, \lambda_2, \lambda_3 \in \mathcal{C}$, $\mu_1, \mu_2, \mu_3 : \mathcal{K} \to \mathcal{C}$ are additive maps, $\nu_1, \nu_2 : \mathcal{K}^2 \to \mathcal{C}$ are biadditive, and $\omega : \mathcal{K}^3 \to \mathcal{C}$ is triadditive. Furthermore, ν_2 and ω are symmetric in x and u since P is symmetric in x and u. We can now insert (7.51) in (7.50), that is, we rewrite $(xyx)^\alpha$ and $(yxy)^\alpha$ in (7.50) according to (7.51). From the resulting identity we may conclude, by using Lemma 4.20, that $\lambda_2 = \lambda_3 = \mu_1 = \mu_2 = \mu_3 = \nu_2 = \omega = 0$ and $\nu_1(x, y) = \nu_1(y, x)$ for all $x, y \in \mathcal{K}$. Thus (7.51) becomes

$$(xyx)^\alpha = \lambda x^\alpha y^\alpha x^\alpha + \nu(x, y) x^\alpha, \tag{7.52}$$

where $\lambda = \lambda_1$ and $\nu = \nu_1$. Linearizing (7.52) we obtain

$$(xyz + zyx)^\alpha = \lambda x^\alpha y^\alpha z^\alpha + \lambda z^\alpha y^\alpha x^\alpha + \nu(x, y) z^\alpha + \nu(z, y) x^\alpha. \tag{7.53}$$

Since

$$[x, y]yx + xy[x, y] + [y, x]xy + yx[y, x] = [x, y](yx - xy) + (xy - yx)[x, y] = 0,$$

it follows that

$$\lambda[x,y]^\alpha y^\alpha x^\alpha + \lambda x^\alpha y^\alpha [x,y]^\alpha + \lambda[y,x]^\alpha x^\alpha y^\alpha$$
$$+ \lambda y^\alpha x^\alpha [y,x]^\alpha + \nu([x,y],y)x^\alpha + \nu([y,x],x)y^\alpha = 0$$

(here we also used that $\nu(x,y) = \nu(y,x)$). This can be rewritten as

$$\lambda[[x^\alpha, y^\alpha], [x,y]^\alpha] + \nu([x,y],y)x^\alpha + \nu([y,x],x)y^\alpha = 0. \qquad (7.54)$$

If $\lambda = 0$, then (7.54) reduces to $\nu([x,y],y)x^\alpha + \nu([y,x],x)y^\alpha = 0$. Using Lemma 4.20 it follows that $\nu([y,x],x) = 0$ for all $x,y \in \mathcal{K}$. But then we infer from (7.52) that

$$[xyx,x]^\alpha = (x[y,x]x)^\alpha = \nu(x,[y,x])x^\alpha = \nu([y,x],x)x^\alpha = 0$$

for all $x,y \in \mathcal{K}$; however, this contradicts our assumption that $[aba,a]^\alpha \ne 0$ for some $a,b \in \mathcal{K}$.

Thus $\lambda \ne 0$. This makes it possible for us to apply Theorem 4.13; indeed, from the linearized form of (7.54) it follows that $[x,y]^\alpha$ is a quasi-polynomial, i.e.,

$$[x,y]^\alpha = \gamma_1 x^\alpha y^\alpha + \gamma_2 y^\alpha x^\alpha + \eta_1(x)y^\alpha + \eta_2(y)x^\alpha + \tau(x,y)$$

for some elements $\gamma_1, \gamma_2 \in \mathcal{C}$, additive maps $\eta_1, \eta_2 : \mathcal{K} \to \mathcal{C}$ and a biadditive map $\tau : \mathcal{K}^2 \to \mathcal{C}$. As $[x,y]^\alpha + [y,x]^\alpha = 0$ we infer from Lemma 4.4 that $\gamma_2 = -\gamma_1$, $\eta_2 = -\eta_1$, and $\tau(x,y) = -\tau(y,x)$ for all $x,y \in \mathcal{K}$. Therefore,

$$[x,y]^\alpha = \gamma_1[x^\alpha, y^\alpha] + \eta_1(x)y^\alpha - \eta_1(y)x^\alpha + \tau(x,y).$$

Substituting this expression in (7.54), we see by applying Lemma 4.20 that $\eta_1 = 0$. So we have

$$[x,y]^\alpha = \gamma_1[x^\alpha, y^\alpha] + \tau(x,y). \qquad (7.55)$$

Using both (7.52) and (7.55) we now see that

$$[x,xyx]^\alpha = \gamma_1[x^\alpha, \lambda x^\alpha y^\alpha x^\alpha + \nu(x,y)x^\alpha] + \tau(x,xyx)$$
$$= \lambda\gamma_1 x^\alpha[x^\alpha, y^\alpha]x^\alpha + \tau(x,xyx).$$

On the other hand,

$$[x,xyx]^\alpha = (x[x,y]x)^\alpha = \lambda x^\alpha[x,y]^\alpha x^\alpha + \nu(x,[x,y])x^\alpha$$
$$= \lambda\gamma_1 x^\alpha[x^\alpha, y^\alpha]x^\alpha + \lambda\tau(x,y)(x^\alpha)^2 + \nu(x,[x,y])x^\alpha.$$

Comparing these two expressions we get

$$\lambda\tau(x,y)(x^\alpha)^2 + \nu(x,[x,y])x^\alpha - \tau(x,xyx) = 0.$$

Since $\mathcal{K}^\alpha$ is in particular 5-free, it follows from Lemma 4.20 that $\lambda\tau(x,y) = 0$, and so $\tau(x,y) = 0$ for all $x,y \in \mathcal{K}$. Now (7.55) reads as $[x,y]^\alpha = \gamma_1[x^\alpha, y^\alpha]$.

Again using our assumption that $[aba, a]^\alpha \neq 0$ it follows that $\gamma_1 \neq 0$. Define $\beta : \mathcal{K} \to \mathcal{Q}$ by $x^\beta = \gamma_1 x^\alpha$. Clearly β is a Lie homomorphism, and $x^\alpha = \gamma x^\beta$ where $\gamma = \gamma_1^{-1}$. Lemma 3.3 (v) tells us that $\mathcal{K}^\beta$ is also an 8-free subset of $\mathcal{Q}$. Now define a Lie homomorphism $\varphi : \mathcal{K} \to \overline{\mathcal{Q}} = \mathcal{Q}/\mathcal{C}$ by $x^\varphi = \overline{x^\beta}$. Note that φ satisfies all assumptions of Theorem 6.15 (the assumption that $\mathcal{C}$ is a direct summand of $\mathcal{Q}$ is automatically fulfilled since $\mathcal{C}$ is a field). Therefore there exists a homomorphism $\sigma : \langle \mathcal{K} \rangle \to \mathcal{Q}$ such that $x^\varphi = \overline{x^\sigma}$ for all $x \in \mathcal{K}$. Thus $x^\beta - x^\sigma \in \mathcal{C}$, and hence $x^\alpha = \gamma x^\sigma + \varepsilon(x)$ for every $x \in \mathcal{K}$, where $\varepsilon : \mathcal{K} \to \mathcal{C}$. Accordingly, (7.50) now reads as

$$\left(\gamma(xyx)^\sigma + \varepsilon(xyx)\right)\left(\gamma y^\sigma + \varepsilon(y)\right) = \left(\gamma x^\sigma + \varepsilon(x)\right)\left(\gamma(yxy)^\sigma + \varepsilon(yxy)\right),$$

which yields

$$\gamma\varepsilon(y)x^\sigma y^\sigma x^\sigma + \gamma\varepsilon(xyx)y^\sigma - \gamma\varepsilon(x)y^\sigma x^\sigma y^\sigma - \gamma\varepsilon(yxy)x^\sigma \in \mathcal{C}.$$

Now convert this identity back to one in terms of α (that is, write $\gamma^{-1}(u^\alpha - \varepsilon(u))$ instead of u^σ), so that the d-freeness assumption can be used again. Accordingly, Lemma 4.20 implies that $\varepsilon = 0$. Thus $x^\alpha = \gamma x^\sigma$.   □

There are some other situations, i.e., other types of sets $\mathcal{S}$, which can be handled. But we will not go any further here. Our aim has been just to indicate yet another applicability of the FI techniques.

Literature and Comments. The problem of describing commutativity preserving linear maps on an algebra $\mathcal{A}$ is one of the most studied linear preserver problems. It originated in linear algebra. In 1976 Watkins [196] characterized (through a version of the condition (7.3)) bijective commutativity preserving linear maps on $\mathcal{A} = M_n(\mathbb{F})$, $n \geq 4$. He obtained this result under some technical assumptions, which, however, have later proved to be unnecessary [14, 179, 180] (in particular, unlike the $n = 2$ case, the $n = 3$ case does not need to be excluded). Somewhat later, in the late 1980s, this result was extended to various infinite dimensional algebras of operators: the case when $\mathcal{A} = \mathcal{B}(H)$, the algebra of all bounded linear operators on a Hilbert space, was treated in [100], the case when $\mathcal{A} = \mathcal{B}(X)$, where X is a Banach space, was treated in [176], and the case when $\mathcal{A}$ is a von Neumann factor was treated in [168] (in each of these papers the assumptions were slightly different). In the proofs the authors have relied heavily on idempotents.

All algebras mentioned in the preceding paragraph are prime and centrally closed. In 1993 Brešar [58] generalized and unified the results from the aforementioned papers by determining bijective commutativity preserving linear maps between centrally closed prime algebras satisfying certain technical restrictions, that is, he obtained a version of Corollary 7.12. The proof was based on the characterization of commuting traces of biadditive maps. This paper was followed by several papers considering commutativity preserving maps (or, more generally, maps satisfying $[(x^2)^\alpha, x^\alpha] = 0$) using FI techniques [12, 18, 19, 21, 27, 36, 48, 68, 79, 145]. The exposition in section 7.1 is close to the last paper in this series [68], which, however, was strongly influenced by [36]. We remark that some of these papers also treat commutativity preservers on certain subsets of rings (Lie ideals, symmetric or skew elements, etc.). The motivation for these more general settings also partially derives from linear algebra and operator theory.

The FI approach to commutativity preserving maps on C^*-algebras is studied in the book by Ara and Mathieu [7].

The literature on commutativity preservers on matrix and operator algebras is really vast; numerous authors have generalized the results by Watkins and others in different directions. One of these directions considers the question of how vital is the bijectivity assumption. In this context FI's have turned out to be indirectly important, that is, not the results and the methods but the philosophy of FI's has proved to be useful. Specifically, the idea to replace the condition that α preserves commutativity by the (weaker) condition $[(x^2)^\alpha, x^\alpha] = 0$ has led to the following result [86]: If $\mathbb{F}$ is an arbitrary field and $\mathcal{A}$ is a finite dimensional central simple $\mathbb{F}$-algebra $\mathcal{A}$ with $\dim_\mathbb{F} \mathcal{A} \neq 4$, then every (not necessarily bijective) commutativity preserving linear map $\alpha : \mathcal{A} \to \mathcal{A}$ (or more generally, a map satisfying $[(x^2)^\alpha, x^\alpha] = 0$) is either of the standard form (7.3) or its range is commutative (for the case when $\mathcal{A} = M_n(\mathbb{F})$ with $\mathbb{F}$ algebraically closed and $\mathrm{char}(\mathbb{F}) = 0$, this was proved before [177] using different methods). The main advantage of the condition $[(x^2)^\alpha, x^\alpha] = 0$ in this context is that, unlike the condition that α preserves commutativity, it makes it possible for one to use the scalar extension argument.

Normality preserving maps on matrix and operator algebras were considered in [100, 131, 181]. The connection betweeen FI's and this topic was first noticed in [82]. Yet this paper also uses some operator theoretic techniques, which have later turned out to be unnecessary, i.e., FI's themselves can produce better results. This was shown in the paper [21] upon which Section 7.2 is based.

The FI approach to zero Jordan product preserving maps was considered in [95] and [98]; Section 7.3 is based on the latter paper. We remark that there are also operator theoretic papers on these maps which use different techniques; see for example [201].

In this context we also mention the paper [96] which uses FI's in a more general problem concerning maps preserving square-zero elements, but only in algebras of matrices over a commutative ring.

Section 7.4 is based on the paper [97].

Further applications of FI's to linear preserver problems can be found in [37, 85, 91, 94, 101].

Chapter 8

Further Applications to Lie Algebras

In this closing chapter we consider three rather unrelated applications of FI's. The common property of all three topics is the Lie algebra framework. But otherwise, each of them has a different background. We shall discuss the motivation and history just briefly (mostly at the end of the chapter), and in each section we will introduce the necessary notions in a very concise manner. It is not our intention to go into the heart of the matter of these topics. Our main goal is to indicate that FI's are hidden behind various mathematical notions, and after tracing them out one can effectively apply the theory presented in Part II.

8.1 Lie-Admissible Algebras

Let $\mathbb{F}$ be a field. Let $\mathcal{A} = (\mathcal{A}, +, *)$ be a nonassociative algebra over $\mathbb{F}$ with addition $+$ and multiplication $*$.

We define an algebra $\mathcal{A}^- = (\mathcal{A}, +, [,])$ as the vector space $\mathcal{A}$ under the product $[x, y] = x * y - y * x$. We say that $\mathcal{A}$ is a *Lie-admissible* algebra if $\mathcal{A}^-$ is a Lie algebra. Further, we say that $*$ is

(i) *flexible* if $(x * y) * x = x * (y * x)$ for all $x, y \in \mathcal{A}$;

(ii) *power-associative* if every subalgebra of $\mathcal{A}$ generated by one element is associative;

(iii) *third power-associative* if $(x * x) * x = x * (x * x)$ for all $x \in \mathcal{A}$;

(iv) *fourth power-associative* if

$$((x*x)*x)*x = (x*(x*x))*x = (x*x)*(x*x) = x*((x*x)*x) = x*(x*(x*x))$$

for all $x \in \mathcal{A}$.

Clearly a flexible algebra is third power-associative. It is interesting to note that if $\mathrm{char}(\mathbb{F}) = 0$, then $\mathcal{A}$ is power-associative if and only if it is third and fourth power-associative (see [2] or [172, Lemma 1.11]).

Let us now consider the following situation. Let $\mathcal{Q}$ be an (associative) unital $\mathbb{F}$-algebra, and let $\mathcal{A}$ be an $\mathbb{F}$-subspace of $\mathcal{Q}$. We suppose furthermore that $\mathcal{A} = (\mathcal{A}, +, *)$ is a nonassociative algebra over $\mathbb{F}$ (where a priori $*$ is just an arbitrary $\mathbb{F}$-bilinear map) with the property that

$$[x, y] = x * y - y * x = [x, y] \quad \text{for all } x, y \in \mathcal{A} \tag{8.1}$$

(here $[x, y]$ denotes $xy - yx$, as usual). Thus $\mathcal{A}$ is a Lie-admissible algebra (and in fact also a Lie subalgebra of $\mathcal{Q}$). A natural problem presents itself: can one describe $*$ in terms of the associative product in $\mathcal{Q}$? The following example shows that in some sense this problem is more general than the problem of describing Lie isomorphisms between associative algebras.

Example 8.1. Let α be a Lie isomorphism from an associative algebra $\mathcal{A}$ onto another associative algebra. Define a new multiplication $* : \mathcal{A}^2 \to \mathcal{A}$ by

$$x * y = (x^\alpha y^\alpha)^{\alpha^{-1}} \quad \text{for all } x, y \in \mathcal{A}.$$

It is straightforward to check that (8.1) holds (that is, $\mathcal{A}^-$ is just $\mathcal{A}$ equipped with the Lie product). Moreover, $*$ is an associative multiplication.

Returning to the general problem that we proposed, we first point out some simple examples of multiplications $*$ satisfying (8.1). The simplest one is that $\mathcal{A}$ is a subalgebra of $\mathcal{Q}$ and $*$ coincides with the ordinary multiplication, so $x * y = xy$. Similarly, one can take $*$ to be defined by $x * y - \quad yx$. A more complex example is given by

$$x * y = \gamma xy + (\gamma - 1)yx + \mu(x)y + \mu(y)x + \tau(x, y)1, \tag{8.2}$$

where $\mu : \mathcal{A} \to \mathbb{F}$ is a linear map, $\gamma \in \mathbb{F}$ and $\tau : \mathcal{A}^2 \to \mathbb{F}$ is a symmetric bilinear map such that $\gamma xy + (\gamma - 1)yx + \tau(x, y)1$ lies in $\mathcal{A}$ for all $x, y \in \mathcal{A}$. Here $\mathcal{A}$ may not be a subalgebra of $\mathcal{Q}$ — for example, one can take $\mathcal{A}$ to be the space of $n \times n$ matrices with trace 0, and set $\gamma = 1$ and $\tau(x, y) = -\frac{1}{n}\mathrm{tr}(xy)1$. We also remark that

$$\gamma xy + (\gamma - 1)yx = \frac{1}{2}\{[x, y] + \lambda x \circ y\}, \quad \text{where } \lambda = 2\gamma - 1.$$

Of course, one cannot expect (8.2) to be the solution of the problem we posed above without further conditions; these we shall presently give in the statement of Theorem 8.2. First, however, it is appropriate at this point to mention that the initial result concerning this problem is due to Benkart and Osborn [46]. They obtained a version of Theorem 8.2 for the case where $\mathcal{A} = \mathcal{Q} = M_n(\mathbb{F})$. We will now show that FI's make it possible for us to consider the problem in a more general context.

Theorem 8.2. *Let* $\mathbb{F}$ *be a field with* $\mathrm{char}(\mathbb{F}) \neq 2$, *and let* $\mathcal{Q}$ *be a central* $\mathbb{F}$-*algebra. Let* $\mathcal{A}$ *be a linear subspace of* $\mathcal{Q}$ *which is a 3-free subset of* $\mathcal{Q}$, *and let* $* : \mathcal{A}^2 \to \mathcal{A}$ *be a multiplication such that* $x * y - y * x = [x, y]$ *for all* $x, y \in \mathcal{A}$.

(a) *The multiplication* $*$ *is third power-associative if and only if there exist an element* $\lambda \in \mathbb{F}$, *a linear map* $\mu : \mathcal{A} \to \mathbb{F}$, *and a symmetric bilinear map* $\tau : \mathcal{A}^2 \to \mathbb{F}$ *such that*

$$x * y = \frac{1}{2}\{[x, y] + \lambda x \circ y + \mu(x)y + \mu(y)x + \tau(x, y)1\} \qquad (8.3)$$

for all $x, y \in \mathcal{A}$.

(b) *The multiplication* $*$ *is flexible if and only if* (8.3) *holds and*

$$\mu([x, y]) = 0 = \tau(x, [x, y]) \quad \text{for all } x, y \in \mathcal{A}. \qquad (8.4)$$

(c) *Suppose that* $\mathcal{A}$ *is a 5-free subset of* $\mathcal{Q}$, $1 \in \mathcal{A}$, *and* (8.3) *holds with* $\lambda \neq 0$. *If* $*$ *is fourth power-associative, then*

$$\lambda\mu(x \circ y) + \mu(x)\mu(y) + (2\lambda + \mu(1))\tau(x, y) = 0 \quad \text{for all } x, y \in \mathcal{A}. \qquad (8.5)$$

One may wonder about the case $\lambda = 0$ in (c), which we do not touch. Then one can consult [172, Theorem 2.14] which basically covers this case. We also remark that a more careful analysis yields more information than just (8.5), that is, we could obtain further formulas involving λ, μ and τ which all together are equivalent to $*$ being fourth power-associative. The reader will easily understand from the proof how these formulas could be derived — the procedure is straightforward, but very tedious.

Proof. Define $B : \mathcal{A}^2 \to \mathcal{A}$ by $B(x, y) = x * y$ for all $x, y \in \mathcal{A}$. Clearly B is bilinear. Since $x * y - y * x = [x, y]$, we have

$$B(x, y) - B(y, x) = [x, y] = [x, y] \quad \text{for all } x, y \in \mathcal{A},$$

and also

$$(x * x) * x - x * (x * x) = [B(x, x), x] = [B(x, x), x] \quad \text{for all } x \in \mathcal{A}. \qquad (8.6)$$

(a) Let $*$ be third power-associative. It follows from (8.6) that $[B(x, x), x] = 0$ for all $x \in \mathcal{A}$. Since $\mathcal{A}$ is a 3-free subset of $\mathcal{Q}$ we are in a position to apply Corollary 4.16. Hence there exist $\lambda \in \mathbb{F}$ and maps $\mu : \mathcal{A} \to \mathbb{F}$, $\gamma : \mathcal{A}^2 \to \mathbb{F}$ such that

$$B(x, x) = \lambda x^2 + \mu(x)x + \gamma(x, x)1 \quad \text{for all } x \in \mathcal{A}. \qquad (8.7)$$

Moreover, μ is linear and γ is bilinear (cf. Remark 4.7). Linearizing (8.7) we see that (8.3) holds with $\tau(x, y) = \gamma(x, y) + \gamma(y, x)$.

Conversely, assume that (8.3) holds. Then

$$x * x = \lambda x^2 + \mu(x)x + \frac{1}{2}\tau(x,x)1.$$

Therefore $(x*x)*x - x*(x*x) = [x*x, x] = 0$ and so $*$ is third power-associative.

(b) Note that (8.3) implies that

$$
\begin{aligned}
(x * y) * x = \frac{1}{2}\Big\{ & \Big[\frac{1}{2}\{[x,y] + \lambda x \circ y + \mu(x)y + \mu(y)x + \tau(x,y)1\}, x\Big] \\
& + \frac{1}{2}\lambda\{[x,y] + \lambda x \circ y + \mu(x)y + \mu(y)x + \tau(x,y)1\} \circ x \\
& + \mu\Big(\frac{1}{2}\{[x,y] + \lambda x \circ y + \mu(x)y + \mu(y)x + \tau(x,y)1\}\Big)x \\
& + \frac{1}{2}\mu(x)\{[x,y] + \lambda x \circ y + \mu(x)y + \mu(y)x + \tau(x,y)1\} \\
& + \tau\Big(\frac{1}{2}\{[x,y] + \lambda x \circ y + \mu(x)y + \mu(y)x + \tau(x,y)1\}, x\Big)\Big\}
\end{aligned}
$$

and

$$
\begin{aligned}
x * (y * x) = \frac{1}{2}\Big\{ & \Big[x, \frac{1}{2}\{[y,x] + \lambda y \circ x + \mu(y)x + \mu(x)y + \tau(y,x)1\}\Big] \\
& + \frac{1}{2}\lambda x \circ \{[y,x] + \lambda y \circ x + \mu(y)x + \mu(x)y + \tau(y,x)1\} \\
& + \frac{1}{2}\mu(x)\{[y,x] + \lambda y \circ x + \mu(y)x + \mu(x)y + \tau(y,x)1\} \\
& + \mu\Big(\frac{1}{2}\{[y,x] + \lambda y \circ x + \mu(y)x + \mu(x)y + \tau(y,x)1\}\Big)x \\
& + \tau\Big(x, \frac{1}{2}\{[y,x] + \lambda y \circ x + \mu(y)x + \mu(x)y + \tau(y,x)1\}\Big)\Big\}.
\end{aligned}
$$

It then follows that

$$(x * y) * x - x * (y * x) = \frac{1}{2}\Big(\mu([x,y])x + \tau([x,y],x)1\Big) \quad \text{for all } x, y \in \mathcal{A}. \quad (8.8)$$

Therefore, if (8.3) and (8.4) are fulfilled, then the algebra $(\mathcal{A}, +, *)$ is flexible.

Conversely, suppose that $(\mathcal{A}, +, *)$ is flexible. In particular, $*$ is third power-associative, and so (8.3) holds. Consequently (8.8) is valid, and we have

$$\frac{1}{2}\Big(\mu([x,y])x + \tau([x,y],x)1\Big) = (x * y) * x - x * (y * x) = 0 \quad \text{for all } x, y \in \mathcal{A}.$$

Fixing y we may now use Corollary 4.21 and hence conclude that (8.4) holds. Therefore (b) is proved.

(c) Suppose that $1 \in \mathcal{A}$, $\mathcal{A}$ is a 5-free subset of $\mathcal{Q}$ and (8.3) holds with $\lambda \neq 0$. In particular

$$x * x = \lambda x^2 + \mu(x)x + \frac{1}{2}\tau(x,x)1 \quad \text{for all } x \in \mathcal{A}.$$

Therefore, $[x * x, x] = 0$, and

$$(x * x) * (x * x) = \lambda \left(\lambda x^2 + \mu(x)x + \frac{1}{2}\tau(x,x)1 \right)^2$$
$$+ \mu(x * x) \left(\lambda x^2 + \mu(x)x + \frac{1}{2}\tau(x,x)1 \right) + \frac{1}{2}\tau(x * x, x * x)1$$
$$= \lambda^3 x^4 + 2\lambda^2 \mu(x)x^3 + \left(\lambda^2 \tau(x,x) + \lambda\mu(x)^2 + \lambda\mu(x * x) \right)x^2$$
$$+ \Gamma_1(x,x,x)x + \Gamma_2(x,x,x,x)1, \qquad (8.9)$$

where $\Gamma_1 : \mathcal{A}^3 \to \mathbb{F}$ and $\Gamma_2 : \mathcal{A}^4 \to \mathbb{F}$ are multilinear maps; we could easily express Γ_1 and Γ_2 in terms of λ, μ and τ, but this is not relevant for our purposes.

Our next goal is to compute $((x * x) * x) * x$. First we note that (8.3) yields

$$(x * x) * x = \frac{1}{2} \left\{ \lambda \left(2\lambda x^2 + 2\mu(x)x + \tau(x,x) \right)x + \mu(x * x)x \right.$$
$$\left. + \mu(x) \left(\lambda x^2 + \mu(x)x + \frac{1}{2}\tau(x,x)1 \right) + \tau(x * x, x)1 \right\}$$
$$= \lambda^2 x^3 + \frac{3}{2}\lambda\mu(x)x^2 + \frac{1}{2}\left(\lambda\tau(x,x) + \mu(x * x) + \mu(x)^2 \right)x$$
$$+ \frac{1}{4}\mu(x)\tau(x,x)1 + \frac{1}{2}\tau(x * x, x)1.$$

Hence $[(x * x) * x, x] = 0$, and

$$((x * x) * x) * x = \frac{1}{2} \left\{ \lambda((x * x) * x) \circ x + \mu((x * x) * x)x \right.$$
$$\left. + \mu(x)((x * x) * x) + \tau((x * x) * x, x)1 \right\}$$
$$= \lambda^3 x^4 + 2\lambda^2 \mu(x)x^3 + \frac{1}{2}\left(\lambda^2 \tau(x,x) + \lambda\mu(x * x) + \frac{5}{2}\lambda\mu(x)^2 \right)x^2$$
$$+ \Psi_1(x,x,x)x + \Psi_2(x,x,x,x)1 \qquad (8.10)$$

for some multilinear maps $\Psi_1 : \mathcal{A}^3 \to \mathbb{F}$, $\Psi_2 : \mathcal{A}^4 \to \mathbb{F}$.

Comparing (8.9) and (8.10), and also using

$$\mu(x * x) = \mu \left(\lambda x^2 + \mu(x)x + \frac{1}{2}\tau(x,x)1 \right) = \lambda\mu(x^2) + \mu(x)^2 + \frac{1}{2}\tau(x,x)\mu(1),$$

we obtain

$$(x * x) * (x * x) - ((x * x) * x) * x$$
$$= \frac{1}{4}\lambda \left(2\lambda\mu(x^2) + \mu(x)^2 + (2\lambda + \mu(1))\tau(x,x) \right)x^2$$
$$+ \Phi_1(x,x,x)x + \Phi_2(x,x,x,x)1, \qquad (8.11)$$

where $\Phi_1 : \mathcal{A}^3 \to \mathbb{F}$ and $\Phi_2 : \mathcal{A}^4 \to \mathbb{F}$ are multilinear.

Now assume that $*$ is fourth power-associative. Then the left-hand side of (8.11) is 0, so we have

$$\Phi_0(x,x)x^2 + \Phi_1(x,x,x)x + \Phi_2(x,x,x,x)1 = 0 \quad \text{for all } x \in \mathcal{A}, \tag{8.12}$$

where

$$\Phi_0(x,y) = \frac{1}{4}\lambda\Big(\lambda\mu(x \circ y) + \mu(x)\mu(y) + \big(2\lambda + \mu(1)\big)\tau(x,y)\Big);$$

so $\Phi_0 : \mathcal{A}^2 \to \mathbb{F}$ is a symmetric bilinear map. Now by Lemma 4.20 we in particular infer that $\Phi_0(x,x) = 0$ for all $x \in \mathcal{A}$, and hence, as Φ_0 is symmetric, also $\Phi_0(x,y) = 0$ for all $x, y \in \mathcal{A}$. Thus (8.5) holds. $\qquad\square$

Some related conditions on $*$, which Theorem 8.2 does not cover, can be considered in a similar fashion. In particular, the case when $*$ is associative can be easily analysed.

The following remark is in order concerning part (a) of Theorem 8.2. If in (8.3) we have $\lambda \neq 0$, then we see by using $\mathcal{A} * \mathcal{A} \subseteq \mathcal{A}$ that $x \circ y$ lies in $\mathcal{A} + \mathbb{F}1$ for all $x, y \in \mathcal{A}$. Since $[x,y]$ is in $\mathcal{A}$, it follows that xy is in $\mathcal{A} + \mathbb{F}1$ for all $x, y \in \mathcal{A}$. Thus, if $1 \in \mathcal{A}$, then $\mathcal{A}$ is a unital subalgebra of $\mathcal{Q}$. On the other hand, the assumption that $\mathcal{A} * \mathcal{A}$ is contained in $\mathcal{A}$ is in fact unnecessary, it was stated only because the problem arose from the theory of Lie-admissible algebras. We could assume that $* : \mathcal{A}^2 \to \mathcal{Q}$ and obtain the same conclusions.

We do not want to bother the reader by stating corollaries of Theorem 8.2 in terms of concrete d-free sets. Let us just recall that one can take $\mathcal{A}$ to be a matrix algebra $M_n(\mathcal{B})$, $n \geq 3$, where $\mathcal{B}$ is any unital algebra (cf. Corollary 2.22), a prime algebra with $\deg(\mathcal{A}) \geq 3$ (cf. Corollary 5.12), or more generally, any set from the list of d-free subsets of prime algebras given in Section 5.2.

8.2 Poisson Algebras

Let $\mathbb{F}$ be a field. An associative $\mathbb{F}$-algebra $(\mathcal{B}, +, \cdot)$ with binary operation $\{\,,\,\} : \mathcal{B}^2 \to \mathcal{B}$ is called a *Poisson algebra* if $(\mathcal{B}, +, \{\,,\,\})$ is a Lie algebra and

$$\{xy, z\} = x\{y, z\} + \{x, z\}y$$

holds for all $x, y, z \in \mathcal{B}$.

A *Dirac map* is a Lie homomorphism from a commutative Poisson algebra $\mathcal{B}$ into $\mathcal{L}(\mathcal{H})$, the algebra of all linear (not necessarily bounded) operators acting on a Hilbert space $\mathcal{H}$. The Dirac problem is to find all such maps. We will demonstrate the applicability of FI's to this problem. The approach we will use makes it possible for us to treat a considerably more general situation than in the original Dirac problem.

Theorem 8.3. *Let $\mathbb{F}$ be a field with $\operatorname{char}(\mathbb{F}) \neq 2$, and let $\mathcal{Q}$ be a central $\mathbb{F}$-algebra. Let $\mathcal{B}$ be a Poisson algebra over $\mathbb{F}$ and let $\alpha : \mathcal{B} \to \mathcal{Q}$ be a linear map such that*

$$\{x, y\}^\alpha = [x^\alpha, y^\alpha] \quad \text{for all } x, y \in \mathcal{B}.$$

If $\mathcal{B}^\alpha$ is a 4-free subset of $\mathcal{Q}$, then there exist $\lambda \in \mathbb{F}$, linear maps $\mu_1, \mu_2 : \mathcal{B} \to \mathbb{F}$ and a bilinear map $\nu : \mathcal{B}^2 \to \mathbb{F}$ such that either

$$(xy)^\alpha = \lambda x^\alpha y^\alpha + \mu_1(x)y^\alpha + \mu_2(y)x^\alpha + \nu(x,y)1$$

or

$$(xy)^\alpha = \lambda y^\alpha x^\alpha + \mu_1(x)y^\alpha + \mu_2(y)x^\alpha + \nu(x,y)1,$$

and

$$\lambda\nu(x,y) = \mu_1(x)\mu_1(y) - \mu_1(xy) = \mu_2(x)\mu_2(y) - \mu_2(xy)$$

for all $x, y \in \mathcal{B}$. Moreover, if $\mathcal{B}$ is a commutative algebra, then $\lambda = 0$, $\mu_1 = \mu_2$, and ν is symmetric.

So, roughly speaking, Theorem 8.3 determines the structure of those Dirac maps whose range is "big enough" inside $\mathcal{Q}$.

Proof. Define $B : \mathcal{B} \times \mathcal{B} \to \mathcal{Q}$ by

$$B(x,y) = (xy)^\alpha.$$

Our first goal is to show that B is a quasi-polynomial (with respect to α). Set

$$J(x,y) = B(x,y) + B(y,x) \quad \text{and} \quad L(x,y) = B(x,y) - B(y,x).$$

As $B(x,y) = \frac{1}{2}\big(J(x,y) + L(x,y)\big)$, it suffices to show that both J and L are quasi-polynomials.

Applying α to $\{xy, z\} = x\{y, z\} + \{x, z\}y$ we get

$$[B(x,y), z^\alpha] = B(x, \{y, z\}) + B(\{x, z\}, y) \quad \text{for all } x, y, z \in \mathcal{B}. \tag{8.13}$$

Since $\{x, x\} = 0$, it follows from (8.13) that $[B(x,x), x^\alpha] = 0$ for all $x \in \mathcal{B}$. Corollary 4.15 now tells us that $B(x,x) = \lambda_0(x^\alpha)^2 + \lambda_1(x)x^\alpha + \lambda_2(x)1$ where $\lambda_0 \in \mathbb{F}$, $\lambda_1, \lambda_2 : \mathcal{B} \to \mathbb{F}$ with λ_1 additive and λ_2 the trace of a biadditive map. A linearization of this identity shows that J is a quasi-polynomial.

Let us now examine L. We begin by noticing that

$$\{[x, y], z\} = \{xy, z\} - \{yx, z\} = x\{y, z\} + \{x, z\}y - y\{x, z\} - \{y, z\}x$$
$$= [x, \{y, z\}] + [\{x, z\}, y].$$

This implies that

$$[L(x,y), z^\alpha] = L(x, \{y, z\}) + L(\{x, z\}, y) \tag{8.14}$$

for all $x, y, z \in \mathcal{B}$. Interchanging the roles of x and z we get

$$[L(z,y), x^\alpha] = L(z, \{y, x\}) + L(\{z, x\}, y). \tag{8.15}$$

Similarly, by interchanging y and z we have

$$[L(x,z),y^\alpha] = L(x,\{z,y\}) + L(\{x,y\},z). \tag{8.16}$$

Since $\{u,v\} = -\{v,u\}$ and $L(u,v) = -L(v,u)$ for all $u,v \in \mathcal{B}$, summing up (8.14)–(8.16) we obtain

$$2L(z,\{y,x\}) = [L(x,y),z^\alpha] + [L(z,y),x^\alpha] + [L(x,z),y^\alpha] \tag{8.17}$$

for all $x,y,z \in \mathcal{B}$.

Substituting $\{u,v\}$ for z in (8.17), we get

$$2L(\{u,v\},\{y,x\})$$
$$= [L(x,y),[u^\alpha,v^\alpha]] + [L(\{u,v\},y),x^\alpha] + [L(x,\{u,v\}),y^\alpha] \tag{8.18}$$

for all $x,y,u,v \in \mathcal{B}$. Now, interchanging the roles of x and u as well as of y and v in (8.18), we obtain

$$2L(\{x,y\},\{v,u\})$$
$$= [L(u,v),[x^\alpha,y^\alpha]] + [L(\{x,y\},v),u^\alpha] + [L(u,\{x,y\}),v^\alpha] \tag{8.19}$$

for all $x,y,u,v \in \mathcal{B}$.

Since $L(\{u,v\},\{y,x\}) = -L(\{x,y\},\{v,u\})$, summing up (8.18) and (8.19) gives

$$[L(x,y),[u^\alpha,v^\alpha]] + [L(\{u,v\},y),x^\alpha] + [L(x,\{u,v\}),y^\alpha]$$
$$+ [L(u,v),[x^\alpha,y^\alpha]] + [L(\{x,y\},v),u^\alpha] + [L(u,\{x,y\}),v^\alpha] = 0. \tag{8.20}$$

Applying the identity $[a,[b,c]] = [[a,b],c] - [[a,c],b]$ to two of the terms of (8.20) we obtain

$$[L(\{x,y\},v) - [L(x,y),v^\alpha],u^\alpha] - [L(\{x,y\},u) - [L(x,y),u^\alpha],v^\alpha]$$
$$+ [L(\{u,v\},y) - [L(u,v),y^\alpha],x^\alpha] - [L(\{u,v\},x) - [L(u,v),x^\alpha],y^\alpha] = 0$$

for all $x,y,u,v \in \mathcal{B}$. That is,

$$[Q(x,y,v),u^\alpha] - [Q(x,y,u),v^\alpha] + [Q(u,v,y),x^\alpha] - [Q(u,v,x),y^\alpha] = 0,$$

where Q is defined as

$$Q(x,y,z) = L(\{x,y\},z) - [L(x,y),z^\alpha].$$

Since $\mathcal{B}^\alpha$ is 4-free, we are now in a position to apply Theorem 4.13 (for $m = 4$, $n = 3$, $P = 0$ and $c = \pm 1$). It follows that Q is a quasi-polynomial. Using

$$L(z,\{y,x\}) = L(\{x,y\},z) = [L(x,y),z^\alpha] + Q(x,y,z)$$

in (8.17), we see that

$$[L(y,x), z^\alpha] + [L(z,y), x^\alpha] + [L(x,z), y^\alpha] = 2Q(x,y,z)$$

for all $x, y, z \in \mathcal{B}$. Again we may use Theorem 4.13 (or, since L is skew-symmetric, Corollary 4.14), this time for $m = 3$, $n = 2$, $P = 2Q$ and $c = \pm 1$ (again exactly 4-freeness is used). Hence L is a quasi-polynomial.

Thus both J and L are quasi-polynomials, so that B is a quasi-polynomial too. Therefore there exist $\lambda_1, \lambda_2 \in \mathbb{F}$, $\mu_1, \mu_2 : \mathcal{B} \to \mathbb{F}$ and $\nu : \mathcal{B}^2 \to \mathbb{F}$ such that

$$B(x,y) = \lambda_1 x^\alpha y^\alpha + \lambda_2 y^\alpha x^\alpha + \mu_1(x)y^\alpha + \mu_2(y)x^\alpha + \nu(x,y)1 \qquad (8.21)$$

for all $x, y \in \mathcal{B}$. As B is bilinear, it follows that μ_1 and μ_2 are linear and ν is bilinear (see Lemma 4.6 and Remark 4.7).

From (8.21) we infer

$$\begin{aligned}
B(xy, z) = {}& \lambda_1(\lambda_1 x^\alpha y^\alpha + \lambda_2 y^\alpha x^\alpha + \mu_1(x)y^\alpha + \mu_2(y)x^\alpha + \nu(x,y)1)z^\alpha \\
&+ \lambda_2 z^\alpha(\lambda_1 x^\alpha y^\alpha + \lambda_2 y^\alpha x^\alpha + \mu_1(x)y^\alpha + \mu_2(y)x^\alpha + \nu(x,y)1) \\
&+ \mu_2(z)(\lambda_1 x^\alpha y^\alpha + \lambda_2 y^\alpha x^\alpha + \mu_1(x)y^\alpha + \mu_2(y)x^\alpha + \nu(x,y)1) \\
&+ \mu_1(xy)z^\alpha + \nu(xy, z)1,
\end{aligned}$$

and also

$$\begin{aligned}
B(x, yz) = {}& \lambda_1 x^\alpha(\lambda_1 y^\alpha z^\alpha + \lambda_2 z^\alpha y^\alpha + \mu_1(y)z^\alpha + \mu_2(z)y^\alpha + \nu(y,z)1) \\
&+ \lambda_2(\lambda_1 y^\alpha z^\alpha + \lambda_2 z^\alpha y^\alpha + \mu_1(y)z^\alpha + \mu_2(z)y^\alpha + \nu(y,z)1)x^\alpha \\
&+ \mu_1(x)(\lambda_1 y^\alpha z^\alpha + \lambda_2 z^\alpha y^\alpha + \mu_1(y)z^\alpha + \mu_2(z)y^\alpha + \nu(y,z)1) \\
&+ \mu_2(yz)x^\alpha + \nu(x, yz)1.
\end{aligned}$$

However, $B(xy, z) = (xyz)^\alpha = B(x, yz)$. Comparing the above results we therefore obtain

$$\begin{aligned}
&\lambda_1\lambda_2(y^\alpha x^\alpha z^\alpha + z^\alpha x^\alpha y^\alpha - x^\alpha z^\alpha y^\alpha - y^\alpha z^\alpha x^\alpha) + \lambda_1(\mu_2(y) - \mu_1(y))x^\alpha z^\alpha \\
&\quad + \lambda_2(\mu_2(y) - \mu_1(y))z^\alpha x^\alpha + (\mu_2(z)\mu_2(y) - (\lambda_1 + \lambda_2)\nu(y,z) - \mu_2(yz))x^\alpha \\
&\qquad + ((\lambda_1 + \lambda_2)\nu(x,y) + \mu_1(xy) - \mu_1(x)\mu_1(y))z^\alpha \in \mathbb{F}1.
\end{aligned}$$

Applying Lemma 4.4 we get, in particular,

$$\lambda_1\lambda_2 = 0,$$
$$\mu_2(z)\mu_2(y) - (\lambda_1 + \lambda_2)\nu(y,z) - \mu_2(yz) = 0,$$
$$(\lambda_1 + \lambda_2)\nu(x,y) + \mu_1(xy) - \mu_1(x)\mu_1(y) = 0.$$

Therefore, $\lambda_1 = 0$ or $\lambda_2 = 0$. It is now obvious that these relations yield the desired conclusion.

Finally, assume that $\mathcal{B}$ is commutative. Then $B(x,y) = B(y,x)$ and (8.21) yields

$$(\lambda_1 - \lambda_2)x^\alpha y^\alpha + (\lambda_2 - \lambda_1)y^\alpha x^\alpha + (\mu_1(x) - \mu_2(x))y^\alpha$$
$$+ (\mu_2(y) - \mu_1(y))x^\alpha + \nu(x,y)1 - \nu(y,x)1 = 0$$

for all $x, y \in \mathcal{B}$. Lemma 4.4 then gives $\lambda_1 = \lambda_2$ (so both λ_1 and λ_2 are 0), $\mu_1 = \mu_2$ and $\nu(x,y) = \nu(y,x)$. $\qquad\square$

Remark 8.4. If $\mathcal{B}$ is a commutative algebra, then the conclusion of Theorem 8.3 can be read as

$$(xy)^\alpha = \mu(x)y^\alpha + \mu(y)x^\alpha + \nu(x,y)1, \qquad (8.22)$$

where μ is a multiplicative linear functional on $\mathcal{B}$. We can add to this that

$$\mu(\{x,y\}) = 0 \quad \text{and} \quad \nu(x, \{y,z\}) + \nu(\{x,z\}, y) = 0 \qquad (8.23)$$

for all $x, y, z \in \mathcal{B}$. Indeed, by (8.22) we have

$$\{xy, z\}^\alpha = [(xy)^\alpha, z^\alpha] = \mu(x)[y^\alpha, z^\alpha] + \mu(y)[x^\alpha, z^\alpha].$$

Since $\{xy, z\} = x\{y, z\} + \{x, z\}y$ we see that, on the other hand,

$$\{xy, z\}^\alpha = (x\{y,z\})^\alpha + (\{x,z\}y)^\alpha$$
$$= \mu(x)[y^\alpha, z^\alpha] + \mu(\{y,z\})x^\alpha + \nu(x, \{y,z\})1$$
$$+ \mu(\{x,z\})y^\alpha + \mu(y)[x^\alpha, z^\alpha] + \nu(\{x,z\}, y)1.$$

Comparing both expressions we get

$$\mu(\{y,z\})x^\alpha + \mu(\{x,z\})y^\alpha + \nu(x, \{y,z\})1 + \nu(\{x,z\}, y)1 = 0,$$

and so Lemma 4.4 gives us (8.23).

8.3   Maps Covariant Under the Action of Lie Algebras

Let $\mathcal{V}$ be a module over a Lie algebra $\mathcal{L}$, and let $f : \mathcal{V}^n \to \mathcal{V}$ be a multiadditive map. We say that f is *covariant under the action of* $\mathcal{L}$ if

$$\ell f(x_1, x_2, \ldots, x_n) = f(\ell x_1, x_2, \ldots, x_n) + f(x_1, \ell x_2, \ldots, x_n)$$
$$+ \cdots + f(x_1, x_2, \ldots, \ell x_n) \qquad (8.24)$$

for all $x_1, \ldots, x_n \in \mathcal{V}$ and $\ell \in \mathcal{L}$. This condition appears in some problems of mathematical physics (see, for example, [174, Chapter 8]).

We will consider the special case when $\mathcal{V} = \mathcal{L}$, so (8.24) now reads as

$$[f(x_1, x_2, \ldots, x_n), y] = f([x_1, y], x_2, \ldots, x_n) + f(x_1, [x_2, y], \ldots, x_n)$$
$$+ \cdots + f(x_1, x_2, \ldots, [x_n, y]). \qquad (8.25)$$

If $\mathcal{L}$ is a Lie subalgebra of an algebra $\mathcal{Q}$, we can easily find examples of such maps. A very simple one is $f(x_1, x_2, \ldots, x_n) = x_1 x_2 \ldots x_n$ (let us just ignore at this point that $f(x_1, x_2, \ldots, x_n)$ may not lie in $\mathcal{L}$). More generally, we can take $f(x_1, x_2, \ldots, x_n) = \lambda x_{\pi(1)} x_{\pi(2)} \cdots x_{\pi(n)}$ where π is a permutation and λ is an element from the center $\mathcal{C}$ of $\mathcal{Q}$. Furthermore, let $\mu : \mathcal{L}^s \to \mathcal{C}$ be such that $\mu(x_1, \ldots, x_s) = 0$ whenever at least one x_i lies in $[\mathcal{L}, \mathcal{L}]$. Then one can check that

$$f(x_1, x_2, \ldots, x_n) = \mu(x_{\pi(1)}, \ldots, x_{\pi(s)}) x_{\pi(s+1)} \cdots x_{\pi(n)}$$

satisfies (8.25). Further examples can be obtained by taking sums. All these lead to the following: quasi-polynomials are natural candidates for maps satisfying (8.25). We shall confine ourselves to the case where f is skew-symmetric, that is,

$$f(x_1, \ldots, x_i, \ldots, x_j, \ldots, x_n) = -f(x_1, \ldots, x_j, \ldots, x_i, \ldots, x_n)$$

whenever $i \neq j$. Together with the assumption that $\mathcal{L}$ is $(n+2)$-free we will be able to show that f must indeed be a quasi-polynomial. The intriguing part of the proof is the construction of an FI involving f. We will do this in the next lemma.

Lemma 8.5. *Let $\mathcal{L}$ be a Lie subring of a ring $\mathcal{Q}$, and let $f : \mathcal{L}^n \to \mathcal{Q}$ be a skew-symmetric multiadditive map satisfying (8.25) for all $x_1, x_2, \ldots, x_n, y \in \mathcal{L}$. Then*

$$\sum_{\substack{\sigma \in A_{n+2} \\ \sigma(1) < \sigma(2) < \cdots < \sigma(n)}} [f(x_{\sigma(1)}, x_{\sigma(2)}, \ldots, x_{\sigma(n)}), [x_{\sigma(n+1)}, x_{\sigma(n+2)}]] = 0 \qquad (8.26)$$

for all $x_1, x_2, \ldots, x_{n+2} \in \mathcal{L}$, where A_{n+2} is the group of even permutations on $\{1, 2, \ldots, n+2\}$.

Proof. We let I denote the set of all $\sigma \in A_{n+2}$ subject to the stringent requirement that

$$\sigma(1) < \sigma(2) < \ldots < \sigma(n). \qquad (8.27)$$

Setting

$$g(x_1, \ldots, x_{n+2}) = \sum_{\sigma \in I} [f(x_{\sigma(1)}, x_{\sigma(2)}, \ldots, x_{\sigma(n)}), [x_{\sigma(n+1)}, x_{\sigma(n+2)}]], \qquad (8.28)$$

we have to prove that $g(x_1, \ldots, x_{n+2}) = 0$. Expanding the right-hand side of (8.28) by using (8.25), we obtain

$$g(x_1, \ldots, x_{n+2}) = \sum_{\sigma \in I} \Big\{ f\big([x_{\sigma(1)}, [x_{\sigma(n+1)}, x_{\sigma(n+2)}]], x_{\sigma(2)}, \ldots, x_{\sigma(n)}\big)$$
$$+ f\big(x_{\sigma(1)}, [x_{\sigma(2)}, [x_{\sigma(n+1)}, x_{\sigma(n+2)}]], \ldots, x_{\sigma(n)}\big) \qquad (8.29)$$
$$+ \ldots$$
$$+ f\big(x_{\sigma(1)}, x_{\sigma(2)}, \ldots, [x_{\sigma(n)}, [x_{\sigma(n+1)}, x_{\sigma(n+2)}]]\big) \Big\}.$$

It suffices to show that for an arbitrary triple $\{i, j, k\}$ the sum of all summands in (8.29) which involve $[x_\alpha, [x_\beta, x_\gamma]]$, where $\{\alpha, \beta, \gamma\} = \{i, j, k\}$, is 0. Without loss of generality we may assume that $i < j < k$.

First we note that there exists a unique $\sigma \in I$ such that $\{\sigma(n+1), \sigma(n+2)\} = \{j, k\}$; this is clear in view of (8.27) and because exactly one of the two choices for $\sigma(n+1)$, $\sigma(n+2)$ will result in σ lying in A_{n+2}. Furthermore the permutation $\tau = (\sigma(1), \sigma(2), \ldots, \sigma(n), j, k)$ is even if and only if $j + k + 1$ is even; this is seen by noting (with (8.27) in mind) that there are $n + 1 - j + n + 2 - k$ transpositions required to transform τ to the identity permutation. This observation allows us to write $[x_{\sigma(n+1)}, x_{\sigma(n+2)}] = (-1)^{j+k+1}[x_j, x_k]$, thus avoiding the need to handle separate cases. Therefore

$$f(x_{\sigma(1)}, \ldots, [x_i, [x_{\sigma(n+1)}, x_{\sigma(n+2)}]], \ldots, x_{\sigma(n)})$$
$$= (-1)^{j+k+1} f(x_{\sigma(1)}, \ldots, [x_i, [x_j, x_k]], \ldots, x_{\sigma(n)}). \quad (8.30)$$

In view of (8.27) it is also clear that $i = \sigma(i)$, so $[x_i, [x_j, x_k]]$ is located at the ith position.

Next, by the same argument (now with $i < k$) there is a unique $\gamma \in I$ such that

$$f(x_{\gamma(1)}, \ldots, [x_j, [x_{\gamma(n+1)}, x_{\gamma(n+2)}]], \ldots, x_{\gamma(n)})$$
$$= (-1)^{i+k+1} f(x_{\gamma(1)}, \ldots, [x_j, [x_i, x_k]], \ldots, x_{\gamma(n)}). \quad (8.31)$$

This time, again making strong use of (8.27), one sees that $[x_j, [x_i, x_k]]$ is located at the $(j-1)$st position.

Thirdly, there is a unique $\rho \in I$ such that

$$f(x_{\rho(1)}, \ldots, [x_k, [x_{\rho(n+1)}, x_{\rho(n+2)}]], \ldots, x_{\rho(n)})$$
$$= (-1)^{l+J+1} f(x_{\rho(1)}, \ldots, [x_k, [x_i, x_j]], \ldots, x_{\rho(n)}) \quad (8.32)$$

with $[x_k, [x_i, x_j]]$ located at the $(k-2)$nd position (again with the help of (8.27)).

Now, since f is skew-symmetric, in each of (8.30), (8.31), (8.32) the "double commutators" can be moved to the first position with appropriate sign change and so we have

$$f(x_{\sigma(1)}, \ldots, [x_i, [x_{\sigma(n+1)}, x_{\sigma(n+2)}]], \ldots, x_{\sigma(n)})$$
$$+ f(x_{\gamma(1)}, \ldots, [x_j, [x_{\gamma(n+1)}, x_{\gamma(n+2)}]], \ldots, x_{\gamma(n)})$$
$$+ f(x_{\rho(1)}, \ldots, [x_k, [x_{\rho(n+1)}, x_{\rho(n+2)}]], \ldots, x_{\rho(n)})$$
$$= (-1)^{i-1}(-1)^{j+k+1} f([x_i, [x_j, x_k]], x_{u_1}, \ldots, x_{u_{n-1}})$$
$$+ (-1)^{j-2}(-1)^{i+k+1} f([x_j, [x_i, x_k]], x_{u_1}, \ldots, x_{u_{n-1}})$$
$$+ (-1)^{k-3}(-1)^{i+j+1} f([x_k, [x_i, x_j]], x_{u_1}, \ldots, x_{u_{n-1}})$$
$$= (-1)^{i+j+k} f([x_i, [x_j, x_k]] + [x_j, [x_k, x_i]] + [x_k, [x_i, x_j]], x_{u_1}, \ldots, x_{u_{n-1}}) = 0$$

in light of the Jacobi identity. This completes the proof. $\qquad\square$

Theorem 8.6. *Let $\mathcal{L}$ be a Lie subring of a unital ring $\mathcal{Q}$, and let $f : \mathcal{L}^n \to \mathcal{Q}$ be a skew-symmetric multiadditive map satisfying (8.25) for all $x_1, \ldots, x_n, y \in \mathcal{L}$. If $\mathcal{L}$ is an $(n + 2)$-free subset of $\mathcal{Q}$, then f is a quasi-polynomial.*

Proof. Lemma 8.5 tells us that f satisfies (8.26). Note that this is an FI for which Corollary 4.14 is applicable. Whence f is a quasi-polynomial. $\qquad\square$

Corollaries 5.16, 5.18 and 5.19 provide a list of Lie algebras which are d-free sets, so Theorem 8.6 can be applied to various concrete situations.

Literature and Comments. The study of Lie-admissible algebras was initiated by Albert [2] in 1949. These algebras arise naturally in various areas of mathematics and physics (see [44, 123, 172, 174, 188]). The classification of flexible Lie-admissible algebras $\mathcal{A}$ in the case when $\mathcal{A}^-$ is semisimple was a well-known problem posed by Albert [2].

A partial solution of Albert's problem was obtained in 1962 by Laufer and Tomber [136]. They classified finite dimensional flexible power-associative Lie-admissible algebras $\mathcal{A}$ over algebraically closed fields of characteristic 0 with $\mathcal{A}^-$ semisimple. Myung [171, 172] obtained a description of finite dimensional flexible power-associative Lie-admissible algebras $\mathcal{A}$ over algebraically closed fields of positive characteristics with $\mathcal{A}^-$ being a classical Lie algebra or a generalized Witt algebra. In 1981 Benkart and Osborn [45] and Okubo and Myung [175] independently classified finite dimensional Lie-admissible flexible algebras $\mathcal{A}$ over algebraically closed fields of characteristic 0 such that $\mathcal{A}^-$ are semisimple. Thus Albert's problem was solved completely.

Benkart and Osborn [46] described power-associative products on matrices and Benkart [44] classified third power-associative Lie-admissible algebras $\mathcal{A}$ such that $\mathcal{A}^-$ is semisimple. Jeong, Kang and Lee [123] classified third power-associative Lie-admissible algebras $\mathcal{A}$ such that $\mathcal{A}^-$ are Kac–Moody algebras.

Applications of FI's to Lie-admissible algebras were obtained in papers [28, 33] and Theorem 8.2 is based on these results.

Poisson algebras originally appeared in differential geometry. They were studied abstractly as algebraic structures in papers [105, 106, 107, 129, 130]. The Dirac problem was studied by Souriau [190], Streater [191] and Joseph [125]. According to [190] Dirac maps provide a possible canonical quantization procedure for Classical systems (see also [125, p. 219]). Theorem 8.3 is due to Beidar and Chebotar. This result has not been published before.

In 1973, Nambu formulated a generalization of classical Hamiltonian mechanics [173]. Motivated by Nambu mechanics, Okubo [174, Chapter 8] considered a more general approach based on a covariant action of Lie algebras on vector spaces. Applications of FI's to such problems were obtained in [93, 127]. It is an interesting problem whether the skew-symmetry assumption on f in Theorem 8.6 can be removed.

Appendix A

Maximal Rings of Quotients

Every commutative domain can be embedded into a field, namely, into its field of fractions. A vast number of more general constructions are known in ring theory. Incidentally, not everything is so simple in the noncommutative context; for example, not every domain can be embedded into a division ring (see e.g., [133]). Thus, a simple minded attempt to take formal inverses of elements may not always work, and more sophisticated approaches are necessary.

It is not our intention to treat the general theory of rings of quotients. We shall confine ourselves to maximal rings of quotients, and even this only for the case when the original rings are semiprime. The main reason for considering these rings of quotients is that they are simply most suitable for our purposes (cf. Sections 5.2 and 5.3). In principle they do have one disadvantage, namely, as their name already suggests, they may be very "big", much bigger than the original rings, and so they do not always reflect well their structure. There are other well-known rings of quotients, which are smaller than the maximal ones (in the literature these rings are often called Martindale rings of quotients, while in [40] the terms symmetric and two-sided rings of quotients are used). However, dealing with any of them in the FI context would lead to serious technical problems. Versions of Corollary A.5 below do not hold for them, and this is basically what causes the main problem. Anyway, concerning concrete applications of FI's to prime (and semiprime) rings (at least those that are known so far), maximal rings of quotients are as good as any others would be. Namely, the unknown functions in FI's arriving from concrete problems as a rule turn out to be quasi-polynomials. But then only the center of the bigger ring matters. And it is a fact that all these rings of quotients have the same center, called the extended centroid of the original ring.

Maximal rings of quotients and extended centroids are studied in many books, for instance in [40, 133, 134, 197], to mention just a few. In our exposition we shall mostly follow [40]. We will present the results in a rigorous fashion, while their proofs will be mostly just outlined, pointing out the main ideas and neglecting technicalities.

Assume from now on that $\mathcal{A}$ *is a semiprime ring*. This assumption is not needed in everything that follows, but for simplicity we restrict out attention to the situation in which we are really interested. A left ideal $\mathcal{L}$ of $\mathcal{A}$ is said to be *dense* if given $a_1, a_2 \in \mathcal{A}$ with $a_1 \neq 0$, there exists $a \in \mathcal{A}$ such that $aa_1 \neq 0$ and $aa_2 \in \mathcal{L}$. If $\mathcal{I}$ is a two-sided ideal, then it is easy to see that $\mathcal{I}$ is dense (as a left ideal) if and only if $\mathcal{I}b \neq 0$ for every nonzero $b \in \mathcal{A}$, which is further equivalent to $b\mathcal{I} \neq 0$ for every nonzero $b \in \mathcal{A}$. Furthermore, such ideals are exactly the *essential* ideals, that is ideals having nonzero intersections with all nonzero ideals. If $\mathcal{A}$ is prime, then every nonzero ideal is essential.

Assume for a moment that $\mathcal{A}$ is a commutative domain and $\mathcal{Q}$ is its field of fractions. Pick $q \in \mathcal{Q}$. Then $q = ba^{-1}$ for some $a, b \in \mathcal{A}$ with $a \neq 0$. Let $\mathcal{L}$ be any nonzero ideal of $\mathcal{A}$ contained in $a\mathcal{A}$. Note that $f(x) = xq$ defines an $\mathcal{A}$-module homomorphism from $\mathcal{L}$ into $\mathcal{A}$. Conversely, every $\mathcal{A}$-module homomorphism g from a nonzero ideal $\mathcal{I}$ of $\mathcal{A}$ into $\mathcal{A}$ is of such a form. Indeed, for all $x, y \in \mathcal{I}$ we have $g(x)y = g(xy) = xg(y)$, and so fixing a nonzero y it follows that $g(x) = xr$ where $r = g(y)y^{-1} \in \mathcal{Q}$.

We now return to an arbitrary semiprime ring $\mathcal{A}$. The only aim of the previous paragraph was to help the reader to understand the ideas hidden behind the construction that follows. Let us now consider the set of all pairs $(f; \mathcal{L})$, where $\mathcal{L}$ is a dense left ideal and $f : \mathcal{L} \to \mathcal{A}$ is a left $\mathcal{A}$-module homomorphism. We define $(f; \mathcal{L}) \sim (g; \mathcal{M})$ if f and g coincide on some dense left ideal contained in $\mathcal{L} \cap \mathcal{M}$. It is easy to see that $\sim$ is an equivalence relation. By $[f; \mathcal{L}]$ we denote the equivalence class determined by $(f; \mathcal{L})$. We define the addition and multiplication of equivalence classes as follows:

$$[f; \mathcal{L}] + [g; \mathcal{M}] = [f + g; \mathcal{L} \cap \mathcal{M}],$$
$$[f; \mathcal{L}][g; \mathcal{M}] = [gf; f^{-1}(\mathcal{M})].$$

So basically the sum of equivalence classes corresponds to the sum of homomorphisms, and the product to their composition. One just has to take care about domains so that everything makes sense. Let us point out that $\mathcal{L} \cap \mathcal{M}$ and $f^{-1}(\mathcal{M})$ (the preimage of $\mathcal{M}$), are indeed dense left ideals, as can be easily checked. One can also check that both operations are well-defined, and that the set of all equivalence classes becomes a ring under these operations. All these require some work, but it is elementary and easy. One can embed $\mathcal{A}$ into this ring via $a \mapsto [R_a; \mathcal{A}]$ where R_a is the right multiplication by $a \in \mathcal{A}$, i.e., $R_a(x) = xa$. Identifying each a with $[R_a; \mathcal{A}]$ we thus have $a[f; \mathcal{L}] = f(a)$ for every $a \in \mathcal{L}$. Using this one can easily show that the ring that we constructed has the properties given in the next theorem.

Theorem A.1. *Let $\mathcal{A}$ be a semiprime ring. Then there exists a ring $\mathcal{Q}_{ml}(\mathcal{A})$ satisfying the following conditions:*

(i) *$\mathcal{A}$ is a subring of $\mathcal{Q}_{ml}(\mathcal{A})$;*

(ii) *For every $q \in \mathcal{Q}_{ml}(\mathcal{A})$ there exists a dense left ideal $\mathcal{L}$ of $\mathcal{A}$ such that $\mathcal{L}q \subseteq \mathcal{A}$;*

(iii) *If $0 \neq q \in \mathcal{Q}_{ml}(\mathcal{A})$, then $\mathcal{L}q \neq 0$ for every dense left ideal $\mathcal{L}$ of $\mathcal{A}$;*

(iv) *If $\mathcal{L}$ is a dense left ideal of $\mathcal{A}$ and $f : \mathcal{L} \to \mathcal{A}$ is a left $\mathcal{A}$-module homomorphism, then there exists $q \in \mathcal{Q}_{ml}(\mathcal{A})$ such that $f(x) = xq$ for all $x \in \mathcal{L}$.*

Moreover, the properties (i)–(iv) characterize $\mathcal{Q}_{ml}(\mathcal{A})$ up to an isomorphism.

The last assertion can also be easily established. Indeed, let $\mathcal{Q}$ be a ring satisfying (i)–(iv). Given $q \in \mathcal{Q}$, by assumption there exists a dense left ideal $\mathcal{L}$ such that $\mathcal{L}q \subseteq \mathcal{A}$. One can check that the map $q \mapsto [R_q; \mathcal{L}]$ is a ring isomorphism from $\mathcal{Q}$ onto $\mathcal{Q}_{ml}(\mathcal{A})$.

The ring $\mathcal{Q}_{ml}(\mathcal{A})$ is called the *maximal left ring of quotients* of $\mathcal{A}$. These rings first appeared in the work by Utumi [193], and in the literature they are sometimes also called Utumi left rings of quotients.

One can similarly introduce and study maximal *right* rings of quotients. We have chosen to deal with the left ones by chance. After all, results on FI's are in principle left-right symmetric.

Let us mention just a couple of concrete examples, in order to give some evidence that the concept of $\mathcal{Q}_{ml}(\mathcal{A})$ is a natural one. If $\mathcal{A}$ is a semiprime left Goldie ring, then $\mathcal{Q}_{ml}(\mathcal{A})$ is just the classical left ring of quotients of $\mathcal{A}$. So, for instance, $\mathcal{Q}_{ml}(M_n(\mathbb{Z})) = M_n(\mathbb{Q})$. Next, let $\mathcal{A}$ be a primitive ring containing an idempotent $e \in \mathcal{A}$ such that $\mathcal{D} = e\mathcal{A}e$ is a division ring (more details about such rings can be found at the end of this appendix and in appendix D). Then $\mathcal{Q}_{ml}(\mathcal{A}) = \text{End}_{\mathcal{D}}(e\mathcal{A})$. For more examples we refer to the aforementioned books; especially [133] has plenty of them.

As already mentioned, the intersection of two, and hence also of finitely many dense left ideals is again a dense left ideal. Therefore (ii) can be strengthened as follows.

Corollary A.2. *For any $q_1, \ldots, q_n \in \mathcal{Q}_{ml}(\mathcal{A})$ there exists a dense left ideal $\mathcal{L}$ of $\mathcal{A}$ such that $\mathcal{L}q_i \subseteq \mathcal{A}$ for every i.*

The next lemma is a very special case of the general theory (cf. [40, Section 6.4]). But as this lemma is all we need, we shall give a simple direct proof. Let us first mention that $\mathcal{Q}_{ml}(\mathcal{A})$ is again a semiprime ring, and moreover it is prime in case $\mathcal{A}$ is prime. This can be easily checked.

Lemma A.3. *Let $a, b \in \mathcal{Q} = \mathcal{Q}_{ml}(\mathcal{A})$, and let $\mathcal{I}$ be an essential ideal of $\mathcal{A}$. If $a\mathcal{I}b = 0$, then $a\mathcal{Q}b = b\mathcal{Q}a = 0$.*

Indeed, from $a\mathcal{I}b = 0$ it follows that $(\mathcal{I}b\mathcal{Q}a)^2 = 0$. Thus $\mathcal{J} = \mathcal{I}b\mathcal{Q}a \cap \mathcal{A}$ is a left ideal of $\mathcal{A}$ such that $\mathcal{J}^2 = 0$. Since $\mathcal{A}$ is semiprime, $\mathcal{J} = 0$. If $b\mathcal{Q}a \neq 0$ pick $q \in \mathcal{Q}$ such that $bqa \neq 0$. Then there exists $r \in \mathcal{A}$ such that $0 \neq rbqa \in \mathcal{A}$. Since $\mathcal{I}$ is essential in $\mathcal{A}$ we arrive at the contradiction $0 \neq \mathcal{I}rbqa \subseteq \mathcal{J}$. Thus $b\mathcal{Q}a = 0$. Accordingly, $(a\mathcal{Q}b)\mathcal{Q}(a\mathcal{Q}b) = 0$, forcing $a\mathcal{Q}b = 0$ since $\mathcal{Q}$ is semiprime.

Theorem A.4. *Let $\mathcal{I}$ be an essential ideal of $\mathcal{A}$ and let $\mathcal{B}$ be any ring such that $\mathcal{I} \subseteq \mathcal{B} \subseteq \mathcal{Q}_{ml}(\mathcal{A})$. Then $\mathcal{Q}_{ml}(\mathcal{B}) = \mathcal{Q}_{ml}(\mathcal{A})$.*

One should first note that $\mathcal{B}$ is semiprime. Now, to prove Theorem A.4 it is enough to show that $\mathcal{Q}_{ml}(\mathcal{A})$ satisfies the properties (i)–(iv) of Theorem A.1 (in which we take $\mathcal{B}$ to play the role of $\mathcal{A}$). Since $\mathcal{B}$ is a subring of $\mathcal{Q}_{ml}(\mathcal{A})$, we get (i) for free. Proving the other three properties is not so trivial, but still elementary. We omit details.

A particular case of Theorem A.4 is of special importance.

Corollary A.5. $\mathcal{Q}_{ml}(\mathcal{Q}_{ml}(\mathcal{A})) = \mathcal{Q}_{ml}(\mathcal{A})$.

The center of $\mathcal{Q}_{ml}(\mathcal{A})$ is called the *extended centroid* of $\mathcal{A}$. This term was introduced in the prime ring context by Martindale who also discovered the basic properties and the usefulness of the extended centroid in the study of Lie homomorphisms [151] and generalized polynomial identities [152]. Somewhat later Amitsur considered the extended centroid of semiprime rings [3].

The extended centroid of $\mathcal{A}$ will be denoted by $\mathcal{C}$. In terms of the construction of $\mathcal{Q}_{ml}(\mathcal{A})$ given earlier, it is easy to see that $\mathcal{C}$ is characterized as the set of all equivalence classes $\lambda = [f; \mathcal{L}]$ where $\mathcal{L}$ is an essential ideal of $\mathcal{A}$ and $f : \mathcal{L} \to \mathcal{A}$ is an $(\mathcal{A}, \mathcal{A})$-bimodule map (thus $\lambda \mathcal{L} \subseteq \mathcal{A}$). It can be shown that $\mathcal{C}$ is a von Neumann regular ring, i.e., for every $\lambda \in \mathcal{C}$ there exists $\mu \in \mathcal{C}$ such that $\lambda^2 \mu = \lambda$. The *centroid* Ω of $\mathcal{A}$ is the subring of $\mathcal{C}$ consisting of all equivalence classes of the form $[f; \mathcal{A}]$. The center $\mathcal{Z}$ of $\mathcal{A}$ is embeddable in Ω, so we have $\mathcal{Z} \subseteq \Omega \subseteq \mathcal{C}$. In case $\mathcal{A}$ is unital, $\mathcal{Z}$ is isomorphic to Ω (in which case there is no need for the notion of the centroid). One can check that the centralizer of $\mathcal{A}$ in $\mathcal{Q}$ is just $\mathcal{C}$; moreover, the same is true for the centralizer of every essential ideal of $\mathcal{A}$ in $\mathcal{Q}$.

The $\mathcal{C}$-subalgebra of $\mathcal{Q}_{ml}(\mathcal{A})$ generated by $\mathcal{A}$ is called the *central closure* of $\mathcal{A}$. It will be denoted by $\mathcal{AC}$. Thus a typical element in $\mathcal{AC}$ is of the form $\sum_i \lambda_i a_i$ with $\lambda_i \in \mathcal{C}$ and $a_i \in \mathcal{A}$. We say that $\mathcal{A}$ is a *centrally closed ring* if it is equal to its own central closure. A centrally closed ring is not necessarily unital. For a unital ring, saying that it is centrally closed is the same as saying that its extended centroid coincides with its center. A centrally closed ring is clearly an algebra over the extended centroid. By a *centrally closed algebra over* $\mathcal{C}$ we shall mean an algebra over a commutative ring $\mathcal{C}$ such that its extended centroid is $\mathcal{C}$.

It is not difficult to show that the central closure is a centrally closed semiprime (and prime if $\mathcal{A}$ is prime) ring. Simple rings are always centrally closed. So, a unital simple ring is a centrally closed algebra over its center. However, one usually refers to these algebras as *central simple algebras* (recall that an algebra over a commutative ring $\mathcal{C}$ is said to be central if $\mathcal{C}$ is its center).

FI's have turned out to be useful in solving some problems in algebras that appear in functional analysis. But we did not consider these topics, in order to avoid making the book too diverse. Let us now make a short digression. If one takes, for example, a semiprime Banach algebra, then its extended centroid of course exists, but it may not have any reasonable topological properties and so it is just a "creature from another planet", apparently useless for the category of Banach algebras. If, however, we restrict ourselves to some special classes of algebras, then this is no longer the case. Let us mention just two nice examples:

primitive (complex) Banach algebras and prime C^*-algebras are centrally closed algebras over $\mathbb{C}$. Therefore, all results that involve prime rings and their extended centroids are directly applicable to these algebras.

The extended centroid plays a particularly important role in prime rings. Here is one of the main reasons:

Theorem A.6. $\mathcal{C}$ *is a field if and only if* $\mathcal{A}$ *is prime.*

Let us sketch the proof. First suppose $\mathcal{A}$ is prime. Let $0 \neq \lambda \in \mathcal{C}$ and let $\mathcal{I}$ be a nonzero ideal of $\mathcal{A}$ such that $\lambda\mathcal{I} \subseteq \mathcal{A}$. Then the inverse of λ is determined by the map $f : \lambda\mathcal{I} \to \mathcal{A}$ given by $f(\lambda x) = x$ (well-definedness follows from λ not being a zero divisor). Conversely, suppose $\mathcal{A}$ is not prime, and accordingly let $\mathcal{I} \neq 0$ be a non-essential ideal of $\mathcal{A}$. Then $\mathcal{J} = \{x \in \mathcal{A} \,|\, x\mathcal{I} = 0\}$ is a nonzero ideal of $\mathcal{A}$ and $\mathcal{K} = \mathcal{I} \oplus \mathcal{J}$ is an essential ideal of $\mathcal{A}$. Define $f, g : \mathcal{K} \to \mathcal{A}$ respectively by $f(x + y) = x$ and $g(x + y) = y$. Then $[f; \mathcal{K}]$ and $[g; \mathcal{K}]$ are nonzero orthogonal idempotents in $\mathcal{C}$, whence $\mathcal{C}$ cannot be a field.

The centroid Ω of a prime ring $\mathcal{A}$ is a commutative unital domain containing the center $\mathcal{Z}$ (note that $\mathcal{Z}$ could well be 0). In general $\mathcal{C}$ need not be the field of fractions of Ω (or of $\mathcal{Z}$), even if Ω (or $\mathcal{Z}$) should be a field itself (cf. Examples 5.29 and 6.10). However, in some cases $\mathcal{C}$ is the field of fractions of $\mathcal{Z}$; e.g., if $\mathcal{A}$ is a prime PI-ring then $\mathcal{Z} \neq 0$ and $\mathcal{C}$ is the field of fractions of $\mathcal{Z}$.

Until further notice $\mathcal{A}$ will be a prime ring. Suppose that $0 \neq a, b \in \mathcal{Q}_{ml}(\mathcal{A})$ are such that $axb = bxa$ for all $x \in \mathcal{A}$. We claim that then a and b are linearly dependent over $\mathcal{C}$, i.e., $b = \lambda a$ for some $\lambda \in \mathcal{C}$. Indeed, pick a dense left ideal $\mathcal{L}$ of $\mathcal{A}$ such that $\mathcal{L}a \subseteq \mathcal{A}$. Then $\mathcal{I} = \mathcal{L}a\mathcal{A}$ is a nonzero (and hence automatically essential) ideal of $\mathcal{A}$. Define $f : \mathcal{I} \to \mathcal{A}$ by $f(\sum_i u_i a x_i) = \sum_i u_i b x_i$. To show that f is well-defined, assume that $\sum_i u_i a x_i = 0$. Then also $(\sum_i u_i a x_i)yb = 0$ for every $y \in \mathcal{A}$. However, according to our assumption we have $ax_i yb = bx_i ya$, and so it follows that $(\sum_i u_i b x_i)ya = 0$. Since $\mathcal{A}$ is prime and $a \neq 0$ this yields $\sum_i u_i b x_i = 0$, as desired. Since f is a left $\mathcal{A}$-module homomorphism we have that $f(y) = y\lambda$ for some $\lambda \in \mathcal{Q}_{ml}(\mathcal{A})$ and all $y \in \mathcal{I}$. But f is clearly also a right $\mathcal{A}$-module homomorphism, from which we easily infer that $\lambda \in \mathcal{C}$. Consequently, $b = \lambda a$.

What we just proved is a very special case of the following result.

Theorem A.7. *Let* $\mathcal{A}$ *be prime, and let* $a_i, b_i, c_j, d_j \in \mathcal{Q}_{ml}(\mathcal{A})$ *be such that*

$$\sum_{i=1}^{n} a_i x b_i = \sum_{j=1}^{m} c_j x d_j \quad \text{for all } x \in \mathcal{A}.$$

If $a_1, \ldots, a_n$ *are linearly independent over* $\mathcal{C}$, *then each* b_i *is a* $\mathcal{C}$-*linear combination of* $d_1, \ldots, d_m$. *Similarly, if* $b_1, \ldots, b_n$ *are linearly independent over* $\mathcal{C}$, *then each* a_i *is a* $\mathcal{C}$-*linear combination of* $c_1, \ldots, c_m$.

The proof of Theorem A.7 can be quite easily reduced to the $axb = bxa$ case that we have just settled. In the first step we reduce the problem to the case where each $c_j = 0$. This is easy. Assume the linear independence of the a_i's, and choose a basis of the linear span of all a_i's and c_j's that contains all a_i's. Then write each c_j as a linear combination of elements from this basis, which gives $\sum_{i=1}^{n} a_i x b_i' - \sum_{j=1}^{k} e_j x f_j = 0$ where the set $\{a_1, \ldots, a_n, e_1, \ldots, e_k\}$ is independent and each b_i' is the sum of b_i and a linear combination of the d_j's. This shows that indeed we may assume that each $c_j = 0$. Now our goal is to prove that every $b_i = 0$. Let $\mathcal{L}$ be a dense left ideal of $\mathcal{A}$ such that $\mathcal{L}b_n \subseteq \mathcal{A}$. For all $u \in \mathcal{L}$ and $y \in \mathcal{A}$ we have

$$\sum_{i=1}^{n-1} a_i u (b_i y b_n - b_n y b_i) = \left(\sum_{i=1}^{n} a_i u b_i y \right) b_n - \sum_{i=1}^{n} a_i (u b_n y) b_i = 0.$$

This makes it possible for one to use induction on n. We already know that $b_i y b_n - b_n y b_i \neq 0$ for some $y \in \mathcal{A}$, unless b_n and b_i are linearly dependent. The rest of the proof is easy.

To prove the second assertion, i.e., the one concerning the case where the b_i's are linearly independent, one can follow the same pattern, although some care is needed since the concept of $\mathcal{Q}_{ml}(\mathcal{A})$ is not left-right symmetric.

The next result is reminiscent of the density theorems.

Theorem A.8. *Let $\mathcal{A}$ be prime, and let $a_1, a_2, \ldots, a_n \in \mathcal{Q}_{ml}(\mathcal{A})$ be such that a_1 does not lie in the C-linear span of $a_2, \ldots, a_n$. Then there exists $\mathcal{E} \in \mathcal{M}(\mathcal{A})$, the multiplication ring of $\mathcal{A}$, such that*

$$\mathcal{E}(a_1) \neq 0 \quad and \quad \mathcal{E}(a_2) = \ldots = \mathcal{E}(a_n) = 0.$$

The proof given in [40] is based on the so-called weak density theorem, while the proof in the original paper [104] is more direct. We will give the proof only for the special case where $n = 2$ and $\mathcal{A}$ is unital, just to indicate why the result is not so surprising. The following simple argument is taken from [76] in which a generalization of Theorem A.8 is proved. Let $a_1, a_2 \in \mathcal{Q}_{ml}(\mathcal{A})$ be linearly independent. As shown above, there exists $x \in \mathcal{A}$ such that $a_1 x a_2 \neq a_2 x a_1$. Accordingly, $\mathcal{E} = {}_1 M_{x a_2} - {}_{a_2 x} M_1 \in \mathcal{M}(\mathcal{A})$ satisfies $\mathcal{E}(a_1) \neq 0$ and $\mathcal{E}(a_2) = 0$.

Our final result in this appendix is of great importance for the theory of (generalized) polynomial identities. It links the concept of the extended centroid with the structure theory of rings. Before stating it we first recall some elementary facts about minimal one-sided ideals.

A nonzero left (resp. right) ideal $\mathcal{I}$ of a ring $\mathcal{A}$ is said to be *minimal* if it does not properly contain a nonzero left (resp. right) ideal of $\mathcal{A}$. Minimal left and right ideals of semiprime rings are generated by idempotents. Indeed, let $\mathcal{I}$ be a minimal left ideal of a semiprime ring $\mathcal{A}$. Then $\mathcal{I}^2 \neq 0$. Picking $a \in \mathcal{I}$ such that $\mathcal{I}a \neq 0$ it follows that $\mathcal{I}a = \mathcal{I}$ by the minimality of $\mathcal{I}$. In particular, $ea = a$ for some $e \in \mathcal{I}$. The set $\mathcal{N} = \{x \in \mathcal{I} \mid xa = 0\}$ is a left ideal of $\mathcal{A}$ and a proper subset of $\mathcal{I}$ as $e \notin \mathcal{N}$. Therefore $\mathcal{N} = 0$. Noting that $e^2 - e \in \mathcal{N}$ it follows that e is an idempotent.

Again using the minimality of $\mathcal{I}$ we get that $\mathcal{I} = \mathcal{A}e$, i.e., $\mathcal{I}$ is generated by an idempotent. Of course, similarly we see that every minimal right ideal of $\mathcal{A}$ is of the form $f\mathcal{A}$ for some idempotent f. But actually the connection between minimal left and right ideals is even closer. For an idempotent e in a semiprime ring $\mathcal{A}$ the following three conditions are equivalent: (a) $\mathcal{A}e$ is a minimal left ideal, (b) $e\mathcal{A}$ is a minimal right ideal, and (c) $e\mathcal{A}e$ is a division ring. The proof is just an exercise. For example, let us show that (a) implies (c). If $\mathcal{A}e$ is a minimal left ideal and $b \in \mathcal{A}$ is such that $ebe \neq 0$, then we have $\mathcal{A}ebe = \mathcal{A}e$ by the minimality condition. Hence there is $c \in \mathcal{A}$ such that $cebe = e$, and hence $ecebe = e$. Thus every nonzero element in $e\mathcal{A}e$ has a left inverse, and so $e\mathcal{A}e$ is a division ring. An idempotent e in a semiprime ring $\mathcal{A}$ is called a *minimal idempotent* if it satisfies the (equivalent) conditions (a)–(c).

Theorem A.9. *Let $\mathcal{A}$ be a centrally closed prime ring. Suppose there exists a nonzero $\mathcal{E} \in \mathcal{M}(\mathcal{A})$ such that its range is finite dimensional over $\mathcal{C}$. Then $\mathcal{A}$ contains a minimal idempotent e such that $\dim_\mathcal{C} e\mathcal{A}e < \infty$.*

The first step of the proof is to show that there exist nonzero elements $b, c \in \mathcal{A}$ such that $\dim_\mathcal{C} b\mathcal{A}c < \infty$. Indeed, we may write $\mathcal{E}(x) = \sum_{i=1}^n a_i x b_i$, $n \geq 1$, with $a_1, \ldots, a_n$ $\mathcal{C}$-independent and $\mathcal{E}(x) \in \mathcal{V}$ where $\mathcal{V}$ is finite dimensional over $\mathcal{C}$. By Theorem A.8 there exists $\mathcal{F} \in \mathcal{M}(\mathcal{A})$, with $\mathcal{F}(x) = \sum_{j=1}^m s_j x t_j$, such that $\mathcal{F}(a_1) \neq 0$ and $\mathcal{F}(a_i) = 0$, $i \geq 2$. Therefore $\mathcal{F}(a_1) x b_1 = \sum_{i=1}^n \sum_{j=1}^m s_j a_i t_j x b_i \in \sum_{j=1}^m s_j \mathcal{V}$, noting that $\sum_{j=1}^m s_j \mathcal{V}$ is finite dimensional over $\mathcal{C}$. So we may take $b = \mathcal{F}(a_1)$ and $c = b_1$. Thus $\mathcal{A}$ contains nonzero left ideals $\mathcal{L}$ and right ideals $\mathcal{R}$ such that $\dim_\mathcal{C} \mathcal{R}\mathcal{L} \leq \infty$. Pick a left ideal $\mathcal{L}_0$ and a right ideal $\mathcal{R}_0$ such that $\mathcal{R}_0\mathcal{L}_0$ has minimal (nonzero) dimension. Set $\mathcal{I} = \mathcal{A}\mathcal{R}_0\mathcal{L}_0$. Suppose that $\mathcal{I}'$ is a left ideal of $\mathcal{A}$ such that $0 \neq \mathcal{I}' \subseteq \mathcal{I}$. Then $\mathcal{I}' \subseteq \mathcal{L}_0$ and hence $\mathcal{R}_0\mathcal{I}' \subseteq \mathcal{R}_0\mathcal{L}_0$, which forces $\mathcal{R}_0\mathcal{I}' = \mathcal{R}_0\mathcal{L}_0$. Consequently, $\mathcal{I}' \supseteq \mathcal{A}\mathcal{R}_0\mathcal{I}' = \mathcal{A}\mathcal{R}_0\mathcal{L}_0 = \mathcal{I}$, so that $\mathcal{I}' = \mathcal{I}$. Thus $\mathcal{I}$ is a minimal left ideal of $\mathcal{A}$, and so there exists a minimal idempotent $e \in \mathcal{A}$ such that $\mathcal{I} = \mathcal{A}e$. Since $e \in \mathcal{I} \subseteq \mathcal{A}\mathcal{R}_0$ and $\mathcal{A}e = \mathcal{I} \subseteq \mathcal{L}_0$ it follows that $e\mathcal{A}e = e \cdot \mathcal{A}e$ is finite dimensional.

Appendix B

The Orthogonal Completion

The theory of orthogonal completions was created by Beidar and Mikhalev in a series of papers [15, 41, 42, 43, 169]. An account of it is given in the book [40]. The material in this appendix is drawn from various parts of [40, Chapters 2 and 3] and is designed to provide the necessary background material for proving d-freeness of semiprime rings (under appropriate conditions) in Section 5.3. We shall do this in a self-contained manner, in particular avoiding using the tools of mathematical logic. Unlike in the other three appendices, in this one we shall give complete proofs.

Throughout this appendix, $\mathcal{A}$ will be a semiprime ring with extended centroid $\mathcal{C}$ and maximal left quotient ring $\mathcal{Q} = \mathcal{Q}_{ml}(\mathcal{A})$. For sets $\mathcal{S}, \mathcal{T} \subseteq \mathcal{Q}$ we let $\ell(\mathcal{T}; \mathcal{S})$ denote the left annihilator of $\mathcal{S}$ in $\mathcal{T}$. The set $\mathcal{B}$ of idempotents in $\mathcal{C}$ will play a key role in the theory we outline in this appendix. Its importance is immediately recognized in view of the following lemma.

Lemma B.1. *For every subset $\mathcal{S} \subseteq \mathcal{Q}$ there exists a unique element $E(\mathcal{S}) \in \mathcal{B}$ such that $\ell(\mathcal{Q}; \mathcal{Q}\mathcal{S}) = (1 - E(\mathcal{S}))\mathcal{Q}$ (and hence $E(\mathcal{S})t = t$ for all $t \in \mathcal{S}$). Further, for every $e \in \mathcal{B}$ we have $E(e\mathcal{S}) = eE(\mathcal{S})$.*

Proof. Let $\mathcal{I}$ be the ideal of $\mathcal{Q}$ generated by $\mathcal{S}$ and let $\mathcal{J} = \ell(\mathcal{Q}; \mathcal{I}) = \ell(\mathcal{Q}; \mathcal{Q}\mathcal{S})$. It is easy to see that $\mathcal{I} \oplus \mathcal{J}$ is an essential ideal of $\mathcal{Q}$. We define a map from $\mathcal{I} \oplus \mathcal{J}$ into $\mathcal{Q}$ via $x + y \mapsto x$ for $x \in \mathcal{I}$ and $y \in \mathcal{J}$. Note that, in view of Theorem A.1 (iv), this map determines an element $f = E(\mathcal{S})$ in $\mathcal{Q}_{ml}(\mathcal{Q}) = \mathcal{Q}$ (see Corollary A.5). One can check that $f^2 = f$ and that f commutes with every element in $\mathcal{Q}$, that is to say, $f \in \mathcal{B}$. Furthermore, $fx = x$ for $x \in \mathcal{I}$ and $fy = 0$ for $y \in \mathcal{J}$. It is then easily seen that $\ell(\mathcal{Q}; \mathcal{Q}\mathcal{S}) = \mathcal{J} = (1 - f)\mathcal{Q}$. Of course f is uniquely determined by this property.

Pick $e \in \mathcal{B}$, and let $q \in \ell(\mathcal{Q}; \mathcal{Q}e\mathcal{S}) = (1 - E(e\mathcal{S}))\mathcal{Q}$. Then $eq \in \ell(\mathcal{Q}; \mathcal{Q}\mathcal{S})$, and so $q = eq + (1 - e)q \in (1 - E(\mathcal{S}))\mathcal{Q} + (1 - e)\mathcal{Q} \subseteq \ell(\mathcal{Q}; \mathcal{Q}e\mathcal{S})$. Thus $\ell(\mathcal{Q}; \mathcal{Q}e\mathcal{S}) = (1 - E(\mathcal{S}))\mathcal{Q} + (1 - e)\mathcal{Q}$. Using the fact that for any $e_1, e_2 \in \mathcal{B}$ we have $e_1\mathcal{Q} + e_2\mathcal{Q} = (e_1 - e_1e_2)\mathcal{Q} \oplus e_2\mathcal{Q} = (e_1 + e_2 - e_1e_2)\mathcal{Q}$, we see that

$\ell(Q; QeS) = (1 - E(S) + 1 - e - (1 - E(S))(1 - e))Q = (1 - eE(S))Q$. Thus $(1 - E(eS))Q = (1 - eE(S))Q$, and so $E(eS) = eE(S)$. □

We shall write $E(s)$ for $E(\{s\})$. Further, for $e \in B$ we set $\mathcal{L}_e = \{x \in \mathcal{A} \mid ex \in \mathcal{A}\}$.

Lemma B.2. $\mathcal{L}_e$ *is an essential ideal of* $\mathcal{A}$ *for every* $e \in B$. *Moreover,* $\mathcal{L}_e e$ *is an ideal of* $\mathcal{A}$ *and* $Q_{ml}(\mathcal{L}_e e) = Qe$.

Proof. Clearly $\mathcal{L}_e$ is an ideal of $\mathcal{A}$. According to Theorem A.1 (ii) it contains a dense left ideal of $\mathcal{A}$, which implies that $\mathcal{L}_e$ is an essential ideal. It is also clear that $\mathcal{L}_e e$ is an ideal of $\mathcal{A}$. Using Theorem A.4 we have

$$Q = Q_{ml}(\mathcal{L}_e) = Q_{ml}(\mathcal{L}_e e \oplus \mathcal{L}_e(1 - e)) = Q_{ml}(\mathcal{L}_e e) \oplus Q_{ml}(\mathcal{L}_e(1 - e)),$$

from which $Q_{ml}(\mathcal{L}_e e) = Qe$ easily follows. □

A subset $\mathcal{U} \subseteq B$ is said to be *dense* if $\ell(Q; \mathcal{U}) = 0$, i.e., $E(\mathcal{U}) = 1$, and $\mathcal{U}$ is *orthogonal* if $uv = 0$ for all $u, v \in \mathcal{U}$ with $u \ne v$. For future reference we record two simple observations concerned with such subsets. The first one is immediate.

Lemma B.3. *If* $\mathcal{U}$ *and* $\mathcal{V}$ *are dense orthogonal subsets of* B, *then* $\mathcal{U}\mathcal{V} = \{uv \mid u \in \mathcal{U}, v \in \mathcal{V}\}$ *is also a dense orthogonal subset of* B.

Lemma B.4. *If* $\mathcal{U}$ *is a dense subset of* B, *then* $\mathcal{I} = \sum_{u \in \mathcal{U}} \mathcal{L}_u u$ *is an essential ideal of* $\mathcal{A}$.

Proof. Lemma B.2 implies that $\mathcal{I}$ is an ideal of $\mathcal{A}$. Let $b \in \mathcal{A}$ be such that $b\mathcal{I} = 0$. Then $(bu)\mathcal{L}_u = b\mathcal{L}_u u = 0$ for every $u \in \mathcal{U}$. Since $\mathcal{L}_u$ is an essential ideal of $\mathcal{A}$ by Lemma B.2, it is easy to see (e.g., by using Theorem A.1) that $\ell(Q; \mathcal{L}_u) = 0$. Therefore $bu = 0$ for every $u \in \mathcal{U}$, and hence $b = 0$ since $\mathcal{U}$ is dense. Thus $\mathcal{I}$ is essential. □

We now make the key definition: A subset $\mathcal{T} \subseteq Q$ is said to be *orthogonally complete* if for any orthogonal dense subset $\mathcal{U} \subseteq B$ and any elements $t_u \in \mathcal{T}$, $u \in \mathcal{U}$, there exists $t \in \mathcal{T}$ such that $tu = t_u u$ for all $u \in \mathcal{U}$. We denote this element t, which is clearly unique, by the suggestive notation

$$t = \sum_{u \in \mathcal{U}}^{\perp} t_u u.$$

To show this is not just an empty concept we have the following

Lemma B.5. Q *is orthogonally complete.*

Proof. Let $\mathcal{U}$ be a dense orthogonal subset of B and let $\{q_u \mid u \in \mathcal{U}\} \subseteq Q$. Note that $\mathcal{D} = \sum_{u \in \mathcal{U}} Qu$ is an essential ideal of Q. We define $f : \mathcal{D} \to Q$ according to $f(\sum x_u u) = \sum x_u q_u u$. We note that f is a well-defined left Q-module homomorphism, and so there exists $q \in Q(= Q_{ml}(Q))$ such that, in particular, for each $u \in \mathcal{U}$ we have $qu = f(u) = q_u u$. □

Lemma B.5 shows that $\sum_{u \in \mathcal{U}}^{\perp} q_u u$ always exists in $\mathcal{Q}$ for any choice of an orthogonal dense subset $\mathcal{U} \subseteq \mathcal{B}$ and any elements $q_u \in \mathcal{Q}$. The above definition can now be rephrased as follows: $\mathcal{T}$ is orthogonally complete if $\sum_{u \in \mathcal{U}}^{\perp} t_u u$ lies in $\mathcal{T}$ whenever every $t_u \in \mathcal{T}$.

We can now define the *orthogonal completion* $\mathcal{O}(\mathcal{T})$ of any subset $\mathcal{T} \subseteq \mathcal{Q}$ to be the intersection of all orthogonally complete subsets containing $\mathcal{T}$. It is straightforward to show that $\mathcal{O}(\mathcal{T})$ is in fact orthogonally complete. This, of course, is not very enlightening as to the nature of $\mathcal{O}(\mathcal{T})$, but fortunately one has the much more tangible characterization of $\mathcal{O}(\mathcal{T})$ given by the following

Lemma B.6. *Let $\mathcal{T} \subseteq \mathcal{Q}$. Then $\mathcal{O}(\mathcal{T})$ consists of all elements of the form $\sum_{u \in \mathcal{U}}^{\perp} t_u u$ where $\mathcal{U}$ is a dense orthogonal subset of $\mathcal{B}$ and $t_u \in \mathcal{T}$ for every $u \in \mathcal{U}$.*

Proof. Let $\mathcal{H}$ denote the set of all elements of the form $\sum_{u \in \mathcal{U}}^{\perp} t_u u$ where $\mathcal{U}$ is a dense orthogonal subset of $\mathcal{B}$ and $t_u \in \mathcal{T}$. Clearly $\mathcal{H} \subseteq \mathcal{O}(\mathcal{T})$; our task is to show that $\mathcal{H}$ itself is orthogonally complete. To this end we let W be a dense orthogonal subset of $\mathcal{B}$ and for each $w \in W$ let $h_w \in \mathcal{H}$. By Lemma B.5 we know that $q = \sum_{w \in W}^{\perp} h_w w$ exists in $\mathcal{Q}$. We have to show that $q \in \mathcal{H}$. Each h_w can be written as $h_w = \sum_{u_w \in \mathcal{U}_w}^{\perp} t_{u_w} u_w$ where $t_{u_w} \in \mathcal{T}$ and $\mathcal{U}_w$ is a dense orthogonal subset of $\mathcal{B}$. Now $\mathcal{V} = \{w u_w \mid w \in W, u_w \in \mathcal{U}_w\}$ is easily seen to be a dense orthogonal subset of $\mathcal{B}$. For $v = w u_w \in \mathcal{V}$ we define $t_v = t_{u_w}$, and set $p = \sum_{v \in \mathcal{V}}^{\perp} t_v v \in \mathcal{H}$. For $v = w u_w \in \mathcal{V}$ we shall show that $pv = qv$. Indeed, $pv = t_v v = t_{u_w} w u_w$, and on the other hand $qv = q w u_w = h_w w u_w = h_w u_w w = t_{u_w} u_w w = t_{u_w} w u_w$. Since $\mathcal{V}$ is dense it follows that $q = p \in \mathcal{H}$. $\square$

We will need the following facts about orthogonally complete subsets.

Lemma B.7. *Let $\mathcal{T}$ be an orthogonally complete set such that $0 \in \mathcal{T}$. Then:*

(i) *$e\mathcal{T} \subseteq \mathcal{T}$ for all $e \in \mathcal{B}$.*

(ii) *There exists $t \in \mathcal{T}$ such that $E(t) = E(\mathcal{T})$.*

Proof. (i) Clearly $\{e, 1 - e\}$ is a dense orthogonal subset of $\mathcal{B}$. Let $t \in \mathcal{T}$, and set $t_e = t$, $t_{1-e} = 0$. Since $\mathcal{T}$ is orthogonally complete there exists $s \in \mathcal{T}$ such that $se = te$ and $s(1 - e) = 0$. Therefore $te = se = s \in \mathcal{T}$.

(ii) Let $W = \{E(t) \mid t \in \mathcal{T}\}$. For $t \in \mathcal{T}$ and $e \in \mathcal{B}$ we know from (i) that $te \in \mathcal{T}$. Therefore by Lemma B.1 we have $eE(t) = E(et) \in W$, i.e., $eW \subseteq W$. By Zorn's Lemma there exists a maximal orthogonal subset $\mathcal{V} \subseteq W$. We set $\mathcal{U} = \mathcal{V} \cup \{1 - E(W)\}$. Clearly $\mathcal{U}$ is an orthogonal subset of $\mathcal{B}$. Suppose $E(\mathcal{U}) \neq 1$. Then $e = 1 - E(\mathcal{U}) \neq 0$ in particular satisfies $e(1 - E(W)) = 0$ and therefore $ew \neq 0$ for some $w \in W$. But $ew \in W$ by what we proved, and so we have that $\mathcal{V} \cup \{ew\}$ is an orthogonal subset of $\mathcal{V}$, in contradiction to the maximality of $\mathcal{V}$. Thus $\mathcal{U}$ is a dense orthogonal subset of $\mathcal{B}$. By definition of W for each $v \in \mathcal{V}$ there exists $t_v \in \mathcal{T}$ such that $E(t_v) = v$. We set $t_{1-E(W)} = 0$. Since $\mathcal{T}$ is orthogonally complete, $t = \sum_{u \in \mathcal{U}}^{\perp} t_u u$ belongs to $\mathcal{T}$. We claim that $E(\mathcal{T}) = E(t)$. Since $t \in \mathcal{T}$ it follows easily that $E(t) = E(t)E(\mathcal{T})$. If $E(\mathcal{T}) \neq E(t)$, then $e = E(\mathcal{T}) - E(t) \neq 0$,

$e \in \mathcal{B}$, $et = 0$ but $e\mathcal{T} \neq 0$. From $et = 0$ we conclude that $0 = E(etv) = E(et_v v)$, which by Lemma B.1 yields $0 = evE(t_v) = ev^2 = ev$ for each $v \in \mathcal{V}$, i.e., $e\mathcal{V} = 0$. On the other hand there exists $x \in \mathcal{T}$ such that $ex \neq 0$, whence $E(ex) \neq 0$. But $E(ex) \in \mathcal{W}$ since $ex \in \mathcal{T}$, and from $E(ex) = eE(x)$ we obtain $E(ex)\mathcal{V} = 0$. This is a contradiction to the maximality of $\mathcal{V}$. □

Given dense orthogonal subsets $\mathcal{U}$ and $\mathcal{V}$ of $\mathcal{B}$, and elements $x = \sum_{u \in \mathcal{U}}^{\perp} x_u u$ and $y = \sum_{v \in \mathcal{V}}^{\perp} y_v v$, it is an easy exercise to show that

$$x \pm y = \sum_{uv \in \mathcal{U}\mathcal{V}}^{\perp} (x_u \pm y_v)uv, \quad xy = \sum_{uv \in \mathcal{U}\mathcal{V}}^{\perp} x_u y_v uv.$$

From these and Lemma B.6 we see that the orthogonal completion of a subring of $\mathcal{Q}$ is again a subring of $\mathcal{Q}$. The ring we are especially interested in is

$$\mathcal{O} = \mathcal{O}(\mathcal{A}),$$

the orthogonal completion of $\mathcal{A}$, so we now direct our attention to $\mathcal{O}$. First we mention an illustrative example. Let $\{\mathcal{A}_i \mid i \in I\}$ be a family of prime rings and let $\mathcal{A} = \oplus_{i \in I} \mathcal{A}_i$ be their direct sum. One can check that in this case $\mathcal{Q} = \prod_{i \in I} \mathcal{Q}_i$, where $\mathcal{Q}_i = \mathcal{Q}_{ml}(\mathcal{A}_i)$, and $\mathcal{O} = \prod_{i \in I} \mathcal{A}_i$.

Now we explore $\mathcal{B}$ in more detail. First, we note that $\mathcal{B}$ becomes a Boolean ring under a new addition $e \oplus f = e + f - 2ef$ but with the same multiplication. Further, $\mathcal{B}$ becomes a partially ordered set by defining $e \leq f$ if $e = ef$. We let $Spec(\mathcal{B})$ denote the collection of maximal ideals of the Boolean ring $\mathcal{B}$. We note that an ideal $\mathcal{M}$ of $\mathcal{B}$ is maximal if and only if for all $e \in \mathcal{B}$ either $e \in \mathcal{M}$ or $1 - e \in \mathcal{M}$ but not both. Corresponding to $\mathcal{M} \in Spec(\mathcal{B})$ is the ideal of $\mathcal{O}$

$$\mathcal{O}\mathcal{M} = \{\sum_i s_i e_i \mid s_i \in \mathcal{O} \text{ and } e_i \in \mathcal{M}\}.$$

An important observation for us is the following

Lemma B.8. *Let $a \in \mathcal{O}$ and let $\mathcal{M} \in Spec(\mathcal{B})$. Then $a \in \mathcal{O}\mathcal{M}$ if and only if $E(a) \in \mathcal{M}$.*

Proof. If $E(a) \in \mathcal{M}$, then $a = aE(a) \in \mathcal{O}\mathcal{M}$. Conversely, suppose $a = \sum_{i=1}^{n} s_i e_i \in \mathcal{O}\mathcal{M}$. For each i we have $1 - e_i \notin \mathcal{M}$ and so $e = \prod_{i=1}^{n}(1 - e_i) \notin \mathcal{M}$. But $ae = 0$, whence $0 = E(ae) = eE(a)$. Since $e \notin \mathcal{M}$ it follows that $E(a) \in \mathcal{M}$. □

One of the key results of this theory is

Theorem B.9. *For $\mathcal{M} \in Spec(\mathcal{B})$, $\mathcal{O}\mathcal{M}$ is a prime ideal of $\mathcal{O}$, i.e., $\mathcal{O}_{\mathcal{M}} = \mathcal{O}/\mathcal{O}\mathcal{M}$ is a prime ring.*

Proof. Suppose $a\mathcal{O}b \subseteq \mathcal{O}\mathcal{M}$ for some $a, b \in \mathcal{O}$, with $b \notin \mathcal{O}\mathcal{M}$. It is easy to see that $a\mathcal{O}b$ is orthogonally complete. By Lemma B.7 (ii) $E(a\mathcal{O}b) = E(t)$ for some $t \in a\mathcal{O}b$.

Consequently, since $E(t) \in \mathcal{M}$ in view of Lemma B.8, $e = 1 - E(a\mathcal{O}b) \notin \mathcal{M}$. From Lemma B.8 we also see that $E(b) \notin \mathcal{M}$. We have $ea\mathcal{O}b = 0$, and hence $a\mathcal{Q}(eb) = 0$ by Lemma A.3. Thus $a \in (1 - E(eb))\mathcal{Q}$, so that $aE(eb) = 0$. Lemma B.1 now shows that $aeE(b) = 0$, and note that this yields $E(a)eE(b) = 0$. Since $eE(b) \notin \mathcal{M}$ this forces $E(a) \in \mathcal{M}$, and so, by Lemma B.8, $a \in \mathcal{O}\mathcal{M}$. $\qquad\square$

The following lemma will prove useful in that it converts a seemingly "infinite" situation into a "finite" one.

Lemma B.10. *For every $\mathcal{M} \in Spec(\mathcal{B})$ let $w_\mathcal{M} \in \mathcal{B} \setminus \mathcal{M}$. Then there exist $\mathcal{M}_1, \mathcal{M}_2,$ $\ldots, \mathcal{M}_q \in Spec(\mathcal{B})$ and orthogonal idempotents $e_1, e_2, \ldots, e_q \in \mathcal{B}$ whose sum is 1 such that $e_p \leq w_{\mathcal{M}_p}$ for $p = 1, 2, \ldots, q$.*

Proof. Let $\mathcal{W}$ be the ideal of the Boolean ring $\mathcal{B}$ generated by all $w_\mathcal{M}$. If $\mathcal{W} \neq \mathcal{B}$ then $\mathcal{W} \subseteq \mathcal{M}$ for some $\mathcal{M} \in Spec(\mathcal{B})$, whence the contradiction that $w_\mathcal{M} \notin \mathcal{M}$. Thus $\mathcal{W} = \mathcal{B}$ and in particular $1 = w_1 b_1 \oplus w_2 b_2 \oplus \ldots \oplus w_q b_q$ (Boolean sum) for some $w_p = w_{\mathcal{M}_p}$ and $b_p \in \mathcal{B}$. From the definition of the Boolean operations it is easy to see that $\mathcal{B} = w_1\mathcal{B} + w_2\mathcal{B} + \ldots + w_q\mathcal{B}$ (usual sum). Set $e_1 = w_1$ and $e_2 = w_2 - w_1 w_2$. Clearly $e_1\mathcal{B} + e_2\mathcal{B} = w_1\mathcal{B} + w_2\mathcal{B}$, with $e_1 e_2 = 0$ and $e_2 \leq w_2$. This is just the first step in the well-known process of replacing idempotents by orthogonal ones, and so we eventually have that $\mathcal{B} = e_1\mathcal{B} + e_2\mathcal{B} + \ldots + e_q\mathcal{B}$ with the e_i's orthogonal and $e_p \leq w_p$. From this it follows that $1 = e_1 + e_2 + \ldots + e_q$ and the lemma is proved. $\qquad\square$

Appendix C

Polynomial Identities

The theory of rings with polynomial identities is well documented in several mono-graphs, for instance in [120], [184] and [187]. We shall survey those elements of the theory that are important for understanding functional identities. We will omit rigorous proofs, but rather try to give some informal evidence for the truthfulness of the results that will be stated.

Let $X = \{x_1, x_2, \ldots\}$ be a countable set, and let $\mathbb{Z}\langle X \rangle$ be the free algebra on X over $\mathbb{Z}$. Let $f = f(x_1, \ldots, x_n) \in \mathbb{Z}\langle X \rangle$ be a polynomial such that at least one of its monomials of highest degree has coefficient 1. Let $\mathcal{R}$ be a nonempty subset of a ring $\mathcal{A}$. We say that f is a *polynomial identity* on $\mathcal{R}$ if $f(r_1, \ldots, r_n) = 0$ for all $r_1, \ldots, r_n \in \mathcal{R}$, that is, if the polynomial function determined by f vanishes on $\mathcal{R}^n$. In this case we also say that $\mathcal{R}$ satisfies the polynomial identity f. In what follows we will only consider the case where $\mathcal{R} = \mathcal{A}$, i.e., we will treat rings satisfying polynomial identities. Such rings are called *PI-rings*.

The simplest examples of PI-rings are commutative rings. Indeed, saying that a ring $\mathcal{A}$ is commutative is the same as saying that $\mathcal{A}$ satisfies the polynomial identity $x_1 x_2 - x_2 x_1$. Similarly, a ring $\mathcal{A}$ is Boolean if and only if it satisfies $x_1^2 - x_1$, and $\mathcal{A}$ is a nilpotent ring if and only if it satisfies $x_1 x_2 \ldots x_n$ for some positive integer n.

As we shall see, PI-rings are rather special. Incidentally, the polynomial func-tion determined by the polynomial $p x_1$ vanishes on every ring with characteristic p, but this does not mean that rings of finite characteristic are necessarily PI-rings. Note that we have required that one of the monomials of highest degree in a polynomial identity should have coefficient 1. We remark that in general PI theory takes place in the framework of algebras $\mathcal{A}$ over a commutative domain $\mathcal{C}$ (e.g., a field), but in this book we only have need of the theory when $\mathcal{C} = \mathbb{Z}$, i.e., $\mathcal{A}$ is just a ring.

A polynomial $f = f(x_1, \ldots, x_n) \in \mathbb{Z}\langle X \rangle$ is said to be multilinear if every x_i, $1 \leq i \leq n$, appears exactly once in each of the monomials of f. Thus f is of the form

$$f(x_1, \ldots, x_n) = \sum_{\pi \in S_n} n_\pi x_{\pi(1)} x_{\pi(2)} \cdots x_{\pi(n)},$$

where S_n is the symmetric group of order n and n_π are integers. If $\mathcal{A}$ satisfies a polynomial identity of degree n, then it also satisfies a multilinear polynomial identity of degree $\leq n$. One can show this by using the standard linearization procedure. A trivial example: the polynomial identity $x_1^2 - x_1$, through which Boolean rings are defined, leads to the polynomial identity $x_1x_2 + x_2x_1$. This suggests that sometimes one can lose some important information when reducing general identities to multilinear ones. But for our purposes such a loss is of no significance. We are interested only in structural properties of a ring that satisfies a polynomial identity of a certain degree, and so we may immediately assume the multilinearity of this identity.

A polynomial of extreme importance in PI theory is

$$St_d = St_d(x_1, \ldots, x_d) = \sum_{\pi \in S_d} (-1)^\pi x_{\pi(1)} x_{\pi(2)} \cdots x_{\pi(d)},$$

which we call the *standard polynomial* of degree d. Here, $(-1)^\pi$ denotes the sign of the permutation π. For example, $St_2 = x_1x_2 - x_2x_1$. It is easy to check that

$$St_d(x_1, \ldots, x_d) = \sum_{i=1}^d (-1)^{i+1} x_i St_{d-1}(x_1, \ldots, x_{i-1}, x_{i+1}, \ldots, x_d).$$

Therefore, if a ring $\mathcal{A}$ satisfies St_d, then it satisfies St_m for every $m \geq d$. Another useful property of St_d is that it vanishes if any two of its arguments are equal, i.e.,

$$St_d(x_1, \ldots x_i, \ldots, x_i, \ldots, x_d) = 0.$$

This has an important consequence: every n-dimensional algebra $\mathcal{A}$ over a field $\mathcal{K}$ satisfies St_{n+1}. So, for example, $M_n(\mathcal{K})$ satisfies St_{n^2+1}. But in fact a much sharper result is true: $M_n(\mathcal{K})$, where $\mathcal{K}$ can be any commutative ring, satisfies St_{2n}. This is the celebrated Amitsur–Levitzki theorem. Various proofs are known, some of them short, but all nontrivial. The following fact gives another light to the meaning of the Amitsur–Levitzki theorem: $M_n(\mathcal{K})$ does not satisfy a polynomial identity of degree $< 2n$. This is easy to prove. Just consider the matrix units $e_{11}, e_{12}, e_{22}, e_{23}, \ldots, e_{n-1\,n}, e_{nn}$; there are $2n - 1$ of them, their product in the given order is e_{1n}, and their product in any other order is 0. Therefore, if f is a multilinear polynomial of degree $2n - 1$ such that its coefficient at $x_1 x_2 \ldots x_{2n-1}$ is 1, then $f(e_{11}, e_{12}, e_{22}, \ldots, e_{nn}) = e_{1n} \neq 0$.

We shall now consider prime PI-rings and begin with

Theorem C.1. *Let $\mathcal{A}$ be a prime ring. Then $\mathcal{A}$ is a PI-ring if and only if its central closure $\mathcal{A}\mathcal{C}$ is a finite dimensional central simple algebra over the extended centroid $\mathcal{C}$ of $\mathcal{A}$.*

This is a partial statement of Posner's theorem [183], one of the cornerstones of PI theory. Actually much more can be said: the center $\mathcal{Z}$ of $\mathcal{A}$ is nonzero and has $\mathcal{C}$ as its field of fractions. Consequently, every element in $\mathcal{AC}$ is of the form $\frac{a}{\lambda}$ where $a \in \mathcal{A}$ and $\lambda \in \mathcal{Z}$. This is a highly nontrivial result whose proof is based on the existence of the so-called central polynomials in matrix algebras.

Let us recall that by the classical Wedderburn theorem, a finite dimensional central simple algebra is up to an isomorphism the same as $M_n(D)$ where D is a finite dimensional division algebra. This is also apparent from the proof of Theorem C.1 which we now sketch.

The "if" part is obvious. Namely, finite dimensional algebras are PI-rings, and subrings of PI-rings are trivially also PI-rings. To prove the converse, assume that $\mathcal{A}$ is a PI-ring. Since $\mathcal{A}$ and $\mathcal{AC}$ clearly satisfy the same multilinear polynomial identities, we may assume without loss of generality that $\mathcal{A}$ is centrally closed. We now invoke a result from the next appendix, namely, Theorem D.1, that considers a more general situation when $\mathcal{A}$ satisfies a generalized polynomial identity, and an outline of whose proof is given. One thereby concludes that $\mathcal{A}$ contains an idempotent e such that $\mathcal{A}e$ is a minimal left ideal of $\mathcal{A}$ and $e\mathcal{A}e$ is a division ring with $\dim_{\mathcal{C}} e\mathcal{A}e < \infty$. We may regard $\mathcal{A}e$ as a faithful simple left $\mathcal{A}$-module (thus $\mathcal{A}$ is a primitive ring). It is easy to see that $\mathrm{End}_{\mathcal{A}}(\mathcal{A}e)$ is antiisomorphic to $e\mathcal{A}e$. According to the well-known corollary to Jacobson's density theorem we have two possibilities: either $\mathcal{A}$ is isomorphic to $M_n(e\mathcal{A}e)$ for some positive integer n or for every $m \in \mathbb{N}$ there exist a subring $\mathcal{A}_m$ of $\mathcal{A}$ and an ideal $\mathcal{I}_m$ of $\mathcal{A}_m$ such that $\mathcal{A}_m/\mathcal{I}_m \cong M_m(e\mathcal{A}e)$. Clearly, if f is a polynomial identity of $\mathcal{A}$, then it is also a polynomial identity of $\mathcal{A}_m/\mathcal{I}_m$. Therefore, in the latter case f would be a polynomial identity of $M_m(e\mathcal{A}e)$ for every $m \in \mathbb{N}$, and hence also of its subring $M_m(\mathcal{C}e) \cong M_m(\mathcal{C})$. However, as noticed above, $M_m(\mathcal{C})$ does not satisfy polynomial identities of degree $2m - 1$, so no polynomial exists that would be a polynomial identity of $M_m(\mathcal{C})$ for every m. Therefore the first possibility occurs, i.e., $\mathcal{A} \cong M_n(e\mathcal{A}e)$ for some $n \in \mathbb{N}$, and hence $\mathcal{A}$ is a finite dimensional central simple algebra over $\mathcal{C}$.

The following theorem gives more detailed information about prime PI-rings. By $\deg(.)$ we denote the degree of algebraicity over $\mathcal{C}$ (cf. Section 5.2).

Theorem C.2. *Let $\mathcal{A}$ be prime ring, and let $n \in \mathbb{N}$. The following conditions are equivalent:*

(i) *$\mathcal{AC}$ is a finite dimensional central simple algebra over $\mathcal{C}$ with $\dim_{\mathcal{C}} \mathcal{AC} = n^2$;*

(ii) *$\mathcal{A}$ satisfies St_{2n} and does not satisfy any polynomial identity of degree $< 2n$;*

(iii) *There exists a field $\mathbb{F}$ such that $\mathcal{A}$ can be embedded into the ring $M_n(\mathbb{F})$, and $M_n(\mathbb{F})$ satisfies the same multilinear polynomial identities as $\mathcal{A}$ (and hence $\mathcal{A}$ cannot be embedded in $M_{n-1}(\mathcal{K})$ for any commutative ring $\mathcal{K}$);*

(iv) *$\deg(\mathcal{A}) = n$.*

Moreover, in this case there exist traces of k-additive maps $\alpha_k : \mathcal{A} \to \mathcal{C}$, $k = 1, \ldots, n$, such that

$$x^n + \alpha_1(x)x^{n-1} + \ldots + \alpha_{n-1}(x)x + \alpha_n(x) = 0$$

for all $x \in \mathcal{A}$. Also, we have $\mathcal{Q}_{ml}(\mathcal{A}) = \mathcal{AC}$.

If $\mathcal{A}$ was the algebra of all square matrices over a field, then the equivalence of (i)–(iv) would be easy to establish. Most of the implications can be proved by reducing the general situation to this simple and tractable one. The idea is to consider the scalar extension of the $\mathcal{C}$-algebra $\mathcal{AC}$ by the algebraic closure $\bar{\mathcal{C}}$ of $\mathcal{C}$ (incidentally, $\mathbb{F}$ in (iii) can be chosen to be $\bar{\mathcal{C}}$). Not everything is entirely obvious. In particular, showing that (iv) implies any of (i)–(iii) requires some more effort since the condition $\deg(\mathcal{A}) = n$ is not a multilinear one, and so it is more difficult to deal with scalar extensions. Anyhow, making use of certain standard tools of PI theory this problem can be handled as well. The last assertion concerning α_i's is based on the Cayley–Hamilton theorem (cf. the discussion following Example 1.2).

Note that Theorem C.2 in particular implies that for every prime PI-ring $\mathcal{A}$ there exists $n \in \mathbb{N}$ such that $\mathcal{A}$ satisfies St_{2n}, but does not satisfy any polynomial identity of degree $< 2n$. So the minimal degree of all polynomial identities of $\mathcal{A}$ is an even number.

If $\mathcal{A}$ is a simple unital ring, then $\mathcal{A}$ is centrally closed and moreover, the extended centroid is just the center $\mathcal{Z}$ of $\mathcal{A}$. Therefore parts of Theorem C.2 can be written in a simpler way. Let us record this.

Corollary C.3. *Let $\mathcal{A}$ be a simple unital ring. Then $\dim_{\mathcal{Z}} \mathcal{A} = n^2$ if and only if $\deg(\mathcal{A}) = n$. Moreover, in this case there exist traces of k-additive maps $\alpha_k : \mathcal{A} \to \mathcal{Z}$, $k = 1, \ldots, n$, such that*

$$x^n + \alpha_1(x)x^{n-1} + \ldots + \alpha_{n-1}(x)x + \alpha_n(x) = 0$$

for all $x \in \mathcal{A}$.

As mentioned earlier, the center of a nonzero prime PI-ring is nonzero; moreover, the same is true for semiprime rings. This is the result by Rowen [186].

Theorem C.4. *A nonzero semiprime PI-ring has a nonzero center.*

The results stated so far could be described as folklore. The next two lemmas are more special. They are taken from papers on FI's ([29, Lemmas 2.1 and 2.2] and [22, Lemma 2.1]), although their connection to FI's is only indirect. The proofs use standard PI theory in order to reduce the general case to the one where $\mathcal{A}$ is the algebra $M_n(\mathbb{F})$ of matrices over a field. Then the problem of course becomes very concrete; using the fact that the set of matrices with zero trace is the only proper noncentral Lie ideal of $M_n(\mathbb{F})$ (in the first lemma), and that the transpose and the symplectic involution are basically (here we are neglecting certain technicalities) the only involutions on $M_n(\mathbb{F})$ (in the second lemma), one then just has to find matrices with a "big" degree of algebraicity in appropriate sets.

Lemma C.5. *If $\mathcal{L}$ is a noncentral Lie ideal of a prime ring $\mathcal{A}$, then $\deg(\mathcal{L}) = \deg(\mathcal{A})$.*

Lemma C.6. *If $\mathcal{A}$ is a prime ring with involution and $\mathrm{char}(\mathcal{A}) \neq 2$, then $\deg(\mathcal{S}(\mathcal{A}) \cup \mathcal{K}(\mathcal{A})) = \deg(\mathcal{A})$. Moreover, if $\deg(\mathcal{A}) \geq 5$, then $\deg(\mathcal{L}) = \deg(\mathcal{A})$ for every noncentral Lie ideal $\mathcal{L}$ of $\mathcal{K}(\mathcal{A})$.*

lie in $\mathcal{Q}_s(\mathcal{A})$). We say that $\mathcal{A}$ is a *GPI-ring* if it satisfies a nonzero generalized polynomial identity.

As in the case of polynomial identities, for most purposes it is enough to consider multilinear generalized polynomial identities (their definition should be self-explanatory). We have dealt with linear generalized polynomial identities in one variable (i.e., elements of the form $\sum_i a_i x_1 b_i$) already in Theorem A.7. In fact, this theorem implies that a prime ring cannot satisfy a nonzero linear generalized polynomial identity in one variable. The next case of multilinear identities in two variables is more interesting, as our initial example clearly suggests. So assume that a prime ring $\mathcal{A}$ satisfies a generalized polynomial identity

$$0 \neq f = f(x_1, x_2) = \sum_{i=1}^{p} a_i x_1 b_i x_2 c_i + \sum_{j=1}^{q} d_j x_2 e_j x_1 f_j.$$

This means that this expression equals 0 if we replace the indeterminates x_1 and x_2 by any two elements in $\mathcal{A}$. Assume for simplicity that all $a_i, b_i, c_i, d_j, e_j, f_j$ lie in $\mathcal{A}$, and also that $\mathcal{A}$ is centrally closed. Further, without loss of generality we may assume that the first summation of f, $\sum_{i=1}^{p} a_i x_1 b_i x_2 c_i$, is also nonzero, and that $\{a_1, \ldots, a_n\}$ is a maximal linearly independent subset of $\{a_1, \ldots, a_p\}$. Note that we can rewrite f as

$$f = \sum_{i=1}^{n} a_i x_1 \mathcal{E}_i(x_2) + \sum_{j=1}^{q} d_j x_2 e_j x_1 f_j,$$

where $\mathcal{E}_i$ lies in $\mathcal{M}(\mathcal{A})$, the multiplication ring of $\mathcal{A}$. If every $\mathcal{E}_i$ was zero (as an element of $\mathcal{M}(\mathcal{A})$), then, by the result on linear identities in one variable, $\sum_{i=1}^{n} a_i x_1 \mathcal{E}_i(x_2) = \sum_{i=1}^{p} a_i x_1 b_i x_2 c_i$ would be 0, contrary to our assumption. So we may assume that $\mathcal{E}_1 \neq 0$. As $f(x, y) = 0$ for all $x, y \in \mathcal{A}$, we are now in a position to apply Theorem A.7. Hence it follows that for every $y \in \mathcal{A}$, $\mathcal{E}_1(y)$ is a $\mathcal{C}$-linear combination of $f_1, \ldots, f_q$, meaning that the range of $\mathcal{E}_1$ is finite dimensional. Now we can use Theorem A.9. Thus $\mathcal{A}$ contains a minimal idempotent e such that $e\mathcal{A}e$ is a finite dimensional division algebra over $\mathcal{C}$.

If $\mathcal{A}$ is not centrally closed, then the above conclusion holds for its central closure $\mathcal{A}\mathcal{C}$. Namely, obviously f is also a generalized polynomial identity of $\mathcal{A}\mathcal{C}$, and $\mathcal{A}\mathcal{C}$ is centrally closed.

If $\mathcal{A}$ satisfies a multilinear identity in three or more variables, the result is the same. At the first glance it may not appear entirely obvious how to extend the above argument concerning the two variables case. Anyway, it turns out that it is possible, and the following theorem, established in [152] by Martindale, holds.

Theorem D.1. *Let $\mathcal{A}$ be a prime ring. Then $\mathcal{A}$ is a GPI-ring if and only if $\mathcal{A}\mathcal{C}$ contains a minimal idempotent e such that $\dim_{\mathcal{C}} e\mathcal{A}\mathcal{C}e < \infty$.*

We have proved the "only if" part for the case when the GPI is of degree ≤ 2. The "if" part is easy. Namely, if $m = \dim_{\mathcal{C}} e\mathcal{A}\mathcal{C}e$, then $\mathcal{A}$ satisfies the generalized polynomial identity $St_{m+1}(ex_1e, ex_2e, \ldots, ex_{m+1}e)$.

It should be pointed out that the existence of an idempotent e satisfying the conditions of Theorem D.1 tells a great deal about the structure of $\mathcal{B} = \mathcal{AC}$ (and hence of $\mathcal{A}$). Apparently this is just a local property, it concerns only one element. But it has global consequences. Already the fact that there exist minimal left ideals in $\mathcal{B}$ is decisive. First of all, $\mathcal{B}$ is then a primitive ring since a minimal left ideal of a prime ring can be considered as a faithful simple module. More importantly, the existence of one minimal left ideal $\mathcal{I}$ implies the existence of "many". Indeed, for every $b \in \mathcal{B}$ we have that either $\mathcal{I}b = 0$ or $\mathcal{I}b$ is again a minimal left ideal; namely, if $\mathcal{I}b \neq 0$, then $\mathcal{I}$ and $\mathcal{I}b$ are obviously isomorphic as left $\mathcal{A}$-modules. This implies that the sum of all minimal left ideals in $\mathcal{B}$ is a two-sided ideal of $\mathcal{B}$. It is called the *socle* of $\mathcal{B}$. One can similarly consider the sum of all right ideals of $\mathcal{B}$, but fortunately we get the same ideal (see remarks about minimal one-sided ideals in appendix A; the sum of all minimal left ideals coincides with the sum of all minimal right ideals as long as the ring in question is semiprime). So, $\mathcal{B}$ contains a nonzero ideal that has a very concrete form: its elements are of the form $a_1 e_1 + \ldots + a_n e_n$ where $a_i \in \mathcal{A}$ and every e_i is a minimal idempotent. When dealing with a prime ring, it is often the case that if one controls a nonzero ideal, then one controls the entire ring. So the ring $\mathcal{B}$ is really tractable. The information that $\dim_{\mathcal{C}} e\mathcal{B}e < \infty$ is also important. One can show that if e and f are two minimal idempotents in $\mathcal{B}$, then the division algebras $e\mathcal{B}e$ and $f\mathcal{B}f$ are isomorphic. So, in particular, $\dim_{\mathcal{C}} e\mathcal{B}e = \dim_{\mathcal{C}} f\mathcal{B}f$.

Actually, even more can be said about prime (and hence primitive) rings with nonzero socle. They can be represented as rings of linear operators on a vector space (over a division ring) which contain "many" finite rank operators. In fact, the socle is equal to the set of all finite rank operators in this ring. See [40, section 4.3] for details.

There is just one technical result that we still have to record. Its statement is somewhat lengthy, but it is exactly what is needed in the proof of Theorem 5.36.

Lemma D.2. *Let $\mathcal{A}$ be a non-GPI prime ring.*

(i) *If $\{q_{i1}, q_{i2}, \ldots, q_{in_i}\} \subseteq \mathcal{Q}_{ml}(\mathcal{A})$, $i = 1, 2, \ldots, p$, is a collection of $\mathcal{C}$-independent sets, then there exists $x \in \mathcal{A}$ such that the set*

$$\{xq_{i1}, xq_{i2}, \ldots, xq_{in_i}\}$$

is $\mathcal{C}$-independent for every $i = 1, 2, \ldots, p$, and moreover each $xq_{ik} \in \mathcal{A}$.

(ii) *If $\{a_{i1}, a_{i2}, \ldots, a_{in_i}\} \subseteq \mathcal{A}$, $i = 1, 2, \ldots, r$, is a collection of $\mathcal{C}$-independent sets, and $0 \neq a \in \mathcal{A}$, then there exists $y \in \mathcal{A}$ such that the set*

$$\{a_{i1}, a_{i2}, \ldots, a_{in_i}, a_{i1}ya, a_{i2}ya, \ldots, a_{in_i}ya\}$$

is $\mathcal{C}$-independent for every $i = 1, 2, \ldots, r$.

(iii) *If $\{b_{j1}, b_{j2}, \ldots, b_{jn_j}\} \subseteq \mathcal{A}$, $j = 1, 2, \ldots, s$, is a collection of $\mathcal{C}$-independent sets, and $0 \neq b \in \mathcal{A}$, then there exists $z \in \mathcal{A}$ such that the set*

$$\{b_{j1}, b_{j2}, \ldots, b_{jn_j}, bzb_{j1}, bzb_{j2}, \ldots, bzb_{jn_j}\}$$

is $\mathcal{C}$-independent for every $j = 1, 2, \ldots, s$.

The proof is based on the fact that the linear dependence of elements can be expressed through a standard polynomial. For example, if the elements

$$a_1, \ldots, a_n, a_1 y a, \ldots, a_n y a$$

are $\mathcal{C}$-dependent for every $y \in \mathcal{A}$, then also

$$a_1 x, \ldots, a_n x, a_1 y a x, \ldots, a_n y a x$$

are $\mathcal{C}$-dependent for every $y \in \mathcal{A}$ and every $x \in \mathcal{A}$, so that

$$St_{2n}(a_1 x, \ldots, a_n x, a_1 y a x, \ldots, a_n y a x) = 0.$$

But this can be interpreted as a nonzero generalized polynomial identity. A more complicated situation involving more sets is just seemingly more difficult; the problem can be resolved by simply multiplying the adequate standard polynomials. We have thereby indicated the idea of the proofs of (ii) and (iii). A modification of this idea, together with Lemma A.2, works for (i) as well. To be honest, the fact that (i) involves $\mathcal{Q}_{ml}(\mathcal{A})$ creates some technical difficulties. However, using [40, Proposition 2.10, Corollary 6.1.7 and Theorem 6.4.4] they can be overcome.

As one could expect, all assertions of the lemma are just special cases of more general phenomena; see [40, Lemma 6.1.8]. We have chosen, however, to avoid stating a more abstract version of the lemma, and rather confine ourselves to what we really need.

Bibliography

[1] J. Alaminos, M. Brešar, A. R. Villena, The strong degree and the structure of Lie and Jordan derivations from von Neumann algebras, *Math. Proc. Camb. Phil. Soc.* **137** (2004), 441–463.

[2] A. A. Albert, Power-associative rings, *Trans. Amer. Math. Soc.* **64** (1948), 318–328.

[3] S. A. Amitsur, On rings of quotients, *Symposia Math.* **8** (1972), 149–164.

[4] G. Ancochea, Le théorème de von Staudt en géometrie projective quaternionienne, *J. Reine Angew. Math.* **184** (1942), 192–198.

[5] G. Ancochea, On semi-automorphisms of division algebras, *Ann. Math.* **48** (1947), 147–153.

[6] P. Ara, M. Mathieu, An application of local multipliers to centralizing mappings of C^*-algebras, *Quart. J. Math.* **44** (1993), 129–138.

[7] P. Ara, M. Mathieu, *Local multipliers of C^*-algebras*, Springer, 2003.

[8] S. A. Ayupov, Anti-automorphisms of factors and Lie operator algebras, *Quart. J. Math.* **46** (1995), 129–140.

[9] S. A. Ayupov, Skew commutators and Lie isomorphisms in real von Neumann algebras, *J. Funct. Anal.* **138** (1996), 170–187.

[10] S. A. Ayupov, N. A. Azamov, Commutators and Lie isomorphisms of skew elements in prime operator algebras, *Comm. Algebra* **24** (1996), 1501–1520.

[11] S. A. Ayupov, A. Rakhimov, S. Usmanov, *Jordan, real and Lie structures in operator algebras*, Kluwer Academic Publishers, Dordrecht–Boston–London, 1997.

[12] R. Banning, M. Mathieu, Commutativity preserving mappings on semiprime rings, *Comm. Algebra* **25** (1997), 247–265.

[13] W. E. Baxter, Lie simplicity of a special class of associative rings II, *Trans. Amer. Math. Soc.* **87** (1958), 63–75.

[14] L. B. Beasley, Linear transformations on matrices: The invariance of commuting pairs of matrices, *Linear and Multilinear Algebra* **6** (1978/79), 179–183.

[15] K. I. Beidar, Rings of quotients of semiprime rings, *Moscow Univ. Math. Bull.* **33** (1978), 29–34.

[16] K. I. Beidar, On functional identities and commuting additive mappings, *Comm. Algebra* **26** (1998), 1819–1850.

[17] K. I. Beidar, M. Brešar, M. A. Chebotar, Generalized functional identities with (anti-)automorphisms and derivations on prime rings, I, *J. Algebra* **215** (1999), 644–665.

[18] K. I. Beidar, M. Brešar, M. A. Chebotar, Functional identities on upper triangular matrix algebras, *J. Math. Sci.* (New York) **102** (2000), 4557–4565.

[19] K. I. Beidar, M. Brešar, M. A. Chebotar, Functional identities revised: the fractional and the strong degree, *Comm. Algebra* **30** (2002), 935–969.

[20] K. I. Beidar, M. Brešar, M. A. Chebotar, Functional identities with r-independent coefficients, *Comm. Algebra* **30** (2002), 5725–5755.

[21] K. I. Beidar, M. Brešar, M. A. Chebotar, Y. Fong, Applying functional identities to some linear preserver problems, *Pacific J. Math.* **204** (2002), 257–271.

[22] K. I. Beidar, M. Brešar, M. A. Chebotar, W. S. Martindale 3rd, On functional identities in prime rings with involution II, *Comm. Algebra* **28** (2000), 3169–3183.

[23] K. I. Beidar, M. Brešar, M. A. Chebotar, W. S. Martindale 3rd, On Herstein's Lie map conjectures, I, *Trans. Amer. Math. Soc.* **353** (2001), 4235–4260.

[24] K. I. Beidar, M. Brešar, M. A. Chebotar, W. S. Martindale 3rd, On Herstein's Lie map conjectures, II, *J. Algebra* **238** (2001), 239–264.

[25] K. I. Beidar, M. Brešar, M. A. Chebotar, W. S. Martindale 3rd, On Herstein's Lie map conjectures, III, *J. Algebra* **249** (2002), 59–94.

[26] K. I. Beidar, M. Brešar, M. A. Chebotar, W. S. Martindale 3rd, Polynomial preserving maps on certain Jordan algebras, *Israel J. Math.* **141** (2004), 285–313.

[27] K. I. Beidar, S.-C. Chang, M. A. Chebotar, Y. Fong, On functional identities in left ideals of prime rings, *Comm. Algebra* **28** (2000), 3041–3058.

[28] K. I. Beidar, M. A. Chebotar, On Lie-admissible algebras whose commutator Lie algebras are Lie subalgebras of prime associative algebras, *J. Algebra* **233** (2000), 675–703.

[29] K. I. Beidar, M. A. Chebotar, On functional identities and *d*-free subsets of rings I, *Comm. Algebra* **28** (2000), 3925–3951.

[30] K. I. Beidar, M. A. Chebotar, On functional identities and *d*-free subsets of rings II, *Comm. Algebra* **28** (2000), 3953–3972.

[31] K. I. Beidar, M. A. Chebotar, On surjective Lie homomorphisms onto Lie ideals of prime rings, *Comm. Algebra* **29** (2001), 4775–4793.

[32] K. I. Beidar, M. A. Chebotar, On Lie derivations of Lie ideals of prime rings, *Israel J. Math.* **123** (2001), 131–148.

[33] K. I. Beidar, M. A. Chebotar, Y. Fong, W.-F. Ke, On some Lie-admissible subalgebras of matrix algebras, *J. Math. Sci.* **131** (2005), 5939–5947.

[34] K. I. Beidar, Y. Fong, On additive isomorphisms of prime rings preserving polynomials, *J. Algebra* **217** (1999), 650–667.

[35] K. I. Beidar, Y. Fong, P.-H. Lee, T.-L. Wong, On additive maps of prime rings satisfying Engel condition, *Comm. Algebra* **25** (1997), 3889–3902.

[36] K. I. Beidar, Y.-F. Lin, On surjective linear maps preserving commutativity, *Proc. Roy. Soc. Edinburgh Sect. A* **134** (2004), 1023–1040.

[37] K. I. Beidar, Y.-F. Lin, Maps characterized by action on Lie zero products, *Comm. Algebra* **33** (2005), 2697–2703.

[38] K. I. Beidar, W. S. Martindale 3rd, On functional identities in prime rings with involution, *J. Algebra* **203** (1998), 491–532.

[39] K. I. Beidar, W. S. Martindale 3rd, A. V. Mikhalev, Lie isomorphisms in prime rings with involution, *J. Algebra* **169** (1994), 304–327.

[40] K. I. Beidar, W. S. Martindale 3rd, A. V. Mikhalev, *Rings with generalized identities*, Marcel Dekker, Inc., 1996.

[41] K. I. Beidar, A. V. Mikhalev, Orthogonal completeness and algebraic systems, *Russian Math. Surveys* **40** (1985), 51–95.

[42] K. I. Beidar, A. V. Mikhalev, Homogeneous boundness almost everywhere for orthogonal complete algebraic systems, *Vestnik Kievskogo Universiteta, Ser. Mat. Mekh.* **27** (1985), 15–17 (Ukrainian).

[43] K. I. Beidar, A. V. Mikhalev, The method of orthogonal completeness in structure theory of rings, *J. Math. Sci.* **73** (1995), 1–44.

[44] G. M. Benkart, Power-associative Lie-admissible algebras, *J. Algebra* **90** (1984), 37–58.

[45] G. M. Benkart, J. M. Osborn, Flexible Lie-admissible algebras, *J. Algebra* **71** (1981), 11–31.

[46] G. M. Benkart, J. M. Osborn, Power-associative products on matrices, *Hadronic J. Math.* **5** (1982), 1859–1892.

[47] D. Benkovič, D. Eremita, Characterizing left centralizers by their action on a polynomial, *Publ. Math.* **64** (2004), 343–351.

[48] D. Benkovič, D. Eremita, Commuting traces and commutativity preserving maps on triangular algebras, *J. Algebra* **280** (2004), 797–824.

[49] M. I. Berenguer, A. R. Villena, Continuity of Lie derivations on Banach algebras, *Proc. Edinburgh Math. Soc.* **41** (1998), 625–630.

[50] M. I. Berenguer, A. R. Villena, Continuity of Lie mappings of the skew elements of Banach algebras with involution, *Proc. Amer. Math. Soc.* **126** (1998), 2717–2720.

[51] M. I. Berenguer, A. R. Villena, Continuity of Lie isomorphisms of Banach algebras, *Bull. London Math. Soc.* **31** (1999), 6–10.

[52] G. M. Bergman, Centralizers in free associative algebras, *Trans. Amer. Math. Soc.* **137** (1969), 327–344.

[53] P. S. Blau, Lie isomorphisms of prime rings satisfying St_4, *Southeast Asian Bull. Math.* **25** (2002), 581–587.

[54] M. Brešar, Centralizing mappings on von Neumann algebras, *Proc. Amer. Math. Soc.* **111** (1991), 501–510.

[55] M. Brešar, On a generalization of the notion of centralizing mappings, *Proc. Amer. Math. Soc.* **114** (1992), 641–649.

[56] M. Brešar, Centralizing mappings and derivations in prime rings, *J. Algebra* **156** (1993), 385–394.

[57] M. Brešar, On skew-commuting mappings of rings, *Bull. Austral. Math. Soc.* **47** (1993), 291–296.

[58] M. Brešar, Commuting traces of biadditive mappings, commutativity-preserving mappings and Lie mappings, *Trans. Amer. Math. Soc.* **335** (1993), 525–546.

[59] M. Brešar, On certain pairs of functions of semiprime rings, *Proc. Amer. Math. Soc.* **120** (1994), 709–713.

[60] M. Brešar, On generalized biderivations and related maps, *J. Algebra* **172** (1995), 764–786.

[61] M. Brešar, Functional identities of degree two, *J. Algebra* **172** (1995), 690–720.

[62] M. Brešar, Applying the theorem on functional identities, *Nova Journal of Mathematics, Game Theory, and Algebra* **4** (1995), 43–54.

[63] M. Brešar, On a certain identity satisfied by a derivation and an arbitrary additive mapping II, *Aequationes Math.* **51** (1996), 83–85.

[64] M. Brešar, Functional identities: A survey, *Contemporary Math.* **259** (2000), 93–109.

[65] M. Brešar, On *d*-free rings, *Comm. Algebra* **31** (2003), 2287–2309.

[66] M. Brešar, Commuting maps: A survey, *Taiwanese J. Math.* **8** (2004), 361–397.

[67] M. Brešar, The range and kernel inclusion of algebraic derivations and commuting maps, *Quart. J. Math.* **56** (2005), 31–41.

[68] M. Brešar, Commutativity preserving maps revisited, *Israel J. Math.*, to appear.

[69] M. Brešar, M. Cabrera, A. R. Villena, Functional identities in Jordan algebras: Associating maps, *Comm. Algebra* **30** (2002), 5241–5252.

[70] M. Brešar, M. Cabrera, M. Fošner, A. R. Villena, Lie triple ideals and continuity of Lie triple isomorphisms on Jordan–Banach algebras, *Studia Math.* **169** (2005), 207–228.

[71] M. Brešar, M. A. Chebotar, On a certain functional identity in prime rings, *Comm. Algebra* **26** (1998), 3765–3782.

[72] M. Brešar, M. A. Chebotar, On a certain functional identity in prime rings, II, *Beitr. Alg. Geom.* **43** (2002), 333–338.

[73] M. Brešar, D. Eremita, A. R. Villena, Functional identities in Jordan algebras: Associating traces and Lie triple isomorphisms, *Comm. Algebra* **31** (2003), 1207–1234.

[74] M. Brešar, B. Hvala, On additive maps of prime rings, *Bull. Austral. Math. Soc.* **51** (1995), 377–381.

[75] M. Brešar, B. Hvala, On additive maps of prime rings, II, *Publ. Math. Debrecen* **54** (1999), 39–54.

[76] M. Brešar, W. S. Martindale 3rd, On the multiplication ring of a prime ring, *Comm. Algebra* **34** (2006), 2195–2203.

[77] M. Brešar, W. S. Martindale 3rd, C. R. Miers, Centralizing maps in prime rings with involution, *J. Algebra* **161** (1993), 342–357.

[78] M. Brešar, W. S. Martindale 3rd, C. R. Miers, Maps preserving n^{th} powers, *Comm. Algebra* **26** (1998), 117–138.

[79] M. Brešar, C. R. Miers, Commutativity preserving mappings of von Neumann algebras, *Canad. J. Math.* **45** (1993), 695–708.

[80] M. Brešar, C. R. Miers, Strong commutativity preserving maps of semiprime rings, *Canad. Math. Bull.* **37** (1994), 457–460.

[81] M. Brešar, C. R. Miers, Commuting maps on Lie ideals, *Comm. Algebra* **23** (1995), 5539–5553.

[82] M. Brešar, P. Šemrl, Normal-preserving linear mappings, *Canad. Math. Bull.* **37** (1994), 306–309.

[83] M. Brešar, P. Šemrl, Linear preservers on $\mathcal{B}(X)$, *Banach Center Publ.* **38** (1997), 49–58.

[84] M. Brešar, P. Šemrl, Commuting traces of biadditive maps revisited, *Comm. Algebra* **31** (2003), 381–388.

[85] M. Brešar, P. Šemrl, Elementary operators as Lie homomorphisms or commutativity preservers, *Proc. Edinb. Math. Soc.* **48** (2005), 37–49.

[86] M. Brešar, P. Šemrl, On bilinear maps on matrices with applications to commutativity preservers, *J. Algebra* **301** (2006), 803–837.

[87] M. A. Chebotar, On Lie automorphisms of simple rings of characteristic 2, *Fundam. Prikl. Mat.* **2** (1996), 1257–1268 (Russian).

[88] M. A. Chebotar, On generalized functional identities in prime rings, *J. Algebra* **202** (1998), 655–670.

[89] M. A. Chebotar, On functional identities of degree 2 in prime rings, *Fundam. Prikl. Mat.* **6** (2000), 923–938.

[90] M. A. Chebotar, On Lie isomorphisms in prime rings with involution, *Comm. Algebra* **27** (1999), 2767–2777.

[91] M. A. Chebotar, Y. Fong, P.-H. Lee, On maps preserving zeros of the polynomial $xy - yx^*$, *Linear Algebra Appl.* **408** (2005), 230–243.

[92] M. A. Chebotar, Y. Fong, L.-S. Shiao, On functional identities involving quasi-polynomials of degree one, *Comm. Algebra* **32** (2004), 3673–3683.

[93] M. A. Chebotar, W.-F. Ke, On skew-symmetric maps on Lie algebras. *Proc. Roy. Soc. Edinburgh Sect. A* **133** (2003), 1273–1281.

[94] M. A. Chebotar, W.-F. Ke, P.-H. Lee, Maps characterized by action on zero products, *Pacific J. Math.* **216** (2004), 217–228.

[95] M. A. Chebotar, W.-F. Ke, P.-H. Lee, Maps preserving zero Jordan products on Hermitian operators, *Illinois J. Math.* **49** (2005), 445–452.

[96] M. A. Chebotar, W.-F. Ke, P.-H. Lee, On maps preserving square-zero matrices, *J. Algebra* **289** (2005), 421–445.

[97] M. A. Chebotar, W.-F. Ke, P.-H. Lee, L.-S. Shiao, On maps preserving products, *Canad. Math. Bull.* **48** (2005), 355–369.

[98] M. A. Chebotar, W.-F. Ke, P.-H. Lee, R.-B. Zhang, On maps preserving zero Jordan products, *Monatshefte Math.* **149** (2006), 91–101.

[99] W.-S. Cheung, Commuting maps of triangular algebras, *J. London Math. Soc.* **63** (2001), 117–127.

[100] M. D. Choi, A. A. Jafarian, H. Radjavi, Linear maps preserving commutativity, *Linear Algebra Appl.* **87** (1987), 227–242.

[101] J. Cui, J. Hou, Linear maps preserving elements annihilated by the polynomial $XY - YX^\dagger$, *Studia Math.* **174** (2006), 183–199.

[102] D. Eremita, A functional identity with an automorphism in semiprime rings, *Algebra Colloq.* **8** (2001), 301–306.

[103] D. Eremita, On some special generalized functional identities, *Taiwanese J. Math.* **8** (2004), 191–202.

[104] T. S. Erickson, W. S. Martindale 3rd, J. M. Osborn, Prime nonassociative algebras, *Pacific J. Math.* **60** (1975), 49–63.

[105] D. R. Farkas, Characterization of Poisson algebras, *Comm. Algebra* **23** (1995), 4669–4686.

[106] D. R. Farkas, Poisson polynomial identities, *Comm. Algebra* **26** (1998), 401–416.

[107] D. R. Farkas, G. Letzter, Ring theory from symplectic geometry, *J. Pure Appl. Algebra* **125** (1998), 155–190.

[108] M. Fošner, On the extended centroid of prime associative superalgebras with applications to superderivations, *Comm. Algebra* **32** (2004), 689–705.

[109] G. Frobenius, Über die darstellung der endlichen gruppen durch lineare substitutionen, *Sitzungsber. Deutsch. Akad. Wiss. Berlin* (1897), 994–1015.

[110] P. de la Harpe, *Classical Banach–Lie algebras and Banach–Lie groups of operators in Hilbert spaces*, Lecture Notes Math. **285**, Springer-Verlag, 1972.

[111] I. N. Herstein, Jordan homomorphisms, *Trans. Amer. Math. Soc.* **81** (1956), 331–341.

[112] I. N. Herstein, Jordan derivations of prime rings, *Proc. Amer. Math. Soc.* **8** (1957), 1104–1110.

[113] I. N. Herstein, Lie and Jordan structures in simple, associative rings, *Bull. Amer. Math. Soc.* **67** (1961), 517–531.

[114] I. N. Herstein, *Topics in ring theory*, The University of Chicago Press, Chicago, 1969.

[115] I. N. Herstein, *Rings with involution*, The University of Chicago Press, Chicago, 1976.

[152] W. S. Martindale 3rd, Prime rings satisfying a generalized polynomial identity, *J. Algebra* **12** (1969), 576–584.

[153] W. S. Martindale 3rd, A note on Lie isomorphisms, *Canad. Math. Bull.* **17** (1974), 243–245.

[154] W. S. Martindale 3rd, Lie isomorphisms of the skew elements of a simple ring with involution, *J. Algebra* **36** (1975), 408–415.

[155] W. S. Martindale 3rd, Lie isomorphisms of the skew elements of a prime ring with involution, *Comm. Algebra* **4** (1976), 927–977.

[156] W. S. Martindale 3rd, The extended center of coproducts, *Canad. Math. Bull.* **25** (1982), 245–248.

[157] W. S. Martindale 3rd, Jordan homomorphisms onto nondegenerate Jordan algebras, *J. Algebra* **133** (1990), 500–511.

[158] M. Mathieu, Where to find the image of a derivation, *Banach Center Publ.* **30** (1994), 237–249.

[159] J. Mayne, Centralizing automorphisms of prime rings, *Canad. Math. Bull.* **19** (1976), 113–115.

[160] K. McCrimmon, The Zelmanov approach to Jordan homomorphisms of associative algebras, *J. Algebra* **123** (1989), 457–477.

[161] C. R. Miers, Lie isomorphisms of factors, *Trans. Amer. Math. Soc.* **147** (1970), 55–63.

[162] C. R. Miers, Lie homomorphisms of operator algebras, *Pacific J. Math.* **38** (1971), 717–735.

[163] C. R. Miers, Derived ring isomorphisms of von Neumann algebras, *Canad. J. Math.* **25** (1973), 1254–1268.

[164] C. R. Miers, Lie derivations of von Neumann algebras, *Duke Math. J.* **40** (1973), 403–409.

[165] C. R. Miers, Lie *-triple homomorphisms into von Neumann algebras, *Proc. Amer. Math. Soc.* **58** (1976), 169–172.

[166] C. R. Miers, Lie triple derivations of von Neumann algebras, *Proc. Amer. Math. Soc.* **71** (1978), 57–61.

[167] C. R. Miers, Centralizing mappings of operator algebras, *J. Algebra* **59** (1979), 56–64.

[168] C. R. Miers, Commutativity preserving maps of factors, *Canad. J. Math.* **40** (1988), 248–256.

[169] A. V. Mikhalev, Orthogonally complete many-sorted systems, *Doklady Akad. Nauk SSSR* **289** (1986), 1304–1308.

[170] S. Montgomery, Constructing simple Lie superalgebras from associative graded algebras, *J. Algebra* **195** (1997), 558–579.

[171] H. C. Myung, Some classes of flexible Lie-admissible algebras, *Trans. Amer. Math. Soc.* **167** (1972), 79–88.

[172] H. C. Myung, *Malcev-admissible algebras*, "Progress in Mathematics", Vol. 64, Birkhäuser, 1986.

[173] Y. Nambu, Generalized Hamiltonian dynamics, *Phys. Rev. D.* **7** (1973), 2405–2412.

[174] S. Okubo, *Introduction to octonion and other non-associative algebras in physics,* Cambridge University Press, New York, 1995.

[175] S. Okubo, H. C. Myung, Adjoint operators in Lie algebras and the classification of simple flexible Lie-admissible algebras, *Trans. Amer. Math. Soc.* **264** (1981), 459–472.

[176] M. Omladič, On operators preserving commutativity, *J. Funct. Anal.* **66** (1986), 105–122.

[177] M. Omladič, H. Radjavi, P. Šemrl, Preserving commutativity, *J. Pure Appl. Algebra* **156** (2001), 309–328.

[178] S. Pierce, et. al., A survey of linear preserver problems, *Linear and Multilinear Algebra* **33** (1992), 1–129.

[179] S. Pierce, W. Watkins, Invariants of linear maps on matrix algebras, *Linear and Multilinear Algebra* **6** (1978/79), 185–200.

[180] V. P. Platonov, D. Ž. Djoković, Linear preserver problems and algebraic groups, *Math. Ann.* **303** (1995), 165–184.

[181] V. P. Platonov, D. Ž. Djoković, Subgroups of $GL(n^2, \mathbb{C})$ containing $PSU(n)$, *Trans. Amer. Math. Soc.* **348** (1996), 141–152.

[182] E. C. Posner, Derivations in prime rings, *Proc. Amer. Math. Soc.* **8** (1957), 1093–1100.

[183] E. C. Posner, Prime rings satisfying a polynomial identity, *Proc. Amer. Math. Soc.* **11** (1960), 180–183.

[184] C. Procesi, *Rings with Polynomial Identities*, Marcel Dekker, 1973.

[185] M. P. Rosen, Lie isomorphisms of a certain class of prime rings, *J. Algebra* **89** (1984), 291–317.

[186] L. H. Rowen, Some results on the center of a ring with polynomial identity, *Bull. Amer. Math. Soc.* **79** (1973), 219–223.

[187] L. H. Rowen, *Polynomial identities in ring theory*, Academic Press, 1980.

[188] R. M. Santilli, Embedding of a Lie algebra in nonassociative structures, *Nuovo Cimento A* **51** (1967), 570–576.

[189] V. G. Skosyrskii, Strongly prime noncommutative Jordan algebras, *Trudy Inst. Mat. (Novosibirsk)* **16** (1989), 131–164 (in Russian).

[190] J. M. Souriau, Quantification géométrique, *Comm. Math. Phys.* **1** (1965/1966), 374–398.

[191] R. F. Streater, Canonical quantization, *Comm. Math. Phys.* **2** (1966), 354–374.

[192] G. A. Swain, Lie derivations of the skew elements of prime rings with involution, *J. Algebra* **184** (1996), 679–704.

[193] Y. Utumi, On quotient rings, *Osaka J. Math.* **8** (1956), 1–18.

[194] A. R. Villena, Lie derivations on Banach algebras, *J. Algebra* **226** (2000), 390–409.

[195] Y. Wang, The ranges of additive maps in generalized functional identities on prime rings, *Comm. Algebra* **30** (2002), 2897–2913.

[196] W. Watkins, Linear maps that preserve commuting pairs of matrices, *Linear Algebra Appl.* **14** (1976), 29–35.

[197] R. Wisbauer, *Modules and algebras: Bimodule structure and group actions on algebras*, Addison Wesley Longman Ltd., 1996.

[198] T.-L. Wong, A special functional identity in prime rings, *Comm. Algebra* **32** (2004), 363–377.

[199] E. I. Zelmanov, On prime Jordan algebras II, *Siberian Math. J.* **24** (1983), 89–104.

[200] K. Zhao, Weyl type algebras from quantum tori, *Commun. Contemp. Math.* **8** (2006), 135–165.

[201] L. Zhao, J. Hou, Jordan zero-product preserving additive maps on operator algebras, *J. Math. Anal. Appl.* **314** (2006), 689–700.

Index